Results and Problems in Cell Differentiation

A Series of Topical Volumes in Developmental Biology

19

Editors

W. Hennig, L. Nover, and U. Scheer

Steven T. Case (Ed.)

Structure, Cellular Synthesis and Assembly of Biopolymers

With 128 Figures and 13 Tables

Springer-Verlag
Berlin Heidelberg New York
London Paris Tokyo
Hong Kong Barcelona
Budapest

Professor Dr. STEVEN T. CASE
Dept. of Biochemistry
The University of Mississippi
Medical Center
2500 North State Street
Jackson, MS 39216-4505, USA

ISBN 3-540-55549-8 Springer-Verlag Berlin Heidelberg NewYork
ISBN 0-387-55549-8 Springer-Verlag NewYork Berlin Heidelberg

Library of Congress Cataloging-in-Publication Data
Structure, cellular synthesis, and assembly of biopolymers / Steven T. Case (ed.).
p. cm. -- (Results and problems in cell differentiation ; 19)
Includes bibliographical references and index.
ISBN 0-387-55549-8 (alk. paper : NewYork). -- ISBN 3-540-55549-8 (alk paper : Berlin)
1. Biopolymers. 2. Developmental biology. I. Case, Steven T., 1949- . II. Series.
QH607.R4 vol. 19 [QP801.B69] 574.87'612 s--dc20 [574.19'24] 92-36490 CIP

Typesetting: Macmillan India Ltd., Bangalore;
Offsetprinting: Saladruck, Berlin; Binding: Lüderitz & Bauer, Berlin
31/3020-5 4 3 2 1 0 – Printed on acid-free paper

This book is dedicated with gratitude and affection to

Professor Irvin M. Gottlieb

in honor of his 70th birthday and retirement as
Professor of Chemistry, Widener University, Chester, PA, USA

Preface

Modern technology has enabled scientists to generate data at an unprecedented rate. This deluge of information can lead to two problems. With the exception of a few dedicated colleagues, the scientific community at large is unable to compile an overall picture of progress because of the volume and complexity of knowledge gained. Concurrently, the investigator may lose a global perspective that relates his or her research to other apparently unrelated fields. During the past decade, I have fallen victim to both problems.

Two valued colleagues, Drs. Donald B. Sittman and Susan A. Gerbi, for quite different reasons, have maintained a strong interest in my research. On several occasions, they suggested that it was time to review and consolidate data obtained in the study of Balbiani ring genes and their encoded secretory proteins. While this task in itself was not problematic, the context in which to place this review was not clear.

Over a span of three and a half years, I was fortunate to participate in a series of stimulating meetings that enabled me to interact with a dissimilar group of scientists sharing a common interest; biopolymers. First, there were the Office of Naval Research/Molecular Biology Program meetings for contractors in the Bioengineering for Materials Application (June 1987) and Biopolymers (October 1989) programs. Next, there was the University/Industry Workshop on Biomolecular Materials sponsored by the National Science Foundation (October 1990). Finally, there was a symposium, "Materials Synthesis Based on Biological Processes," at the annual meeting of the Materials Research Society (November 1990). Among these groups of scientists was a subset that studied a variety of naturally occurring biomolecular materials with compositions and properties as diverse as the organisms that fabricated them. In each instance the synthesis, assembly and processing of these biopolymers were accomplished by a tissue composed of highly differentiated cells. Thus, with the encouragement of Dr. Wolfgang Hennig, this emerged as the theme of a volume in the series, *Results and Problems in Cell Differentiation.*

The subject of this volume is an unlikely combination of biopolymers. These natural biomaterials have unique structures and physical or mechanical properties which may be applicable to the design of novel biomolecular materials. The intent of this volume is to sample, in depth, several classes of biopolymers (nanolaminated ceramics, adhesive proteins, eggshell proteins and protein fibers) that are synthesized by a variety of organisms (mollusks, parasitic trematodes, fish and insects) and studied by an equally diverse group of scientists (materials scientists, zoologists, biochemists, molecular biologists, biophysicists and mechanical engineers). The reader will be introduced to the biological systems in order to understand Nature's use of each biopolymer. In each instance, highly differentiated cells are responsible for the synthesis of biopolymer precursors. Available data will be summarized according to their structure, assembly and processing. The reader will also sample some of the microscopic and micromechanical methods used to measure properties of biopolymers. Since the study of each biopolymer has an historically unique origin and approach, our comprehension of their structures and mechanisms of assembly is disparate. Our greatest opportunity to decipher and exploit Nature's skills in biopolymer engineering is through the combined expertise of interdisciplinary teams of scientists such as the contributors to this volume.

I am extremely grateful to the authors for their enthusiasm and cooperation in the preparation and timely submission of manuscripts. They made my task surprisingly pleasant. I wish to express my special thanks to Mrs. Alice Matthews for expert secretarial assistance in compiling the index for this volume.

Last, but certainly not least, I wish to acknowledge the loving support of my wife, Gay Lynn, son, Chad Erik, and daughter, Jill Lynn. Many hours that belonged to them were sacrificed in my persuit of this endeavor.

Jackson, Mississippi, August 1992 Steven T. Case

Contents

1 Nacre of Abalone Shell: a Natural Multifunctional Nanolaminated Ceramic-Polymer Composite Material (With 12 Figures)

MEHMET SARIKAYA and ILHAN A. AKSAY

2 The Formation of Mussel Byssus: Anatomy of a Natural Manufacturing Process (With 8 Figures)

J. HERBERT WAITE

3 Reflections on the Structure of Mussel Adhesive Proteins (With 6 Figures)

RICHARD A. LAURSEN

4 Composition and Design of *Fasciola hepatica* Eggshells (With 3 Figures)

Allison C. Rice-Ficht

5 The Cell and Molecular Biology of Eggshell Formation in *Schistosoma mansoni* (With 6 Figures)

Kim E. Wells and John S. Cordingley

6 Molecular Architecture of Helicoidal Proteinaceous Eggshells (With 38 Figures)

STAVROS J. HAMODRAKAS

7 Secretory Proteins of *Chironomus* Salivary Glands: Structural Motifs and Assembly Characteristics of a Novel Biopolymer

(With 20 Figures)

STEVEN T. CASE and LARS WIESLÄNDER

8 Spider Silk: a Mystery Starting to Unravel
(With 20 Figures)

MIKE HINMAN, ZHENGYU DONG, MING XU, and RANDOLPH V. LEWIS

9 The Nature and Role of Liquid Crystalline Order in Silk Secretions (With 9 Figures)

CHRISTOPHER VINEY

10 Micromechanics of Natural Composites
(With 6 Figures)

JOSEPH E. SALIBA

Chapter 1

Nacre of Abalone Shell: a Natural Multifunctional Nanolaminated Ceramic-Polymer Composite Material

Mehmet Sarikaya[1] and Ilhan A. Aksay[1,2]

1 Introduction: Nanocomposite Materials and Biological Composites

When materials are manufactured with an emphasis on tailoring their properties through microstructural control, the extent of this control is generally at a specific length scale. For instance, the mechanical properties of most metallic materials are controlled through the manipulation of dislocation dynamics at the nanometer length scale (Cottrell 1953), whereas the mechanical properties of ceramic materials are controlled through the propagation of cracks that are initiated from defects of micrometer length scale (Evans and Marshall 1989).

In contrast, the approach used by organisms in processing materials is in many ways more controlled than synthetic methods because biological materials are dynamic systems (Lowenstam and Weiner 1989; Mann et al. 1989; Simkiss and Wilbur 1989). In the formation of biological materials, organisms efficiently design and produce complex and hierarchical microstructures with unique properties at spatial dimensions from the molecular to the centimeter, and with greater structural control. The dynamism of these systems allows the collection and transport of the raw constituents; the nucleation, configuration, and growth of new structures (self-assembly) (see, for instance, Wasserman et al. 1989); and the repair and replacement of old or damaged components. These materials include all organic components, such as spiders' webs (Gosline et al. 1986) and insect cuticles (Bouligand 1965); inorganic-organic composites, such as seashells (Currey 1987) and bones (Glimcher 1981); all ceramic composites, such as sea urchin teeth and spines (Berman et al. 1990); and inorganic, ultrafine particles, such as magnetic (Blakemore 1975; Frankel et al. 1979; Blakemore 1982; Frankel and Blakemore 1991) and semiconducting (see, for instance, in biological systems, Dameron et al. 1989; and in synthetic systems, Fendler 1987) particles produced by bacteria and algae, respectively. In addition, in certain cases, byproducts, such as enzymes (Alper 1991), proteins (see, for instance, Haggin 1988), and other macromolecules (Crueger

[1] Department of Materials Science and Engineering, Roberts Hall, FB-10, University of Washington, Seattle WA 98195, USA and Advanced Materials Technology Center, Washington Technology Center, University of Washington, Seattle, WA 98195, USA
[2] Current address: Department of Chemical Engineering, Princeton University, Princeton, NJ 08544-0563 USA

Results and Problems in Cell Differentiation 19
Biopolymers
Case, S.T. (Ed.)

and Crueger 1982), have chemical or physical properties superior to their synthetic counterparts.

Biological systems can be a source of inspiration for design and processing concepts for novel synthetic materials where structural control can be established through a continuous length scale. An approach based on biological systems can be divided into two categories (Aksay and Sarikaya 1991). First, by studying the structures of biocrystals using various microscopy techniques at all scales of spatial resolution, the fundamentals of their unique structural designs can be acquired and then mimicked by techniques that are currently available to materials scientists — an approach that we will refer to as *biomimicking*. Second, by learning the molecular synthesis and processing mechanisms of biomaterials and using these hitherto unknown methodologies, new technological materials superior to those presently available can be produced — an approach we call *bioduplication*. The bioduplication approach is much more involved and requires a long-term commitment (probably tens of years) to learn not only the intricacies of bioprocessing used by organisms but also to develop new strategies to synthetically process materials at the molecular level with the same size, shape, multifunctional and hierarchical complexity as the biomaterials. The first approach, i.e., biomimicking, can be relatively short term (10 years or less), although by no means simple. Biomimicking involves exploring the structures of biomaterials, which are often hierarchical with each level having a different functionality, and correlating that functionality with the unique microstructural design and multifunctional properties of the biomaterial. Although our research is directed towards both categories, this chapter describes some of our findings on the design of the nacre of abalone shell and answers several questions surrounding this unique, but relatively simple, microstructure, particularly the organization of the inorganic component and its possible relation to the organic matrix.

In this chapter, specifically, the mechanical properties and structure of nacre, a composite of ceramic and proteins, are discussed. The significance of nacre as a structural material is that its mechanical properties, such as fracture toughness and strength, are unprecedentedly high and comparable to those of the high-technology structural ceramics (Sarikaya et al. 1990a,b; Yasrebi et al. 1990). Yet in nacre, the structure is mostly composed of $CaCO_3$, a material that has limited engineering value for materials applications. The unique structure of nacre, which is composed of alternating nanometer-scale laminated layers of thin biomacromolecules and $CaCO_3$ platelets, is highly organized to produce an excellent multifunctional material (armor) for the organism (Currey 1974, 1976; Jackson et al. 1988; Sarikaya et al. 1990a). It has recently been shown that crystallographic defect formation, in particular twins, in the inorganic phase takes place in a hierarchical fashion and can be attributed to the ability of the organism to control defect structures in a hierarchical manner which results in a composite structure with excellent properties (Sarikaya et al. 1992).

Other known facts about nacre are: (1) the overall shell composite is over 95 vol% inorganic material ($CaCO_3$ in the form of aragonite) (Currey 1987); (2) the composite has a "brick and mortar" microarchitecture (Currey 1987; Jackson et al. 1988; Sarikaya et al. 1990a,b) with aragonite forming thin, hexagonally shaped bricks within an organic matrix which is thought to be composed of three distinct layers (Watabe 1965; Bevelander and Nakahara 1968; Towe and Hamilton 1968; Mutvei 1970; Nakahara et al. 1982; Weiner and Traub 1984; Weiner 1986); and (3) the inorganic material consists of highly oriented aragonite platelets (Weiner and Traub 1984).

Despite these findings, however, much more needs to be known about the microstructure if the desired synthetically laminated materials are to be produced through biomimicking nacre. Some of these issues are: (1) the nature of the crystallographic defects and their distribution in the inorganic phase; (2) the crystallographic relationship between aragonite platelets to explain the high degree of organization and structural integrity of nacre; (3) the direct analysis of the possible stereochemistry between the organic and inorganic phases; (4) the nature of the organic phase(s) in the thin film matrix between the aragonite platelets, their size and distribution throughout nacre; and (5) the influence of the aragonite platelets on the formation and resultant functionality of the organic matrix in the overall behavior of the composite. Answers to these questions will shed more light on the mechanisms of formation and particularly the degree of control that the organism has over the growth of the highly ordered biocomposite.

This paper specifically concentrates on points (1) and (2) raised above, and indirectly discusses points (3), (4), and (5). Therefore, we first present a detailed analysis of the mechanical properties of nacre. Second, from the analysis of the toughening and strengthening mechanisms in nacre, we derive a set of criteria for the structural design of synthetic laminated materials via biomimetics. These are followed by a description of our current understanding of the structure of nacre in terms of morphology, composition, and crystallography, its hierarchy (in both the hard and the soft tissues), and possible structural relationships between the organic and inorganic components. Finally, we discuss possible directions for future studies that will lead to a better understanding of the biological and chemical basis of the formation of the relatively simple but highly efficient nacre structure and its significance in terms of designing and processing nanolaminated multifunctional materials for engineering applications.

2 Mechanical Properties of Nacre

2.1 Toughness and Strength in Nacre

The nacre structure is found in many families of mollusks, such as the red abalone (*Haliotis rufescens*) of the gastropod family, cephalopods such as nautilus (*Nautilus pompilius*), and bivalves such as black-lipped pearl oysters (*Pinctada margaritifera*) (Nicol 1960; Grégoire 1972; Morton 1979; Currey 1987). We used red abalone in our studies, since the diameter and the thickness of the shell containing the nacre in this species were large enough to perform standard mechanical tests (Brown and Srawley 1966), allowing direct comparison of nacre with engineering structural ceramics and ceramic-based materials. Mechanical properties were evaluated in terms of fracture toughness, K_{IC}, and fracture strength, σ_F, in three-point straight notched and four-point bend bars, respectively, in the transverse direction. This is the direction perpendicular to the laminated layers through which a crack is normally expected to propagate in the shell of the organism in its natural environment (Currey 1977). Red abalones were collected in Baja California, Mexico, and we estimated that the specimens were about 10–12 years old with a shell diameter of 25–30 cm and a thickness of about 1.5 cm, including the prismatic and nacreous layers. Although it was inevitable that the samples had curved layers, the test samples satisfied the size requirements for American Standards for Testing of Materials (ASTM) standards and allowed for the comparison of test results between nacre and structural ceramics, such as Al_2O_3, which were prepared for this purpose (Gunnison 1991). Hardness indentations were made with a Vickers microhardness indentor with a load of 1 kg.

The results of the mechanical tests of nacre and some of the well-studied ceramics and ceramic-based composites (cermets) are plotted in Fig. 1 in terms of K_{IC} versus σ_F, specific strength (Sarikaya et al. 1990a,b; Yasrebi et al. 1990; Aksay and Sarikaya 1991).

The average values for the toughness and strength of nacre are 7 ± 3 MPa-$m^{1/2}$, and 180 ± 20 MPa, respectively. As seen from the figure, fracture strength and, especially, fracture toughness values show a high degree of scatter. This may be due to the presence of yearly growth layers that are intrinsic to abalone nacre (Currey 1987). The yearly growth layers are absent in other types of nacre, such as pearl oyster; however, the samples are too thin to perform standard tests, requiring tests with undersized samples (Jackson et al. 1988).

Nevertheless, *P. margaritifera* nacre is also commonly used for mechanical tests (Currey 1977; Jackson et al. 1988; Sawyer S, Sarikaya M, Aksay IA, unpubl.

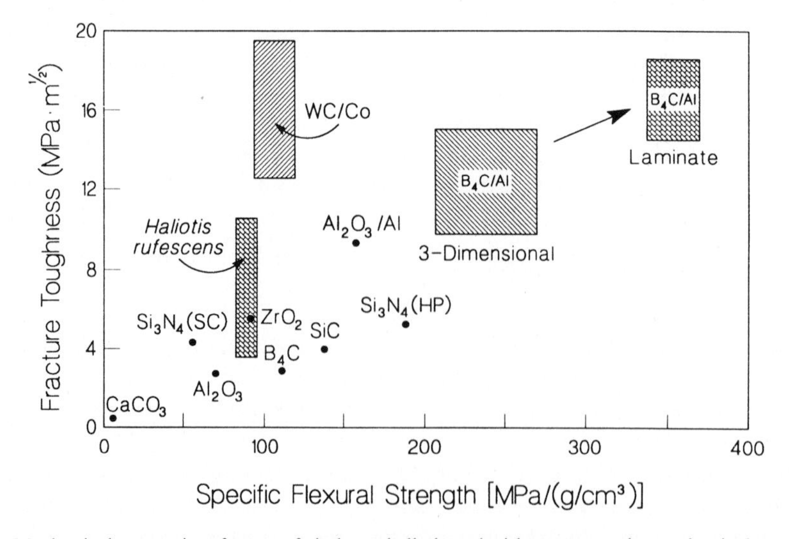

Fig. 1. Mechanical properties of nacre of abalone shell plotted with respect to the mechanical properties of ceramics and ceramic-metal composites

research). The test results from our laboratory on *P. margaritifera* nacre produced values for K_{IC} and σ_F of 10 ± 6 MPa-m$^{1/2}$ and 220 ± 60 MPa, respectively (Sawyer S, Sarikaya M, Aksay IA, unpubl. research). It has been reported that mechanical tests performed on "dry" samples produced much lower results (Currey 1977; Jackson et al. 1988). In the present case, there was no intentional drying, and the samples, kept under laboratory conditions, were tested within 1 to 3 months after the removal of the animal from the ocean. Therefore, the results given above may be slightly higher than those reported in the literature; however, these results should be more representative of the true properties of nacre since they are from relatively fresh samples and are more similar to the organism in its natural environment.

2.2 Toughening Mechanisms

The crack propagation behavior in nacre was studied by microindentation in the edge-on configuration to understand possible toughening mechanisms (Sarikaya et al. 1990a,b). It was revealed that there is a high degree of tortuosity not seen in the more traditional brittle ceramics, such as Al_2O_3, or in high toughness ceramics, such as ZrO_2 (see, for example, Tressler and Bradt 1984; Evans 1988). A microstructure showing the tortuous crack propagation in the nacre is displayed in Fig. 2. The most apparent features of the crack propagation seen in the scanning electron microscopy (SEM) images are crack blunting/branching and microcrack formation. A closer examination of these images reveals that microcracks advance both on planes parallel to the $CaCO_3$ layers and on those perpendicular to them. It is not clear, however, whether the cracks propagate inside the organic layer or through the interface plane between the organic and the inorganic components. The surfaces of fractured samples indicate that a major crack has meandered around the $CaCO_3$ layers, exposing them through the organic surroundings, resulting in a very rough fractured surface that is similar to that seen in fiber-reinforced ceramic composites where a pull-out mechanism operates (Evans and Marshall 1989). In micrographs recorded at much higher magnifications, separation of the $CaCO_3$ layers in both *x* and *y* directions (in-layer plane) is clearly seen; see, for instance, Fig. 3a. The "bricks" have been left intact, indicating they slide on the organic layer (Sarikaya et al. 1990a,b). The micrograph in Fig. 3b, on the other hand, shows organic ligaments stretched in the perpendicular direction, i.e., the *z* direction, between the $CaCO_3$ platelets. This stretching indicates that the interface between the organic and the inorganic phases is strong and that the organic phase acts as a strong binder (Jackson et al. 1988; Sarikaya et al. 1990a).

By a close analysis of many fractured surfaces and indentation cracks, several toughening mechanisms may be proposed (Sarikaya et al. 1990a,b): (1) crack blunting/branching, (2) microcrack formation, (3) plate pull-out, (4) crack bridging (ligament formation), and (5) sliding of $CaCO_3$ layers. In general, it would be desirable to have all of these energy-absorbing mechanisms operative in a composite material in order to increase the overall fracture toughness of the brittle

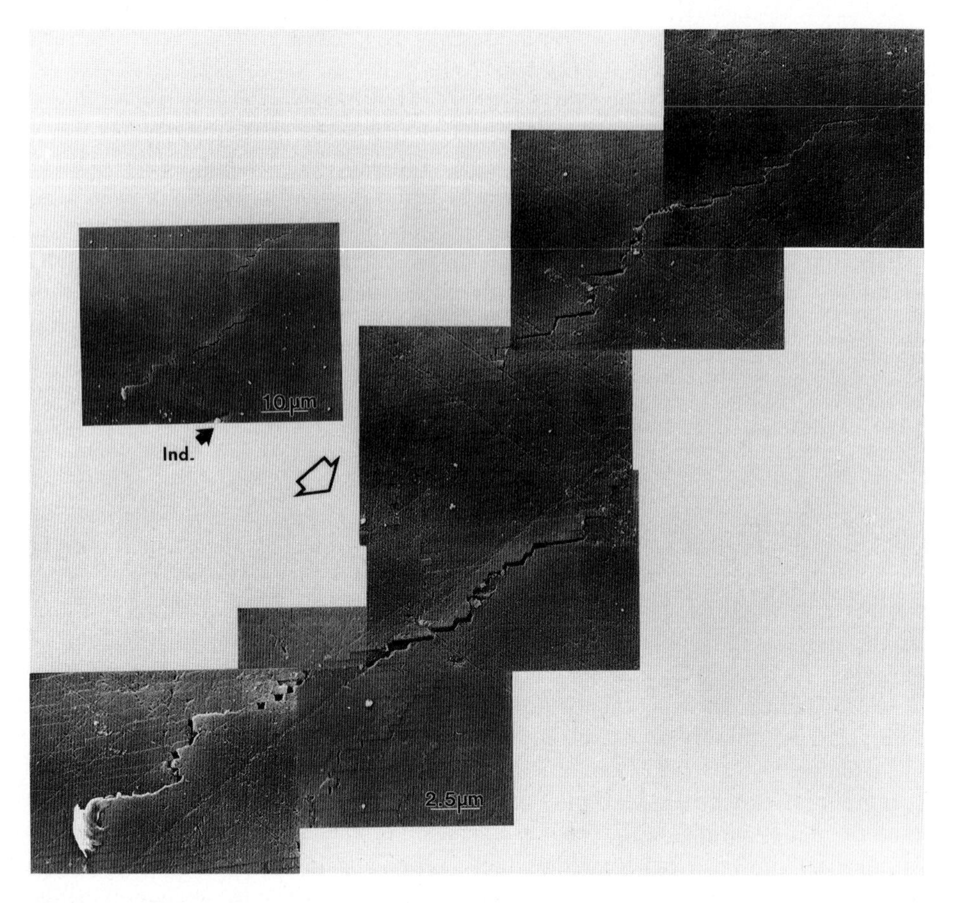

Fig. 2. Secondary electron (SE) image of crack propagation behavior in nacre in edge-on orientation (indentation is indicated by *Ind* in the *inset*)

material (Evans and Marshall 1989). The high degree of tortuosity seen in crack propagation is due mainly to crack blunting and branching. However, tortuosity alone is not a major toughening mechanism in these composites because it cannot account for the many orders of magnitude increase seen in toughness. Our studies showed that the linear tortuosity, determined from a number of indentation cracks, indicates only a 30–50% toughening value of a polycrystalline pure $CaCO_3$ material. The major toughening mechanisms are, therefore, thought to be sliding and ligament formation (Sarikaya et al. 1990a,b).

With reference to the sliding mechanism, similarities exist between the deformation of a nacreous seashell and a metal, and these may be stated as follows: (1) Around the periphery of the indentation, as seen in Fig. 3, the material exhibits deformation features, such as sliding of the $CaCO_3$ layers to accommodate the strain caused by the indentor. These deformation features resemble Luders bands (or slip bands) encountered in metals (Honeycombe 1968; Engel and Klingele 1981), such as in face-centered cubic Ni and Al, in which the lattice is plastically

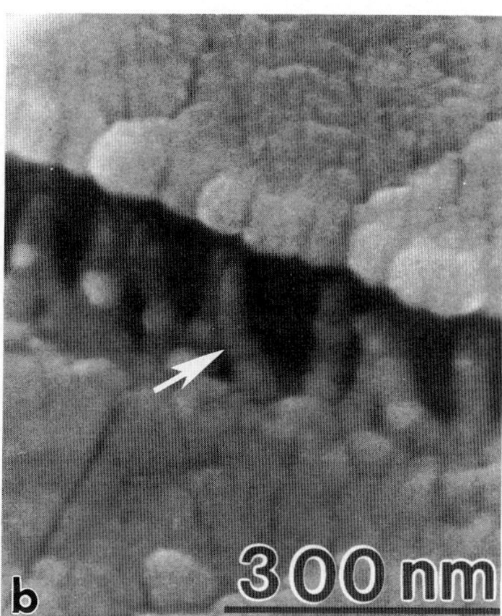

Fig. 3a, b. SEM images illustrate **a** sliding of the platelets (direction of *arrows*) and **b** pull-out (in the direction of *arrows*) of the organic matrix during crack propagation. (Sarikaya et al. 1990a)

deformed from the dislocation movement in order to accommodate the applied stress by slipping on high atomic density planes such as $\{111\}$[1]. It is also noted that the amount of sliding near the indentation is higher and decreases with distance from the indentation. (2) When sliding occurs in nacre, there is no cracking through the layers except for microcrack formation. This is again similar to the process of crack formation in metals, where deformation is associated with a high degree of dislocation activity which causes an increase in dislocation density in local regions ahead of the major crack, which then becomes sites of microcracks. The microcracks eventually line up and grow to form the major crack. This phenomenon indicates that the biocomposite is not brittle and the strain accommodation mostly takes place between $CaCO_3$ layers, not through the ceramic layers. We do not yet know whether the interface is ruptured or the rupturing takes place within the biopolymer itself. Nonetheless, it is likely that this complex deformation mode, similar to that in metals, may be the main mechanism of energy absorption during the application of stress and may be responsible for the many orders of magnitude increase in the toughness of the composite.

Sliding takes place when there is a resolved shear stress acting in the plane of the layers. However, if the resolved shear stress has a tensile component, i.e.,

[1] $\{hkl\}$ refers to a family of crystallographic planes, hkl being the indices of these planes. (111) indicates a specific plane. [UVW] refers to a specific direction, $\langle UVW \rangle$ refers to a family of directions.

normal to the layers, then the $CaCO_3$ layers are forced to separate. This separation is resisted by the formation of organic ligaments that stretch across the organic layer to form bridges, anchored to the platelets, between the inorganics layers. It has been suggested that the organic matrix has a chitin component in the center (Watabe 1965; Weiner 1986); if so, chitin would have a plywood or an accordion-type structure which would then allow stretching during the bridging process (Sarikaya et al. 1990a,b; Gunnison 1991). The fact that there is more than 1000% stretching of the ligaments and that the organic layer is still intact is a strong indication that the interface between the organic and inorganic components of nacre is strong. Still another significant factor is that the organic layer allows either sliding or ligament formation, depending on the condition of the resolved applied stress (Sarikaya et al. 1990a,b), indicating that the ultra-architecture of the organic layer is a critical design parameter in the making of this ceramic-polymer composite and thus a subject of detailed studies (Crenshaw 1972; Weiner and Hood 1975; Weiner and Traub 1980; Greenfield et al. 1984; Addadi and Weiner 1985; Mann 1988).

2.3 Strengthening Mechanisms

The strength of nacre may be related to several factors, including the size and structure of the aragonite platelets and the interfaces between the inorganic and the organic components. If the size of the aragonite platelets is compared to the overall toughness of the nacre, it can be seen that from the limited thickness of the largest flaw, i.e., the thickness of the platelet, 0.5 μm, (Griffith 1920) the increase in the fracture strength of aragonite would be < 50 MPa. This value is not high enough to account for the fracture strength of 185 MPa of nacre measured in our studies. Nor can the rule of mixtures account for the value of the strength measurement (Hull 1981).

The best explanation that we can presently put forward for the high strength of nacre is that the tensile stress applied to nacre is transferred to compressive stress during loading. This may be evident from the inward bulging of the edges of the microindentation (Laraia and Heuer 1990; Gunnison 1991). This energy stored in the organic component results in dilation during load release. The failure of the composite, therefore, seems to take place during unloading. The overall cycle of loading/unloading is schematically represented in Fig. 4 (Gunnison 1991). During loading of the indentor, the sample is placed in the edge-on configuration and, hence, the applied stress is distributed in the directions indicated in the figure: One component, $\vec{\sigma}_z$, in the direction parallel to the platelets and the other, $\vec{\sigma}_x$, perpendicular to them. The former is dissipated by the sliding of the aragonite platelets on the organic matrix, as depicted in Fig. 4c, d and f. The perpendicular stress is absorbed by the compression of the organic matrix, and hence, reduces the actual stress applied to the platelets. The compression of the organic matrix acts as an energy storage mechanism during the application of the force, and the energy is released during the removal of the indentor, thus displacing the aragonite crystals

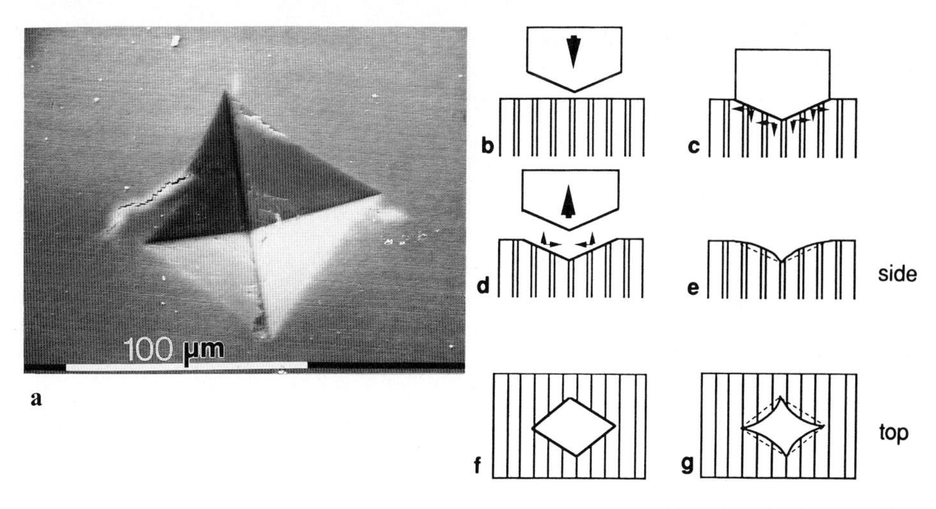

Fig. 4. **a** SE image of indentation in nacre in edge-on configuration, displaying inward bulging, and **b–g** schematic illustration of the structural changes that might occur in nacre during indentation. (Courtesy of K.E. Gunnison)

towards the indentation opening (Fig. 4d and e). This would then result in inward bulging of the sides of the indentation (Fig. 4a and g), at which time failure might occur. As seen in Fig. 4a, the side of the indentation which displays the largest bulging is also the side which possesses the largest crack. In the present system, this failure mechanism could effectively increase the overall strength of the composite many orders of magnitude as the major stress applied to the inorganic component is actually compressive. Such increases are a common occurence in the strength of ceramics if the applied stress is compressive (Dawridge 1979). To ensure the transferrability of stress, further studies are necessary to elucidate the exact mechanism of strengthening in nacre.

3 Design Guidelines for Processing Biomimetic Laminated Composites

The unusual mechanical properties of nacre as compared to synthetic ceramics and composites may be due to: (1) the intrinsic properties of the constituent phases (brittle inorganic and soft organic phases); (2) the highly ordered organization of ceramic and biopolymer layers, and their detailed structures, including the interface structures and properties; and (3) the sizes of the ceramic and biopolymer layers. Despite the fact that many of the structural features in abalone, especially in the soft tissue, are not yet known, from our knowledge of their excellent mechanical properties and our knowledge of their structural architecture at the micro- and submicron levels, some guidelines may be drawn to aid in the processing of synthetic microstructures via biomimicking. These biomimicked synthetic laminated composites are expected to have superior properties over those that were achieved in synthetic composites prepared by traditional processing approaches. Based on studies of nacre (Jackson et al. 1988; Sarikaya et al. 1990a,b), the design of high-fracture/high-toughness synthetic composites should incorporate the following:

1. a laminate thickness of the hard and brittle component of less than 1 μm and a soft component of less than 1000 Å, with an approximate ratio between 5:1 and 10:1;
2. a highly plastic soft phase (deformability >100%);
3. strong interfaces between the soft and the hard phases (so that the interface does not fail during crack propagation); and
4. the ability of the soft phase to bind to the surfaces of the hard component (strong interfacial bonding) and provide either plasticity to the overall composite structure (for sliding and, hence, metal-like deformation behavior) or form ligaments to constrain crack opening (through crack bridging), depending on the resolved applied stress.

For the practical processing of laminated composites, these guidelines may be difficult to apply. For example, there is no practical way of producing structural laminates with laminate layer thicknesses thinner than about 10 μm (Yasrebi et al. 1990). Also, it is difficult, if not impossible, to control the thickness ratio of the hard and soft components at these small dimensions. However, the guidelines stated above are still useful in the sense that they present the ultimate structural features that should be achieved for the submicron and nanolaminate design. In the design of synthetic laminates, high hardness ceramics, such as BN, B_4C, TiC, ZrO_2, and Al_2O_3, can be used as the brittle component. Highly plastic (superplastic) metals and alloys, such as Al and Cu (and their alloys), or organic polymers, such as polyethylene and polypropylene, may be good candidates for the soft phases. Those constituent phases which give the best combination of properties, with emphasis on strong interfaces, can be used provided that they can be processed.

Some of the property requirements stated above were met to a certain degree in the B_4C-Al (Yasrebi et al. 1990; Aksay and Sarikaya 1991; Yasrebi M, Milius DL, Aksay IA, unpubl. research) and B_4C-polymer (Khanuja 1991; Aksay and Sarikaya 1991) laminated systems designed to be used as impact-resistant materials. In one synthesis strategy for processing of B_4C-Al composites, porous B_4C layers, with thicknesses below 100 μm and as low as 15 μm, were tape cast with thin Al sheets in between. The stacks were then heated to induce infiltration and to allow bonding between Al and B_4C without an excessive reaction. The resulting composite displays a structure where both the Al-infiltrated B_4C layers and pure Al layers in between are in the form of continuous films (Yasrebi et al. 1990). The overall architecture provides alternating layers of hard and soft components and strong interfaces. As a result, mechanical properties in terms of fracture toughness and fracture strength both show a 30–40% increase over the monolithic B_4C-Al composite with the same Al composition (Fig. 1), and a structure where Al and B_4C have a three-dimensional interpenetrating network (Halverson et al. 1989). The properties in B_4C-Al laminates are currently the best that can be achieved among the present cermet systems.

As seen from the above example, despite the fact that improvements were achieved in the mechanical properties of laminated composites based on biomimetic architecture, these improvements have not been as extraordinary as when nacre is compared to monolithic $CaCO_3$. This may be due mainly to the fact that the laminate layers are not thin enough; thicknesses below 1 μm in the inorganic layers and below 100 nm in the layers of the organic or the soft phase are needed. Second, as will be clear from the discussion in the following section, both the inorganic and organic layers have complex structures, in terms of their crystallography, substructures, and morphology. In particular, it has been suggested that the organic matrix might actually have a sandwich structure containing three organic sublayers, each having its own unique composition and molecular conformation (Watabe and Wilbur 1960; Weiner and Traub 1981, 1984; Nakahara et al. 1982). Neither the composition of these layers nor their structure have yet been clearly identified (Lowenstam and Weiner 1989; Mann et al. 1989; Simkiss and Wilbur 1989). In addition, identification of the structural relationship between the organic and the inorganic layers is far from complete (Greenfield et al. 1984; Addadi and Weiner 1985; Addadi et al. 1986; Mann 1988). Therefore, a part of our research has been directed towards the study of the structure of nacre at the nanometer scale in order to answer some of these questions (Sarikaya et al. 1992). In the following sections, we summarize the current understanding of the structural relationships in nacre. First, we present a summary of the crystallographic and morphological relationships between the building blocks which make up the inorganic component of nacre. From this study, we construct a possible structure of the underlying organic matrix, which we believe will shed more light on the areas where future studies should be focused in order to understand the structural relationships at the molecular scale.

4 Structure of the Red Abalone Shell: Prismatic and Nacreous Layers

4.1 Structure of the Shell at the Macro Level

Red abalone; *H. rufescens*, belongs to the mollusk species in the family of gastropods. Gastropods first evolved about 500 million years ago and have very successfully survived to the present time with little change (abalone is thus considered a "living fossil") (Wilbur and Yonge 1964; Morton 1979). The shell of the abalone is ear-shaped, with a large opening and a small spire. The dorsal surface is convex and exhibits growth rings. The inner surface has two scars at the position of muscle attachment and is irridescent as it consists of nacreous layers. The shell has respiratory pores (about 20 in the adult) but only the last four to five are open at a given age. The organism has a large foot with which it grabs onto

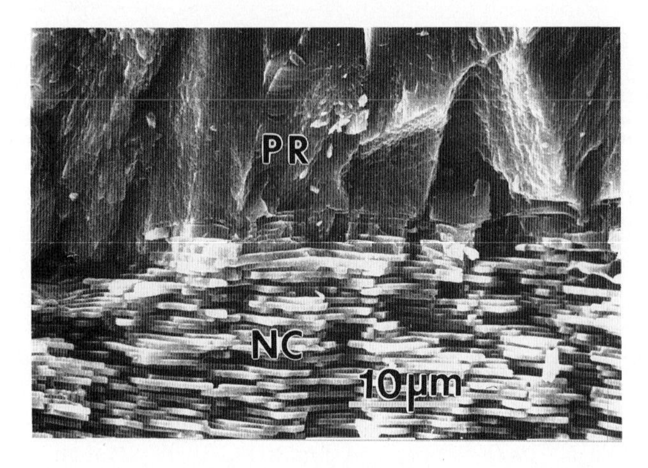

Fig. 5. SE image of a cross-section of abalone reveals the outer prismatic (*PR*) and inner nacreous (*NC*) layers

a rock when it is juvenile and forages as a bottom-dwelling mollusk on algae and plants. The shell grows about 2.5 cm per year and the diameter of the shell can reach 30 cm, after which the shell only thickens (more than 1 cm).

A longitudinal cross-section of the red abalone shell displays two types of microstructures: an outer prismatic layer and inner nacreous layer (Fig. 5). Two forms of $CaCO_3$, calcite (rhombohedral, R $\bar{3}$ c) and aragonite (orthorhombic, Pmnc), constitute the inorganic phase of the composite in the prismatic and nacreous layers, respectively. The structure and properties of the nacreous layer are described in this chapter, since it is this part of the shell that displays an excellent combination of mechanical properties as a result of its highly ordered hierarchical structure (Currey 1987; Jackson et al. 1988; Sarikaya et al. 1990a,b). Furthermore, the structure of nacre is comparatively simple, having alternating layers of organic and inorganic components. This microarchitecture, especially the highly ordered inorganic component, i.e., the aragonite platelets, is more straightforward for studying crystallography with the aid of transmission electron microscopy (TEM).

4.2 Structure of the Nacre at the Nanometer Level

4.2.1 Morphology of Inorganic Phase: Aragonite

The general structure of nacre is seen in Fig. 5, which shows a secondary electron image of an obliquely broken sample recorded by SEM. The nacre is composed of stacked platelets throughout the thickness of the sample. Of the two types, columnar or sheet forms, nacre in abalone belongs to the former (Wilbur and Simkiss 1968; Wise 1970; Currey 1987). The stacking is not random and in a fully grown specimen resembles a "brick and mortar" microarchitecture with the aragonite phase in the form of platelets forming the bricks and the organic matrix

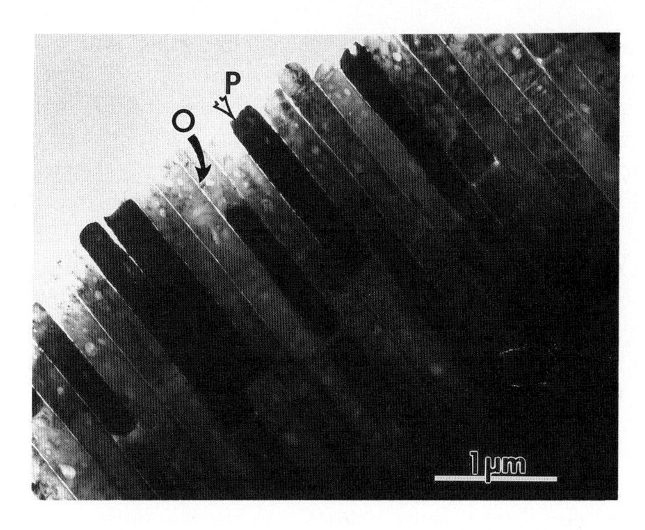

Fig. 6. A TEM image showing edge-on view of the nacre section of red abalone shell with aragonite ($CaCO_3$) platelets, *P*, and thin film organic matrix, *O*

forming the mortar in between. As illustrated by the TEM image recorded in an edge-on configuration in Fig. 6, the thickness of the platelets is about 0.25 µm and the organic matrix is between 10 and 50 nm in thickness, depending on the site of the shell from which the sample is extracted. As described by other researchers (Nakahara et al. 1982; Greenfield et al. 1984; Weiner and Traub 1984), the organic matrix is thought to have a sandwich form, with chitin in the middle, acidic macromolecules surrounding it, and soluble macromolecules as the outer layers adjacent to the platelets. There is some speculation on the conformation of the organic matrix and its composition (Crenshaw 1972; Addadi and Weiner 1985); however, neither direct evidence of the conformation of the macromolecules nor a quantitative analysis of the macromolecules has been possible to date.

4.2.2 Crystallography of Aragonite Platelets

Fundamental knowledge of the overall structural relationships between the organic and inorganic components in the closely knit structure of nacre (and in other hard tissues) is essential from the point of view of understanding, first, its mechanical properties, and second, its mechanism of formation (see, for example, Degens 1976; Mann 1988; Weissbuch et al. 1991). A more direct way to achieve such an understanding is to decipher the crystallographic relationships between the inorganic phase and organic matrix if the organic matrix also forms a crystalline lattice. Crystallinity of the organic matrix has been postulated in previous studies from the fact that it contains chitin and silk fibrion-like proteins as the outer and middle sublayers in the sandwich structure and are known to self-assemble in crystalline arrangements (Nakahara et al. 1982; Weiner and Traub 1984). In the organic matrix, the outermost sublayer, the one in contact with $CaCO_3$, consists mostly of soluble charged macromolecules (proteins) and presumably acts as the binder to the inorganic phase. In this scheme, it is suggested that active sites on the matrix align with those on the $CaCO_3$, i.e., either Ca^{2+} ions (Weiner 1986) or CO_3^{2-} sites (ionotropy mechanism) (Greenfield et al. 1984). To date, it has been impossible to study both the organic and inorganic crystals simultaneously, and the structural relationships between the components of the nacre exist only as a conjecture (Watabe and Wilbur 1960; Meenakshi et al. 1974; Weiner and Hood 1975; Weiner and Traub 1980, 1984; Nakahara et al. 1982). Bulk studies performed by X-ray diffraction on the nacre revealed that the aragonite platelets are organized with their [001] axis perpendicular to the layers (Crenshaw 1972; Weiner and Traub 1984; Addadi and Weiner 1985). It has been postulated that the $\vec{a}$ and $\vec{b}$ axes, within the layer plane, are randomly oriented in each platelet. Furthermore, from this scheme, it was assumed that each aragonite platelet grows on the crystallographically related organic template, which itself has a local random orientation (Addadi and Weiner 1985; Weiner 1986). From the composition of the insoluble fraction of the organic matrix, i.e., the inner crystalline sublayers, which contain a high fraction of aspartic and glutamic acid, it might be possible to deduce a self-assembled structure that would be related epitaxially to the aragonite lattice along the [001] projection (Weiner et al. 1983; Weiner and Traub 1984).

In our studies, we followed an approach in which each aragonite crystallite was analyzed separately and its crystallographic orientation relationship was established with respect to its neighbors, both on the same layer and across the thickness of the nacre (Sarikaya et al. 1992). This was accomplished by high spatial resolution electron microdiffraction in a TEM which allowed isolated diffraction patterns from individual aragonite crystals (electron probe size as small as 50 nm diam). By such an approach, the overall crystallography of the inorganic component of the composite was established in three directions. Assuming that there is a structural relationship between the inorganic and organic components, the organic matrix, which surrounds the aragonite platelets on all sides throughout the nacre, may provide a framework by retaining all the crystallographic relationships. This scheme would then imply a lattice structure within the organic layer itself that retains the crystallographic relationship throughout the nacre structure (Sarikaya et al. 1992). To establish the crystallographic orientation relationship between the adjacent aragonite platelets, microdiffraction studies were performed both in the edge-on and face-on configurations of nacre (Sarikaya et al. unpubl.). The diffraction patterns shown in Fig. 7b–e are from regions circled in the bright field (BF) image in Fig. 7a which was recorded from nacre in the edge-on configuration. It can be seen that all the patterns belong to the same zone axis, or electron beam direction, i.e., [010] of the orthorhombic crystal. However, there is a slight orientation change between them, ranging from 0 to 5°. This misalignment is with respect to the [001] axis which is perpendicular to the flat faces of the platelets, as shown by the arrow in Fig. 7a. Therefore, the platelets are slightly rotated with respect to the [001] axis through the centers of their flat faces. It should be noted that this rotation does not follow a regular pattern but it is random, following a zigzag pattern as depicted in the schematic illustration in Fig. 7f–g.

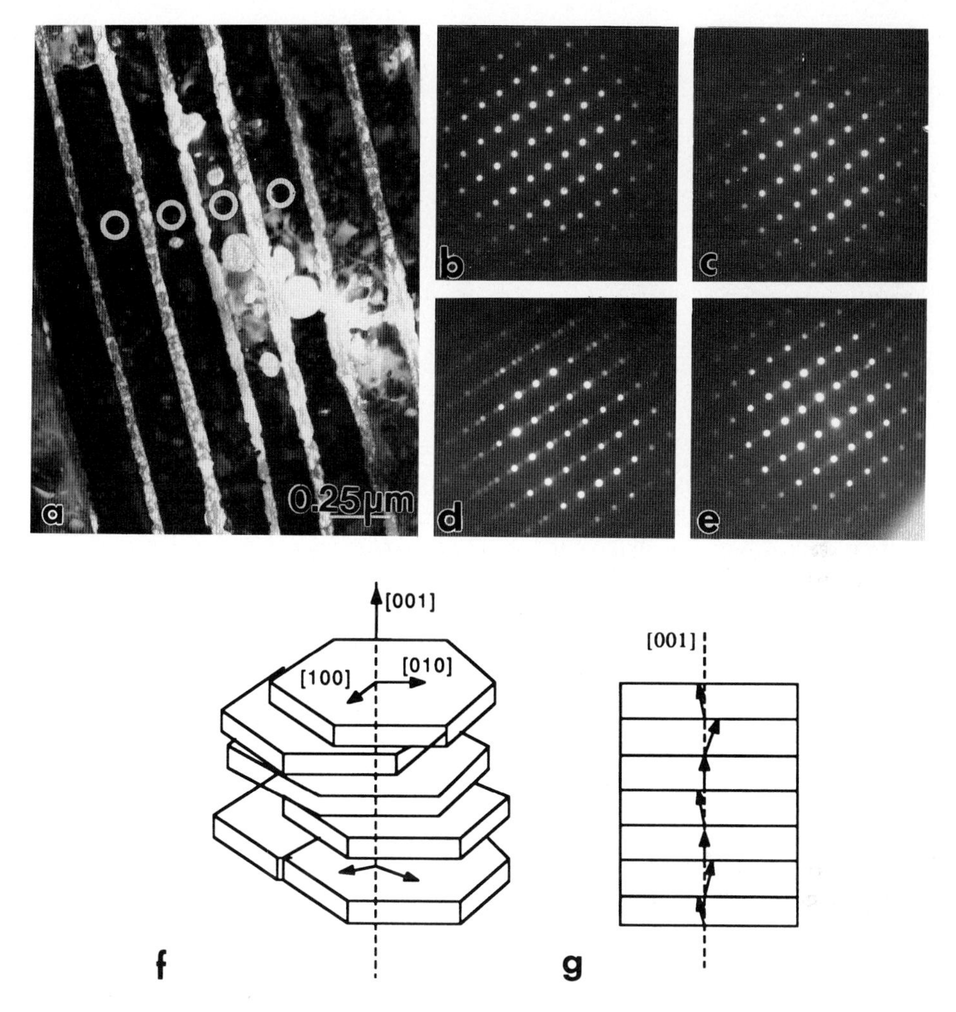

Fig. 7. Electron microdiffraction patterns (**b–e**) taken from each of the aragonite platelets in edge-on configuration shown in **a. f** and **g** are schematic illustrations of the misorientation of each platelet along the [001] direction

In the face-on view, each layer of the nacre is composed of closely packed platelets having four, five, or six edges, which may be curved rather than straight. Our study focused on the microstructure of these platelets and their crystallographic and spatial organization with respect to each other on the same layer of the nacre to draw a conclusion from their crystallographic relationships about the nature of the underlying organic matrix.

As analyzed above, adjacent platelets belong to the same [001] zone axis, but there is a slight rotation among the platelets about this axis with respect to each other. The question remains whether there is any crystallographic relationship between the $\vec{a}$ and $\vec{b}$ axes in each of the platelets on the same layer. The

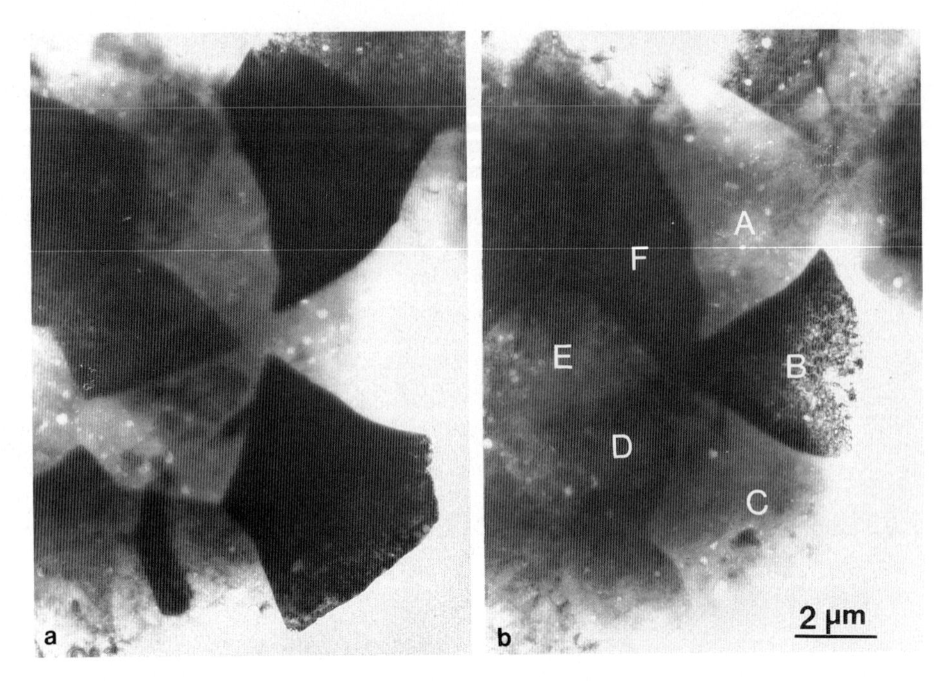

Fig. 8a, b. Face-on image of nacre showing twin-related (*A* to *F*) aragonite platelets before (**a**) and after (**b**) slight tilting. (Sarikaya et al. 1992)

crystallographic relationship among the adjacent platelets in the face-on configuration was also analyzed by electron diffraction, which revealed that each platelet is actually twin related to the one next to it. An example is presented in Fig. 8, where six platelets from a hexagonal arrangement with an approximate 60° angle between each pair. The analysis of the electron diffraction patterns taken from each platelet indicated that they all have approximately the same zone axis. The diffraction patterns taken from the boundaries between each pair of platelets, however, reveal that each platelet is twin-related to the other in a given pair and that the twin plane is of the {110} type. Therefore, in this figure, the platelets A, C, and E are twin-related with platelets B, D, and F, respectively. The images in Fig. 8a and b were taken by slightly tilting the sample (a few degrees) to bring each of the three platelets into a strongly diffracting condition so that they give a dark diffraction contrast, respectively, in each image. This brings about the platelets which are all in the same orientation, and the remaining three in twin orientation. This arrangement would indicate that all the platelets on the same layer are twin-related to each other whether they share a boundary or not. Twinning among the platelets is called *first generation twins* since this twinning takes place at the largest spatial scale.

Close analysis indicates that each platelet actually consists of several domains which are crystallographically oriented specifically with respect to each other, as shown in the BF images in Fig. 9a and b, which were recorded before and after slightly tilting the sample about the [001] zone axis. Electron diffraction patterns

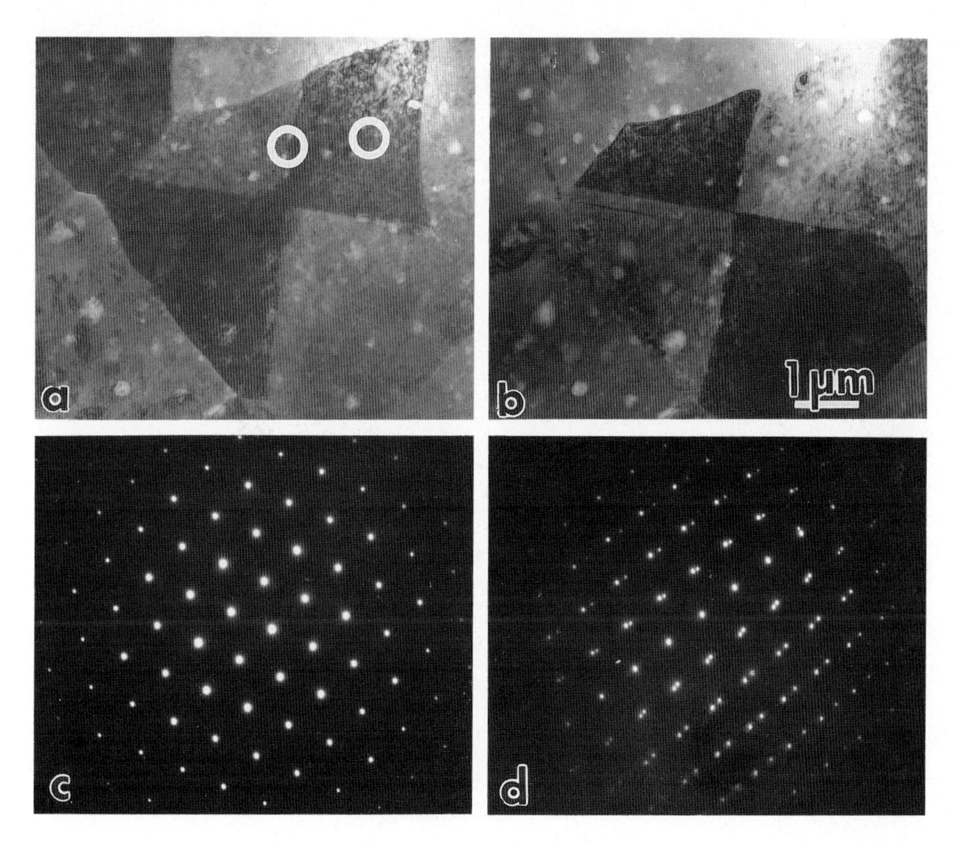

Fig. 9a and **b.** TEM BF images illustrating twin formation between 90° domains in a platelet. **c** and **d** electron diffraction patterns from the interior and boundary, respectively, of a domain. (Sarikaya et al. 1992)

taken from the interior and from the interfaces of the domains in a platelet are given in Fig. 9c and d. The single crystalline pattern in Fig. 9c is from the interior of a domain and is indexed to be in the [001] electron beam direction. On the other hand, the pattern in Fig. 9d, which was taken from the boundary between two domains as outlined by the circle in the figure, indicates that it has two superimposed patterns. Analysis of this selected area diffraction (SAD) pattern reveals that the two superimposed patterns can be correlated to each other with a twin relationship with the twin plane being $\{110\}$ parallel to [001] direction of the unit cell, i.e., either (110) or $(1\bar{1}0)$. In fact, the SAD pattern recorded from all of the domain boundaries shows the same twin reflections, indicating that each domain is related to the one next to it by a $\{110\}$ twin relation. We call these twins *second generation twins*. The analysis of the twin relationship indicates that the $\{110\}$ twin planes are parallel to the outer edges of the platelets.

The number of domains in each platelet can be either six or four. Where there are six domains in a platelet, the boundaries between each pair of domains do not follow any specific crystallographic plane and are incoherent. Four-domained platelets are also frequently observed, as shown in Fig. 9; we call these platelets

90°-domained, as opposed to 60°-domained platelets. In the case of 90°-domained platelets, opposite domains are in the same orientation and are twin-related to the remaining two, with two of the four boundaries being atomically coherent.

In an ideal hexagonally shaped platelet having six twin-related domains, the angle between each pair of domains must be 60°, with six domains completing 360° for the whole platelet, as schematically illustrated in Fig. 10a. This is not possible since the outer edges of the platelets are parallel to {110} planes and the angle between each pair of planes, for example between (110) and $(1\bar{1}0)$, is 63.5° (Fig. 10b). This leaves 3.5° unaccounted for and, as this induces strain into the aragonite matrix, therefore it must be accommodated by some structural deformation. Further imaging of the microstructure at higher magnifications shows that each domain actually contains two sets of nanometer scale twins, each forming on (110) or $(1\bar{1}0)$ planes.

These ultrafine twins, similar to the growth twins in geological minerals, are shown in a BF image in Fig. 11a and in the atomic resolution image in Fig. 11b forming on some crystallographic planes making an angle of about 63.5°. Based on a structural model, the accommodation of the 3.5°-strain is possible by the formation of these nanometer-scale defects on {110} planes. If so, the lattice in each domain will be plastically deformed outwards from the boundaries, i.e., towards the

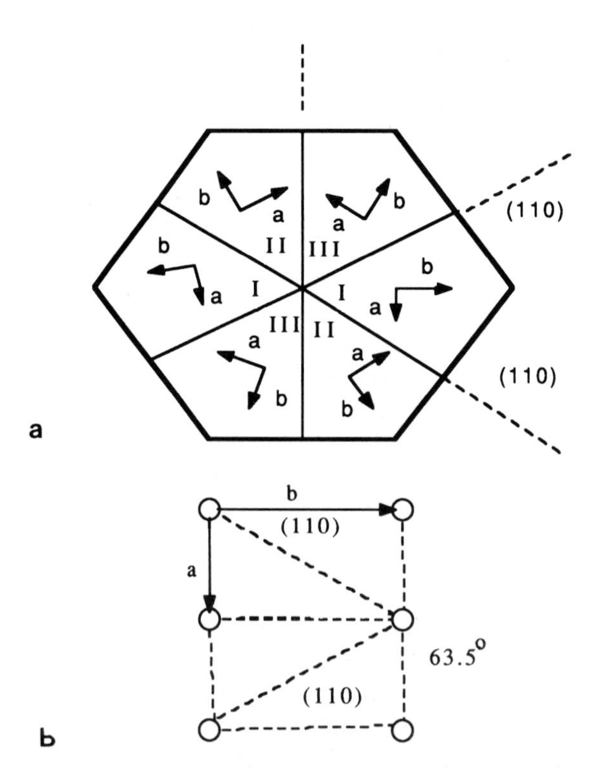

Fig. 10a, b. Schematic illustration of 60° domains in an aragonite platelet, *a* and *b*

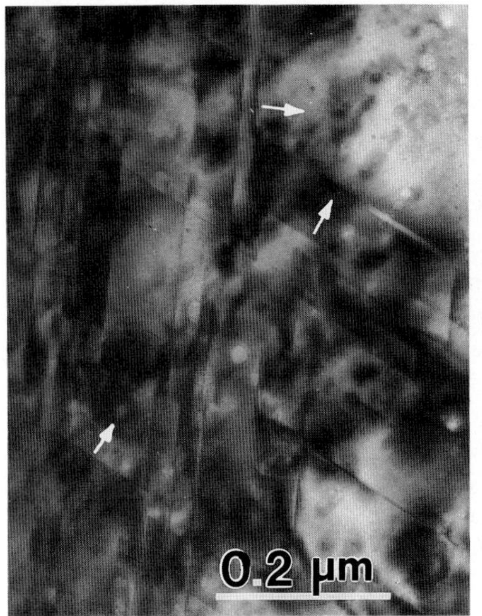

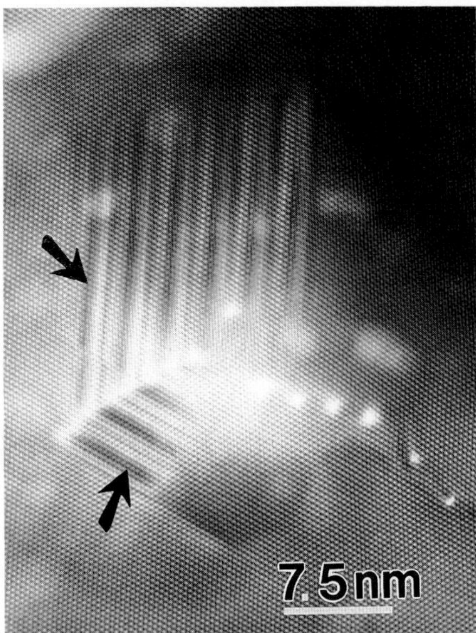

Fig. 11. a TEM image of domain in an aragonite platelet, showing third-generation twins. **b** Atomic resolution electron microscopy image of third-generation nanometer scale twins within a domain

periphery of the platelet. In fact, in most cases, the outer edges of the domains, are curved, with the apex in the middle of the outer edge of the domains, whereas the region at the edge where two domains meet is inwardly curved. We call these ultrafine defects *third generation twins* since they occur at the smallest scale. It must be noted that some of the lattice stress created by the 3.5°-strain can also be accommodated by the misalignment of adjacent domains, as observed. However, this cannot account for all the strain accommodation, as the interfaces between the domains show a high degree of coherency. The above analysis done on 60°-twinned platelets is also correct for 90°-twinned platelets and that the stress created due to the misorientation of domains is accommodated by the formation of similar ultrafine third generation twins.

In summary, we find that there are three scales of twins in the face-on configuration of the nacre section of red abalone shell (Sarikaya et al. 1992): (1) first generation twins among platelets having incoherent boundaries, (2) second generation twins between domains having coherent boundaries within a given platelet, and (3) nanometer-scale third generation twins within the domains. These three twin structures cover a size scale of six orders of magnitude, from the nanometer to the submillimeter, and reveal for the first time a hierarchical structure for hard tissue in a biological material. In the following section, we analyze the relationship of these hierarchical twins and their organization in the form of a space-filling tiling system, a system which will further support our hypothesis of a conformation of the underlying organic matrix that takes into account all the observed crystallographic and geometrical relationships among the aragonite platelets both in the face-on and in the edge-on configurations.

5 Structure of the Organic Matrix and its Relationship to the Inorganic Phase

As is evident from the results of the analyses in Section 4, the aragonite phase in nacre can be crystallographically correlated through three levels of twin relationships in the structural components of the nacre inorganic phase, beginning at the nanometer scale and proceeding through the millimeter scale. Although sixfold twin structures also occur in geological aragonite (Wenk et al. 1983), the hierarchical arrangement in nacre is unique in the sense that each platelet is separate from the others during the early stage of growth (Liu J, Sarikaya M, and Aksay IA; unpubl. research). On the other hand, in geological aragonite the mimetic twin domains grow, in contact, one after another and are influenced by the presence of each other (Bragg 1924). Even after crystallization is complete, all the platelets in nacre are still separated from each other by an organic membrane (Nakahara et al. 1982; Sarikaya et al. 1990a,b). The fact that separate platelets grow simultaneously and yet have a definite crystallographic orientation relationship suggests that the growth process might be mediated by the organic template below the individual crystals, as proposed earlier (see, for example, Weiner et al. 1983; Mann 1988; Mann et al. 1988; Lowenstam and Weiner 1989; Simkiss and Wilbur 1989). In the following paragraphs, we discuss how these hierarchical twins might originate and the implication this has for the structure of the organic template on which the aragonite crystals are grown.

The interaction between the organic species and inorganic crystals must include geometrical, electrostatic, and stereochemical interactions (Addadi and Weiner 1985; Greenfield et al. 1984; Mann 1988). Therefore, the nucleation and growth of the crystals will be influenced by both the nearest and higher order interactions. The aragonite crystal structure belongs to the space group Pmcn (No. 62) with lattice parameters $a = 4.94$ Å, $b = 7.94$ Å, and $c = 5.72$ Å. With the number of formula units per unit cell being four, Ca^{2+} ions occupy positions 1/4, 0.08, 0.25, C at 0.25, 0.75, 0.08 and O(1) and O(2) positions at 0.25, 0.59, 0.08 and 0.02, 0.82, 0.08, respectively. If the origin is taken as the center of symmetry, the mirror plane, the c-glide plane, and n-glide plane are at $= 1/4$, $y = 1/4$ and $z = 1/4$, respectively. In the [001] projection, the Ca^{2+} ions would give the crystal a pseudo-hexagonal symmetry, as shown in Fig. 12a, and the CO_3^{2-} groups reduce the symmetry to an orthorhombic form (Sarikaya et al. unpubl.). This is physically an important characteristic in terms of the crystallographic relationships of the hierarchical twinned components of the nacre aragonite and the stereochemical relationship that might exist between the aragonite and the macromolecules in the organic matrix. The nucleation and growth of the aragonite crystals may involve both Ca^{2+} and CO_3^{2-} ions, but for simplicity, only the arrangement of Ca^{2+} ions will be illustrated in this chapter. As outlined in Fig. 12, if the array Ca^{2+} ions are

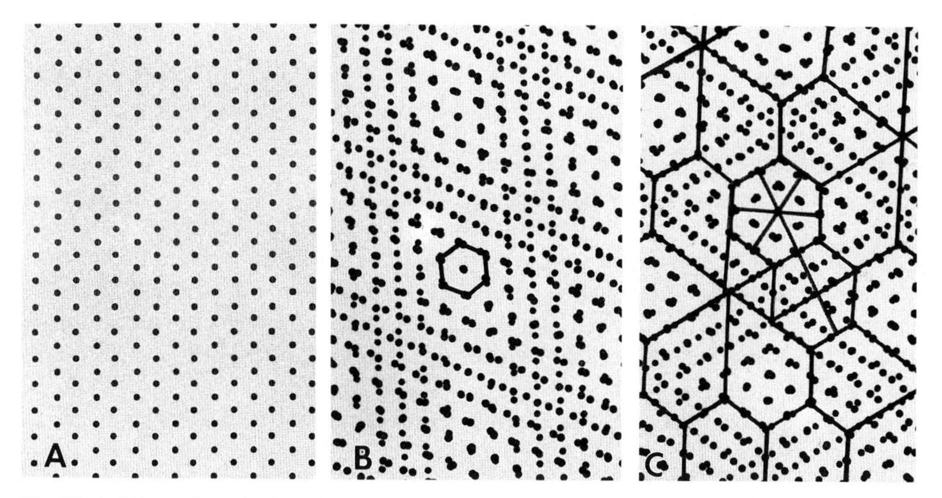

Fig. 12. A Schematic projection of the aragonite lattice in [001] orientation with Ca^{2+} ions highlighted. **B** Schematic illustration of the possible formation of the superstructure on the organic matrix formed by the superimposition of three aragonite lattices in [001] projection. **C** Possible formation of aragonite platelets (tiles) on the superstructure

considered only, they are in an approximate hexagonal closed-packed formation. True hexagonal closed-packed formation would require an axial ratio of $b/a = 1.73$; however, the observed value is 1.69. In the actual aragonite lattice, the Ca^{2+} ions are not in contact but are separated by O ions, and the pseudo-hexagonal arrangement refers to the centers of Ca^{2+} ions rather than their actual packing. Since the same {110} twinning takes place at all length scales, superimposition of the lattices on all three possible twins with a 63.5° angle with respect to each other generates a new superlattice structure, Fig. 12B (Sarikaya et al. 1992). We call this new lattice a *superstructure*, which, in this projection, has a pseudo-hexagonal configuration of fixed lattice points. Taking the actual distance between the Ca^{2+} ions as 5.0 Å, the distance between these lattice points would be about 30 Å. If the nucleation and growth of the aragonite platelets take place on the underlying organic matrix, the geometrical configuration of the active sites for the binding of the Ca^{2+} ions on the organic matrix must accommodate this superlattice and, thus, all twins in the nacre. One possible solution is to assume that the binding sites on the template form a single crystalline pseudo-hexagonal lattice, or integer multiples of it, which is a reasonable assumption since many two-dimensional membranes tend to form hexagonal lattices during self-assembly (Unwig and Henderson 1985). The hypothesis that the organic matrix may be a single crystalline is an essential feature of the structure for the formation of the highly organized tiles in nacre. A local crystalline organization of the matrix, as proposed earlier (Weiner and Traub 1984; Addadi and Weiner 1985, 1990), with no relationship between the neighboring areas and, hence, without long-range order, would result in the formation of aragonite crystals without any definite crystallographic orientation relationship among them.

By tracing along the possible twin boundaries, one can see that the superlattice shown in Fig. 12C allows the generation of the overall hierarchical twin structure. One construction is illustrated in Fig. 12, which contains all the shapes, geometry, and crystallography-related features discussed in this chapter, such as five-edged platelets with 90°-domains, sixfold symmetry of platelets, and six- and three-edged domains (Sarikaya et al. 1992). The fact that one can generate all the possible configurations in this way again illustrates that a pseudo-hexagonal template structure might be a possible solution for the structure of an organic matrix that can accommodate all the twin relationships, rather than the lattice of a single domain. The ultrastructure in the organic matrix would not only be single crystalline on the flat surface but also through the transverse direction (through-thickness direction) in nacre.

The geometrical and crystallographic model of aragonite platelets proposed in this chapter is referred to as multiple tiling in mathematics (see, for example, Grünbaum 1987). It appears that in nacre, nature utilized this mathematical technique to form a highly ordered structure that is compatible with both soft tissue and the crystalline structural constraints of hard tissue. Tiling may also play an important role in providing the overall shape of the nacre and its mechanical properties. Many mollusks, such as gastropods, cephalopods, and bivalves, have aragonite platelets as the fundamental building blocks in nacre, but have grossly different overall shapes (Nicol 1960; Wilbur and Yonge 1964; Morton 1979). For example, in red abalone, the shell is quite flat and thick; in nautilus, the shell is round and thin, forming an elegant chambered structure in which even the separation walls of the chambers are made of nacre. In all these nacre structures, aragonite platelets more or less have the same dimensions, about 0.2–0.5 μm thick and about a 5–10 μm edge length (Sarikaya et al. unpubl.). Yet the multiple tiling of the platelets and the crystallographic relationship between the platelets may be different due to slightly different structures and compositions in their underlying organic matrices. Further studies are required on various species of these organisms, both on the crystallography of the mineral component and on the structural and compositional analyses of the organic matrices, in order to decipher their structures and the unifying, underlying principles for the organization and formation of the various shapes of shell containing nacre structures. It should be clear that a fundamental understanding of these structures is essential for the possible formation of synthetic composites through biomimetics.

6 Summary and Future Directions

We have reviewed the results of most recent studies on the mechanical properties and structure of the nacre section of red abalone. Nacre, which is about 95 vol% aragonite and about 5 vol% organic macromolecules, has fracture toughness and fracture strength properties that are orders of magnitude higher than those of monolithic aragonite. From analysis of the toughening and strengthening mechanisms, one can conclude that the unique structure of nacre in which the inorganic and organic components have a high degree of organization not encountered in synthetic materials is responsible for these superior properties. However, our current understanding of the mechanisms for toughening and strengthening is far from complete.

Morphological and crystallographical analysis by electron microdiffraction of the inorganic phase, aragonite, indicates that the individual aragonite platelets form a multiple tiling system in nacre in which twins form hierarchical defect structures which control the overall structural order. Assuming that crystal-matrix recognition is applicable, a model is forwarded for the structural conformation of the active sites in the organic matrix which may then act as the template for the formation of nacre. This model, called the superstructure, explains all the experimentally observed crystallographic and morphological relationships in the aragonite phase.

This new approach, in which the structure of an organic material is indirectly determined from detailed knowledge of the structural relationships among the subcomponents of the inorganic phase, may prove to be a viable approach for other biocomposites in which the structure of the inorganic is highly organized. This approach would not only serve as a new methodology for studying how the overall shape is determined in various species of mollusks but also for understanding general biomineralization concepts in single and multicell organisms.

Despite considerable effort in the field, our understanding of the mechanisms that operate in nacre to make it a tough and strong composite, and our understanding of the composition and structure of the organic matrix and its structural relationship with the inorganic phase are still limited. Some of the major issues from which an agenda for future research may be formed are as follows:

1. *On Mechanical Properties.* A complete understanding of the toughening and strengthening mechanisms in nacre is necessary because of their dependence upon the structural relationships between the organic and inorganic phases. Also, mechanical property evaluation of the overall shell, particularly under the dynamic conditions in which the organism lives and makes use of its multifunctional characteristics, is essential for the design of multifunctional materials via biomimicking.
2. *On the Organic Matrix.* The protein compositions, conformations, and structural relationships of the organic matrix and its relationship to the inorganic aragonite phase are all necessary parameters for consideration in the design of synthetic materials. The relationship between the organic matrix and calcite in the prismatic layer should also be investigated.

   Items (1) and (2) will lead to structural design rules for biomimicking new materials.
3. *On the Mechanism of Growth.* An understanding of the nucleation of the inorganic phase in the presence of an organic matrix, beginning with the embryonic and juvenile stage of the abalone, and the degree to which the organic matrix controls growth, will assist in understanding the overall formation and shaping of the shell.
4. *On Bioduplication.* Self-assembly studies involving the in vitro isolation, purification, and assembly of various components of the organic matrix, separately

and in combination, for the mineralization of $CaCO_3$ and other minerals under controlled conditions will lead to a fundamental understanding of how the synthesis might be controlled for the structures of hard biological components.

Designing and processing novel materials similar to the multifunctional, nanolaminated composites inspired by nacre through biomimicking approaches, and eventually through bioduplication, will have to wait until we answer the crucial questions regarding the structure and function of this system.

Acknowledgments. We acknowledge the technical input by Dr. Jun Liu. This work was performed under the sponsorship of Air Force Office of Scientific Research under Grant Numbers AFOSR-89-0496, AFOSR-91-0040, and AFOSR-91-0281.

References

Addadi L, Weiner S (1985) Interaction between acidic proteins and crystals: stereochemical requirement in biomineralization. Proc Natl Acad Sci USA 82:4110–4114

Addadi L, Weiner S (1990) Interaction between acidic macromolecules and structured crystal surfaces, stereochemistry and biomineralization. Mol Cryst Liq Cryst 13:305–322

Addadi L, Berkovitch-Yellin Z, Weissbuch I, Lahav M, Leiserowitz L (1986) A link between macroscopic phenomena and molecular chirality: crystals as probes for the direct assignment of absolute configuration of chiral molecules. In: Topics in stereochemistry, vol 16. Wiley, New York, pp 1–85

Aksay IA, Sarikaya M (1991) Bioinspired processing of composite materials. In: Soga S, Kato A (eds) Ceramics: toward the 21st century, Centennial Int Symp. Japanese Ceramic Society, Tokyo, pp 136–149

Alper M (1991) Enzymatic synthesis of materials: an overview. In: Alper M, Rieke PC, Frankel R, Calvert PD, Tirrell DA (eds) Materials synthesis based on biological process: proceedings of a symposium series. vol 218. Materials Research Society, Pittsburgh, pp 3–6

Berman A, Addadi L, Leiserowitz L, Weiner S, Nelson M, Kvik A (1990) A synchrotron X-ray study of a unique protein-calcite composite material. Science 250:664–667

Bevelander G, Nakahara H (1968) An electron microscopy study of the formation of the nacreaous layers in the shells of certain bivalve molluscs. Calcif Tiss Res 3:84–92

Blakemore RP (1975) Magnetotactic bacteria. Science 190:377–379

Blakemore RP (1982) Magnetotactic bacteria. Annu Rev Microbiol 36:217–238

Bouligand Y (1965) Sur une architecture torsadée repandue dans de nombreuses cuticules d'arthropodes. CR Hebd Sceances Acad Sci 261(12):3665–3668

Bragg WL (1924) The structures of aragonite. Proc R Soc Lond Ser A 17

Brown WF Jr, Srawley JE (1966) Plain strain fracture toughness testing of high strength metallic materials, ASTM technical publ, No 410. American Society for Testing and Materials, Philadelphia

Cottrell AH (1953) Dislocations and plastic flow in crystals. Oxford Univ Press, Oxford

Crenshaw MA (1972) Mechanism of normal biological mineralization of calcium carbonates. In: Nancollas GH (ed) Biological mineralization and demineralization. Springer, Berlin Heidelberg New York, pp 243–257

Crueger W, Crueger A (1982) Biotechnology. Sinauer Associates, Sutherland, Massachusetts

Currey JD (1974) The mechanical properties of some molluscan hard tissues. J Zool (Lond) 173:39–406

Currey JD (1976) Further studies on the mechanical properties of mollusc shell material. J Zool (Lond) 180:445–453

Currey JD (1977) Mechanical properties of mother of pearl in tension. Proc R Soc Lond B196:443–463

Currey JD (1987) Biological composites. J Mat Edu 9:[1–2] 118–296

Dameron CT, Reese RN, Mehra RK, Kortan AA, Carroll PJ, Steigerwald ML, Brus LE, Winge DR (1989) Biosynthesis of cadmium sulfite quantum semiconductor crystallites. Nature 338:596–597

Dawridge RW (1979) Mechanical behavior of ceramics. Cambridge Univ Press, Cambridge, p 15

Degens ET (1976) Molecular mechanisms on carbonate, phosphate, and silica deposition in the living cell. Top Curr Chem 64:1–112

Engel L, Klingele H (1981) An atlas of metal damage. Wolfe Science Books, Munich, Germany
Evans AG (1988) High toughness ceramics. J Mat Sci Eng A105/106:65–75
Evans AG, Marshall DA (1989) The mechanical behavior of ceramic matrix composites. Acta Met 37(10):2567–2583
Fendler JH (1987) Atomic and molecular clusters in membrane mimetic chemistry. Chem Rev 87:887–899
Frankel RB, Blakemore RP (1991) Iron biominerals. Plenum, New York
Frankel RB, Blakemore RP, Wolfe RS (1979) Magnetite in freshwater magnetotactic bacteria. Science 203:1355–1356
Glimcher MJ (1981) On the form and function of bone: from molecules to organs. In: Veis A (ed) The chemistry and biology of mineralized biological tissues. Wolff's Law revisited. Elsevier, New York Amsterdam, pp 617–673
Gosline JM, DuMont ME, Denny MW (1986) Structure and properties of spider silk. Endeavour 10(1):37–43
Greenfield EM, Wilson DC, Crenshaw MA (1984) Ionotropic nucleation of calcium carbonate by molluscan matrix. Am Zool 24:925–932
Grégoire C (1972) Structure of the molluscan shell In: Florkin M, Scheer B. (eds.) Chem Zool Academic Press, New York, pp 45–102
Griffith AA (1920) The phenomena of rupture and flow in solids. Philos Trans CCXXI-A:163–198
Grünbaum B (1987) Pattern and tiling. Plenum, New York
Gunnison KE (1991) Structure-mechanical property relationships in a biological ceramic-polymer composite: nacre. MS Thesis, University of Washington, Seattle
Haggin J (1988) Membranes play growing role in small-scale industrial processing. Chem Eng News 66(28):25–32
Halverson DC, Pyzik AJ, Aksay IA, Snowden WE (1989) Processing of boron-carbide/aluminum composites. J Am Ceram Soc 72(5):775–780
Honeycombe RWK (1968) The plastic deformation of metals. St. Martin's Press, New York
Hull D (1981) An introduction to composite materials. Cambridge University Press, Cambridge, UK, pp 81–100
Jackson AP, Vincent JFV, Turner RM (1988) The mechanical design of nacre. Proc R Soc Lond B234:415–440
Khanuja S (1991) Processing of laminated B_4C-polymer laminated composites. MS Thesis, University of Washington, Seattle
Laraia VJ, Heuer AH (1990) The microindentation behavior of several mollusc shells. In: Rieke PC, Calvert PD, Alper M (eds) Materials synthesis utilizing biological processes: proceedings of a symposium series, vol 174. Materials Research Society, Pittsburgh, pp 125–131
Lowenstam HA, Weiner S (1989) On biomineralization. Oxford University Press, New York
Mann S (1988) Molecular recognition in biomineralization. Nature 332:119–123
Mann S, Heywood BR, Rajam S, Birchall D (1988) Controlled crystallization of $CaCO_3$ under stearic acid monolayer. Science 334:692–695
Mann S, Webb J, Williams RJ (eds) (1989) Biomineralization: chemical and biochemical perspectives. VCH publ, Weinheim
Meenakshi VR, Donnay G, Blackwelder PL, Wilbur KM (1974) The influence of substrata on calcification patterns in molluscan shell. Calcif Tissue Res 15:31–44
Morton JE (1979) Molluscs, 5 edn. Hutchinson, London
Mutvei H (1970) Ultrastructure of the mineral and organic components of molluscan nacreous layers. Biomineralization 2:48–72
Nakahara H, Bevelander G, Kakei M (1982) Electron microscopic and amino acid studies on the outer and inner shell layers of *Haliotis rufescens*, VENUS (Jpn J Malac) 41(1):33–46
Nicol JA (1960) The biology of marine animals. Interscience, New York
Sarikaya M, Gunnison KE, Yasrebi M, Aksay IA (1990a) Mechanical property-microstructural relationships in abalone shell. In: Reieke PC, Calvert PD, Alper M (eds) Materials synthesis utilizing biological processes: proceedings of a symposium series, vol 174. Materials Research Society, Pittsburgh, pp 109–116
Sarikaya M, Gunnison KE, Yasrebi M, Milius DL, Aksay IA (1990b) Seashells as a natural model to study laminated composites. In: Proc Am Soc of Composites, Fifth Tech Conf Technomic Publ Co, Lancaster, Pennsylvania, pp 176–183
Sarikaya M, Liu J, Aksay IA (1992) Multiple tiling with hierarchical twins: TEM study of the nacre structure in red abalone. (submitted)
Simkiss K, Wilbur KM (1989) Biomineralization: cell biology and mineral deposition. Academic Press, New York

Towe MK, Hamilton GH (1968) Ultrastructure and inferred calcification of the mature and developing nacre in bivalve mollusks. Calcif Tissue Res 1:306–318

Tressler RI, Bradt RC (eds) (1984) Deformation of ceramic materials II. Materials science research, vol 18. Plenum, New York

Unwig N, Henderson R (1984) The structure of proteins in biological membranes. Sci Am 250(2):78–94

Wasserman SR, Whitesides GM, Tidswell IM, Ocko BM, Pershan PS, Axe JD (1989) The structure of self-assembled monolayers of alkylsiloxanes on silicon: a comparison of results from ellipsometry and low angle X-ray reflectivity. J Am Chem Soc 111:5852–5861

Watabe N (1965) Studies in shell formation: crystal-matrix relationships in the inner layers of molluscan shells. J Ultrastruct Res 12:351–370

Watabe N, Wilbur KM (1960) Influence of the organic matrix on crystal type in molluscs. Nature 188:334–336

Weiner S (1986) Organization of extracellularly mineralized tissues: a comparative study of biological crystal growth. CRC Crit Rev Biochem 20(4):365–380

Weiner S, Hood L, (1975) Soluble protein of the organic matrix of mollusc shells: a potential template for shell formation. Science 190:987–989

Weiner S, Traub W (1980) X-ray diffraction study of the insoluble organic matrix of mollusc shells. FEBS Lett 111(2):311–316

Weiner S, Traub W (1981) Organic matrix-mineral relationships in mollusc shell nacreous layers. In: Balaban M, Sussman JL, Traub W, Yonath A (eds) Structural aspects of recognition and assembly in biological macromolecules. Balaban ISS, Yehevot Philadelphia, pp 462–487

Weiner S, Traub W (1984) Macromolecules in mollusc shells and their functions in biomineralization. Philos Trans R Soc Lond B304:425–434

Weiner S, Talmon Y, Traub W (1983) Electron diffraction studies of molluscan shell organic matrices and their relationship to the mineral phase. Int J Biol Macromol 5:325–328

Weissbuch I, Addadi L, Lahav M, Leiserowitz L (1991) Molecular recognition at crystal interfaces. Science 253:637–645

Wenk HR, Barber DJ, Reeder RJ (1983) Microstructures in carbonates. In: Reeder RJ (ed) Reviews in mineralogy, vol 11. Miner Soc Am, Washington DC, pp 301–367

Wilbur KM, Simkiss K (1968) Calcified shells. Compr Biochem 26A:229–295

Wilbur K, Yonge CM (eds) (1964) Physiology of Mollusca. Academic Press, New York

Wise SW (1970) Microarchitecture and mode of formation of nacre (mother-of-pearl) in pelecypods, gastropods, and cephalopods. Eclogae Geol Helv 63(3):775–797

Yasrebi M, Kim GH, Milius DL, Sarikaya M, Aksay IA (1990) Biomimetic processing of ceramics and ceramic-based composites. In: Brinker CJ, Clark DE, Ulrich DR, Zelinski BJJ (eds) Better ceramics through chemistry IV: proceedings of a symposium series, vol 180. Materials Research Society, Pittsburgh, pp 625–635

Chapter 2

The Formation of Mussel Byssus: Anatomy of a Natural Manufacturing Process

J. Herbert Waite[1]

1 Introduction

The byssus is an extraorganismic polymeric structure in marine mussels and generally employed as a holdfast or tethering device. Like man-made plastics, the byssus is robust, tough, devoid of living cells, and disposable. Unlike plastics, however, it is ultimately biodegradable. Structural and mechanical analysis of the byssus reveals an exquisitely complex design at every level from the microcellular solid in the plaques to the fiber gradients in the thread core to the interpenetrating polymer networks of the byssal varnish. Such fine tuning of materials properties deserves closer scrutiny and, perhaps, imitation. In this vein, the formation of byssus can be caricatured as a series of manufacturing processes including injection and extrusion molding, calendering, and sizing. The biological relevance of each of these caricatures is explored in this chapter.

1.1 Variations of Nature-Envy

Human imagination has a penchant for admiring, even coveting, Nature's achievements. In earliest times this envy was often reflected in the deification of an animal possessing some desirable quality, e.g., Ganapati the Elephant (Hindu god of removing of obstacles), Horus the Falcon (Egyptian god of benign attentiveness), Arachne the Spider (Roman patroness of weaving), and so on. Envious awe of Nature persists in the form of more contemporary comic strip heroes such as Batman, Spiderman, the Green Hornet, etc. The dawn of modern biotechnology has aroused less vicarious ambitions. Since biotechnology promises that even preciously rare natural products can be made in vitro by the cloning and expression of appropriate genes, it has precipitated expectations that perhaps some of Nature's tricks will be directly imitated (Vincent 1990; Amato 1991).

1.2 Mussels: Bearded Models for Industry

Even with their considerable natural talents, marine mussels, to my knowledge, have largely escaped the attention of popular culture. There have been no mussel-

[1] College of Marine Studies and Department of Chemistry, University of Delaware, Lewes, DE 19958, USA

Results and Problems in Cell Differentiation 19
Biopolymers
Case, S.T. (Ed.)

gods or mussel-heroes. The arcane and bizarre paintings of Hieronymus Bosch (Fraenger 1983) are perhaps a singular exception to this, although the pivotal role of mussels in *The Earthly Paradise* (Prado Museum) is not widely appreciated. Nevertheless, mussels have many traits that are avidly coveted by industry. Perhaps the most visible of these is the formation of the beard or byssus in mussels (Strausberg and Link 1990). The byssus mediates the mussel's ability to bond to surfaces underwater. The presence of moisture is the commonest subverter of the performance and durability of man-made bonds (Comyn 1981). This is usually attributed to the four effects of water on adhesive bonds: (1) as a weak boundary layer on the surface, (2) by invasion of the interface by wicking/crazing, (3) adhesive erosion by hydrolysis, and (4) adhesive swelling by water absorption. Industry's attempts to circumvent and defeat these effects are complicated, technically demanding, expensive, and very imperfect (Eagland 1988). As a result there is much room for improvement in the design of adhesives that are environmentally durable. Since mussels and other sessile marine invertebrates routinely bond a wide variety of surfaces underwater, and these bonds are strong and durable, it stands to reason that they should be envied. But there other compelling reasons to study the mussel byssus. The byssus mediates mussel fouling, which is significant on marine structures particularly in the North Atlantic (Southgate and Myers 1985). A better understanding of byssus formation could lead to the development of antifouling strategies that might lessen our dependence on toxic paint additives such as tributyl tin. In addition, the byssus is a scavenger for transition and heavy metals in the marine environment (Hamilton 1980). Its talents in this regard have already been exploited by the US Environmental Protection Agency's "Mussel Watch" program (Koide et al. 1982). Perhaps the metal-binding capacity of some of the precursor proteins are responsible for this (Waite 1990a) and can be utilized to monitor environmental contamination independent of the byssus. Lastly, the byssus represents a robust yet biodegradable set of fibers with a functional lifetime of 2–3 years. As such, the byssus serves as an intriguing model for the future design of degradable consumer materials. Can the byssus or its functional parts be imitated by biotechnology? I hope to establish in this essay that the formation of an adhesive holdfast such as the byssus is going to take much more than the cloning and production of an assortment of proteins.

There is certainly no want of reviews on the subject of the byssus. A few of the recent ones include Young and Crisp (1982), Price (1983), Waite (1983; 1987), and Harris (1990). The aim here is not to rehash reviews but rather to examine the *processes leading to the formation* of the byssus in adult mussels. This has not been previously addressed and for good reason: there is little information on the subject. My aim here is to assemble the known bits and pieces about the formation of byssus, and then, with a dollop of conjecture, to concoct a detailed scenario of the process. Despite the inherent cheekiness of such an approach, it is no less important to the growth of knowledge than the reconstruction of a skull of *Zinjanthropus* from a handful of fragments. My primary intent, then, is not to describe a process for its own sake. Rather it is to arouse excitement and controversy; to expose the possible and the probable; to engage imaginations to find the where-with-all to replace conjecture with fact, and finally to demystify an exquisite natural process without misrepresenting it.

1.3 When and Why Mussels Sprouted Beards

The beard or byssus is an extracorporeal structure produced in all bivalve mollusks at one or more stages during their life history (Carriker 1990). All pediveliger forms appear to produce a byssal thread as a post-larval holdfast (Yonge 1962; Lane and Nott 1975) but in some species of *Mytilus*, it may also serve as a rudder in dispersal by drifting (Lane et al. 1985). It possibly originates from a modification of one of the pedal glands in the Gastropoda primitively concerned with the secretion of mucus (Yonge 1962). Indeed, at least one known occurrence of a byssus-like structure in extant Gastropoda is the thread formed by the foot of *Boonea impressa* (Ward 1985). Adult manifestations of the mussel byssus are thought to have first appeared in the Devonian period (350–400 million years ago) as bivalves that were buried in muck began to emerge to occupy more turbulent epibenthic habitats (Stanley 1972). The progressive transition from buried to exposed is well demonstrated in such byssated species as *Brachidontes citrinus*, *Geukensia demissa*, *Modiolus modiolus*, and *Mytilus edulis* (Stanley 1972). Why should early bivalve mollusks have left the muck that they were so well suited to inhabit? By entering the epibenthic and ultimately intertidal zones, bivalves enhanced or radically changed their diets (from detritus to filter feeding), accelerated exchange of gases and removal of waste products, and extended the range for the dispersal of their young. But why use a byssus? There are, after all, other effective attachment strategies such as cementation (Yonge 1979) and pedal suction (Smith 1991). The advantages of byssal attachment over alternative strategies must include the liberty to close the shells, jettison the holdfast, modulate the tension of attachment, select substrata opportunistically, and regenerate a holdfast on short notice. Whilst the functional byssus of most bivalve groups particularly mussels serves as a holdfast, additional uses have been secondarily evolved. These include byssal tents for protection of nonretractable tentacles in *Lima* (Yonge and Thompson 1976) and byssal nests for the brooding of young in *Musculus* (Ockelmann 1983), byssal tethers for immobilizing predators (Day et al. 1991), byssal hairs and barbs for the protection of the mantle in *Modiolus* (Ockelmann 1983), and byssal pivots for rock borers such as *Botula* (Yonge 1955).

1.4 Byssus Is Tough Stuff

Mussel byssus is the best studied of bivalve byssi. There are four distinct regions in mussel byssus (Fig. 1): root, stem, thread, and plaque (Brown 1952). All but the root are cell-free and exposed to sea water. The root is imbedded in muscular tissue at the base of the foot; the stem is a "telescopic" or brush-like structure each increment of which has a thread attached by way of a cuff or collar; the threads are

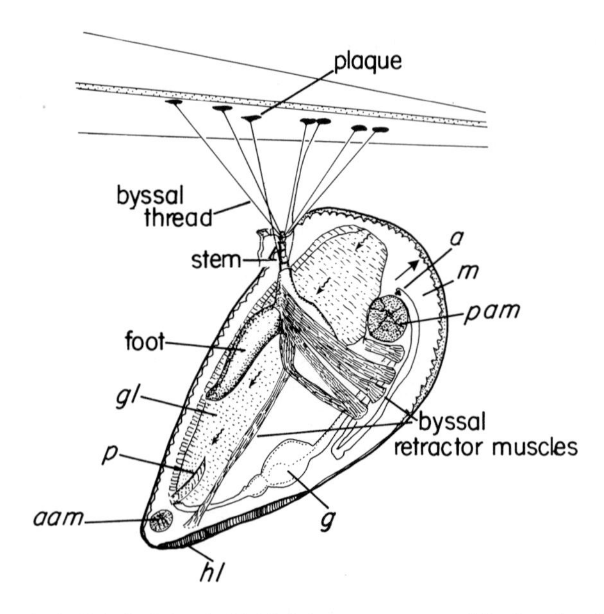

Fig. 1. Mussel on the half-shell (*left side*) highlighting components of the functional byssus. Abbreviations are as follows: *aam* anterior adductor muscle; *a* anus; *g* gut; *gl* gill lamella; *hl* hinge ligament; *pam* posterior adductor muscle; *p* palp. *Arrows* indicate the approximate positions of the exhalant (*solid*) and inhalant (*wavy*) siphons

fibrous extensions of the stem ranging from a few millimeters in length to decimeters. Each thread is tipped by a plaque at which attachment or bonding to the substratum takes place. Dimensions of byssus are highly variable and dependent on age as well as the species in question. In adult *M. edulis*, for example, threads average 2 to 4 cm in length (average diameter 0.2–0.3 mm) and are attached to plaques some 2 to 3 mm in diameter (Allen et al. 1976). The ribbed mussel *G. demissa* has threads of similar length whilst the diameters of both threads and plaques are narrower (van Winkle 1969). In contrast, threads in *Atrina* and *Pinna* have diameters similar to those in *G. demissa*, although lengths can be as long as 12 cm (Turner and Rosewater 1958).

The byssus is widely regarded as a robust natural material in both a chemical and physical sense. Once hardened, it is insoluble in water, detergents, protein denaturants, organic solvents, cold dilute or concentrated hydrochloric, nitric, sulfuric acids, and 3N NaOH; it is hydrolytically dissolved by fused caustic soda, hot concentrated acids, and sodium hypochlorite. Like other quinone-tanned structures, it has low or minimal antigenicity and is highly resistant to proteases (Brown 1950; Waite 1990b). Curiously, only the protective byssal varnish exhibits a resistance to chemical and enzymatic degradation whilst the load-bearing fibrils in the core (such as collagen) are readily degraded once exposed (Benedict and Waite 1986). As with most fibrous materials in Nature and industry, the function of the byssus is to resist and damp forces applied in the direction of the fiber axes. Threads seem quite adequate at doing this and, in the case of *Mytilus* sp., compare favorably with the mechanical properties of tendon (Smeathers and Vincent 1979). Only three byssi have been biomechanically examined in any detail. These are from *M. edulis*,

G. demissa and *Pinna nobilis* (which served for some time as a textile fiber) (Lucas et al. 1955; van Winkle 1969; Smeathers and Vincent 1979). Ultimate strain (= extension at failure) varies greatly between the three species (0.4 for *M. edulis* (Smeathers and Vincent 1979), 0.2 for *G. demissa* (Van Winkle 1969), and 0.5 for *P. nobilis* (Lucas et al. 1955). Perhaps such variation is to be expected in animals coming from different habitats; however, no attempt has been made to probe the interplay between mechanical performance of the byssus and physical environment. The holdfast strength of diverse byssi including *M. californianus* (Harger 1970; Witman and Suchanek 1984), *M. edulis* (Harger 1970), *Arca zebra* (Bowen et al. 1974), *Septifer bifurcatus* (Harger 1970), *Mytilopsis sallei* (Udhayakumar and Karande 1989) has been measured and previously reviewed (Waite 1983). If it seems strange that an adult mussel (say *M. californianus*) which typically weighs 30 to 100 g (i.e., exerting a force of 0.3 to 1 N under gravity) has attachment strengths ranging from 200 to 400 N, it is because some compensation must be made for drag especially when the shells are fouled by epizoites and kelp (Witman and Suchanek 1984). Mussels probably make adjustments in the strength of their holdfast by adding new threads to the byssus.

The testing of holdfast strength has not always been in keeping with biological precepts. Commonly, attachment is measured by connecting a strain gauge to the mussel and applying a force orthogonal to the plane of the surface (Harger 1970; Bowen et al. 1974; Witman and Suchanek 1984; Udhayakumar and Karande 1989). In the natural habitat of the mussel, there is indeed an orthogonal element called lift but it is not the chief of the forces working to detach the animal. This distinction belongs to shear and is derived from the ebb and flow of waves, tides and storms (Witman and Suchanek 1984; Denny 1988). Mussels improve their chances against dislodgement by consolidating arrangement of their threads into bundles each facing a point source of flow, e.g., mussels experiencing mainly bidirectional drag will deposit their threads in two bundles, one facing ebb and the other flow. Only one of these will be challenged in a particular regime. With this in mind, it is uncertain what significance might be attributed to orthogonal force tests. Perhaps a hungry seagull or crab could apply forces perceived by the mussel as pure lift. Whatever the force, mussels have several options in coping with the threat of dislodgement. They can produce more anchoring threads, cluster their threads in the direction of applied forces, and if all else fails, jettison the byssus to seek a more sheltered niche. In fact, all of these options are dynamically exhibited in a natural mussel community (Williamson 1907).

2 The Production Site in *M. edulis*

2.1 Mussel Anatomy

Marine mussels are members of the family Mytilidae in the class Bivalvia and the phylum Mollusca. Convergently evolved forms occur in the family Dreisseni-dae but these are arguably not marine. Mytilids are typically anisomyarian, that is, the shell adductor muscles are unequal in mass such that the anterior muscle is smaller than the posterior one. Mytilid shells are skewed anterio-posteriorly to reflect the muscle mass distribution (Cox 1969). Anatomically, mussels all share the wedge-shaped bivalved shell which encloses a neatly packaged soft body (Fig. 1). The mantle serves as the wrapping for the soft body that is physically contiguous with the shell along the mantle/shell margin, the hinge, and at the adductor and byssal retractor muscle attachment sites. The inhalant and exhalant siphons are located posteriorly in the mantle and function to direct the flow of water to and from the gill lamellae as well as the palps and anus, respectively. In this way, incoming water currents provide dissolved oxygen to the gills and suspended nutrients to the palps; outgoing water currents carry CO_2 and waste products such as feces, pseudofeces, and ammonia. Most mussels are facultative anaerobes with a rather surprising capacity for anaerobiosis. This is especially evident in intertidal species during emersion (McMahon 1988).

2.2 Mussel Foot

The mussel foot is a pigmented appendage in *M. edulis* and located in the center of the soft body mass (Fig. 1). In the pendent pose, it is evident that the byssal stem is rooted in the anterior and posterior byssal retractor muscles at the base of the foot. By contracting and relaxing these muscles, the animal can adjust its proximity to the substratum and modulate byssal tension. Indeed, in juvenile mussels the alternation of byssal thread deposition, contraction of the byssal retractor muscles, and byssal release (jettisoning) is exploited to scale vertical substrata (Tamarin et al. 1976).

The term "foot" is presumably derived from the post-larval mussel's use of the appendage to "walk" along surfaces during exploration and settlement. This is but one of its many functions, and it could as appropriately have been termed "tongue", "sensor", "thread caster", or "byssal factory". Indeed, the extraordinary versatility of the mussel foot cannot be enough emphasized. Imagine a device that can survey its surroundings, distinguish polished glass from Plexiglas, clean a surface, adhere temporarily to surfaces, and cast threads that range from one to seven times its resting length (Waite 1983). The mussel foot is a miniature facility dedicated to producing a continual supply of byssal threads as needed, and is self-reliant with regard to most aspects of the synthesis, assembly, deposition, performance, and finally, disposal of the byssal threads. This self-reliance is derived from the ingenious design and compartmentalization of the foot (Allen et al. 1976; Waite 1983). The foot houses a number of glands that specialize in *holocrine* activities, that is, to prepare products for secretion *outside* the organism.

The schematic drawing in Fig. 2 summarizes the main glands contributing to byssus formation; mucous gland, phenol (purple) gland, collagen (white) gland, accessory (enzyme) gland, and the stem (byssus) gland. The parenthetic appellations persist from early histological studies (see Brown 1952; Pujol 1967). Each one is dedicated to the synthesis and stockpile of specific precursors of the

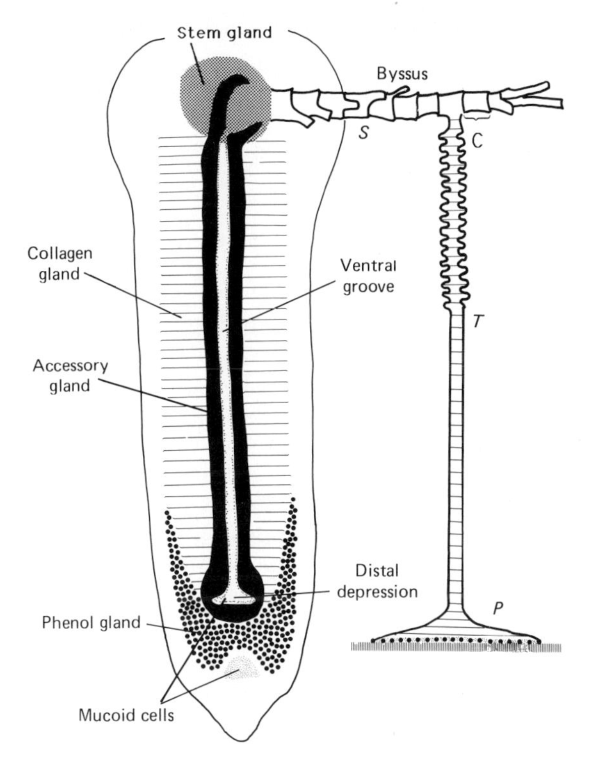

Fig. 2. Schematic view of the byssus-secreting glands in the foot (ventral surface) of *Mytilus* sp. including the byssus with stem (*S*), cuff (*C*), thread (*T*), and plaque (*P*). (Waite 1983)

composite byssus. So far it has not been possible to pinpoint unequivocally the glandular origin of any byssal precursors, although the following associations are widely accepted (Waite 1983): The mucous gland secretes acidic glycosaminoglycans and glycoproteins, the phenol gland contains proteins rich in *o*-diphenols, i.e., 3,4-dihydroxyphenylalanine or DOPA, the collagen gland contains collagenous and elastic proteins, the accessory gland stockpiles both catecholoxidase activity as well as *o*-diphenolic precursors, and the stem gland secretes the lamellar proteins of the stem (Brown 1952; Smyth 1954; Gerzeli 1961; Pujol 1967; Banu et al. 1980). The mucous glands are located near the tip of the foot and along the margins of the ventral groove. The phenol gland is rocket-shaped with the leading tip of the "rocket" just behind the mucous gland and two trailing "fins" running along either side of the ventral groove to about the middle of the foot (Fig. 2). Both the collagen and accessory glands straddle the ventral groove from the distal depression to the stem gland; the last-mentioned is bulbous and restricted to the base of the foot.

The few researchers who have studied the musculature and innervation of the mussel foot have waxed at the intricate complexity of both of these. Not only does every gland exhibit extensive innervation by monaminergic junctions (Vitellaro-Zucarrello et al. 1983a), but distinct muscles are present to extend and rotate the foot (up to 180°), open and shut the lips of the ventral groove, and flatten or cup the distal depression to mention a few (Williamson 1907). The most important questions regarding the coordination of the activities of all these glands and muscles will in all probability be technically off-limits for some time to come.

3 Building Blocks

3.1 Byssal Collagen

The byssus is typically a hydrated proteinaceous structure with a lesser content of carbohydrates and lipids. Given the thoroughly cross-linked fabric of mature byssus, however, few discrete molecules are extractable for characterization. This has posed serious obstacles for the study of structure-function relationships since, in order to accomplish this, it is essential to determine the identity and distribution of various byssal precursors. Of the various proteins in byssus of *M. edulis*, the existence of some was suggested by morphological, histochemical, and X-ray diffraction studies as long ago as 40 years (Mercer 1952; Smyth 1954; Rudall 1955; Gerzeli 1961; Pujol 1967; Banu et al. 1980). Chief among these are the fibrous proteins (such as collagen), DOPA-containing proteins, and enzymes.

At least three types of fibrous proteins (collagen, elastic proteins, and fibrous banded elements) are present in the core of byssal threads (Bairati and Vitellaro 1973; Vitellaro-Zucarrello et al. 1983b). Based on amino acid analysis and low angle X-ray diffraction, collagen prevails in the distal portion of thread (Gathercole and Keller 1975; Mascolo and Waite 1986). Fibrous banded elements are scattered throughout the matrix of the core and have diameters of 25 µm and an average period of 1000 Å (two narrow dark bands separated by a wide lighter area) (Bairati and Vitellaro 1973). The composition of this material is unknown. The elastic protein is best observed in the proximal portion of the thread where it has a wavy, crimped appearance when sectioned parallel to the thread axis (Gathercole and Keller 1975). Only the presence of collagen has a biochemical basis so far. A pepsin-resistant fragment (55 kDa) of this collagen can be extracted from threads (Pikkarainen et al. 1968; Benedict and Waite 1986). It ostensibly has all the hallmarks of a typical collagen: glycine at every third residue, 20% proline/hydroxyproline, collagenase-lability, and metachromasy following staining by Coomassie Blue R-250 (Benedict and Waite 1986b; Qin X, Waite JH, unpubl. results). Like type I–III collagens, the primary sequence-{X-Gly-Y}-possesses imino acids in both the X and Y positions although only prolines at X appear to undergo hydroxylation to 4-*trans*-hydroxyproline. This resembles interstitial collagens of many eukaryotes and contrasts with the collagens of annelid cuticles (Adams 1978).

Byssal collagen is derived from granules secreted by the collagen gland in which precursors appear to be stored in a condensed preassembled state. At least two molecular species are stored within these granules: a microfilamentous material (length 7–9 nm) and supercoiled filaments with periodicities of 150 Å (Vitellaro-

Zucarrello 1980). Collagenase-lability of the microfilaments suggest these to contain collagenous domains (Vitellaro-Zucarrello 1980). What the pepsin-labile sequences in collagen are, or how the granule contents are assembled into cohesive fibers remains a mystery. The coiled filaments may be elastic protein. Whatever they are, the 150 Å periodicities, so clear in the sectioned granules, are lost upon secretion and assembly of the filaments in the thread (Vitellaro-Zucarrello 1980). Perhaps the oddest feature about the collagenous and elastic proteins is the apparent complementarity of their distribution in *Mytilus* threads (Gathercole and Keller 1975; Mascolo and Waite 1986). Collagen, which is rather inextensible, prevails in the distal thread near the plaque (Rudall 1955; Mascolo and Waite 1986) whilst the elastic protein seems to predominate in the extensible proximal thread (Benedict and Waite 1986a, b; Fig. 2). Does this gradient reflect the programmed glandular secretion of granules which are already titrated with different proportions of the two proteins or a secretion of generic composition which is mechanically massaged into complementary gradients of collagen and elastic protein? The former seems the more probable and efficient, but both mechanisms are possible in principle.

3.2 DOPA-Containing Proteins

M. edulis foot proteins (MeFP) are known by virtue of their DOPA content to be destined for the byssus (Mascolo and Waite 1986). MeFP-1, formerly dubbed the polyphenolic protein, has a molecular weight of 125 kDa, is distinguished by a high isoelectric point (> 10) and by the presence of tandemly repeated hexa- and decapeptides such as AKPTYK and AKPSYP*P*TY*K, where P* and Y* denote 4-*trans*- or 3-*trans*-hydroxyproline and DOPA, respectively (Waite et al. 1985). The nearly complete primary structure deduced from cDNA sequences (Filpula et al. 1990) is presented by Laursen (this Vol.). I also defer discussion of the conformation of polyphenolic proteins to Laursen. The consensus repeats from the polyphenolic proteins of other mytilid species are summarized in Table 1. All show a high proportion of DOPA, lysine, and hydroxylated amino acids. The DOPA and lysine may form a covalent cross-link following the oxidation of the former to an *o*-quinone (Waite 1990b), but characterization of specific cross-links has been elusive.

The second DOPA-protein, so-called MeFP-2, has only recently come to light (Rzepecki LM and Waite JH, unpubl. results). It has an apparent molecular weight of 42–45 kDa and a pI of about 8. Like the polyphenolic protein, it has a high proportion of lysine and imino acids but no hydroxyproline. DOPA is generally low (2–3 mol%), and cystine (7 mol%), aspartate/asparagine (15 mol%), and glycine (18 mol%) levels are high. Although the native protein has shown a remarkable resistance towards trypsin, chymotrypsin, and other proteases, prolonged proteolysis (24–48 h) does produce peptide fragments some of which have been sequenced. Reductive alkylation of disulfides also results in a greater lability to proteolysis. In stark contrast to MeFP-1, which has proteolytically derived peptide

Table 1. Distribution and sequence of some DOPA-containing adhesives and varnishes among the Invertebrata

Species	MW	pI	Consensus Peptides (# repeats)	References
Order Mytilacea				
Mytilus edulis L. 1758 (Blue mussel)	130 kDa	10.	A-K-P-S-Y*-**P***-**P***-T-**Y***-K (76) A-K-P-T-Y*-K (13)	[1]
Mytilus californianus Conrad, 1837 (Pacific mussel)	85 kDa	10.	I-T-Y*-**P***-**P***-T-**Y***-K-P*-K R-K-P-S-Y*-**P***-**P***-T-**Y***-K (30)	[2]
Septifer bifurcatus (Conard, 1837) (Bifurcate mussel)	130 kDa	10.	Y-P*-A-K-P-T-S-Y*-G-T-G-**Y***-K (70)	[3]
Geukensia demissa (Dillwyn, 1817) (Ribbed mussel)	130 kDa	8.	Q-T-G-**Y***-X-G-Y*-K (60) X = SA, VP, DP, VL	[4]
Modiolus modiolus squamosus Beauperthuy, 1967 (False tulip mussel)	100 kDa	>10.	T/S-Y-**Y***-P*-K/Q-G-**Y***-G-K (91)	[3]
Brachidontes exustus (Linne 1758) (Scorched mussel)	90 kDa	8.	same as *G. demissa* (60)	
Mytella guyanensis (Soot Ryen) (Ecuadorean mussel)	130 kDa	10.	A/S-H-K-P-Y*-T-G-**Y***-**K** (> 70)	
Aulacomya ater (Shoe mussel)	100 kDa	?	A-G-Y*-G-G-X-K* (?) X = V, L	
Trichomya hirsuta (Lamarck 1819)	120 kDa	> 10.	X-Y*-**Y***-P*-K (50) X = S, T, G	
			G-Y*-G-X-K (25) X = S, A	

X_1 = variable; X_2 = usually D, A, E, G; X_3 = usually S, D, G; X_4 = usually G.

1. Waite et al. (1985), 2. Waite (1986), 3. Rzepecki et al. (1991), 4. Waite et al. (1989)

* indicates occasional hydroxylation, whereas the bold* denotes complete hydroxylation. Numbers in parentheses after the consensus sequence represent the estimated number of repeats. Consensus repeats are represented using the single-letter abbreviations for amino acids; Y* = DOPA.

repeats each with a composition resembling that of the whole protein, in MeFP-2, there are proline-, aspartic acid-, and lysine-rich fragments, the distribution of which has not been determined in the native protein. Nothing is known about the conformation of this protein nor its complete primary sequence.

There is still great confusion about the glandular source of these proteins as well as their distribution in the byssus. Early results suggested that the richest source of the MeFP-1 is the region of the phenol gland near the tip of the foot (Tamarin and Keller 1972). This gland is reputed to contribute the bulk of precursor for the plaque (Tamarin et al. 1976; Waite and Tanzer 1983). More recent studies suggest that it can also be extracted from the accessory gland in regions of the foot uncomplicated by the presence of the phenol gland (Rzepecki LM and Waite JH, unpubl. results). MeFP-2, in contrast, is detectable only in extracts of the foot tip. Conclusive immunohistochemical studies localizing each of the proteins in vivo and in bysso have yet to be done, and they are not likely to be trivial. In a previous study, polyclonal antibodies prepared against MeFP-1 cross-reacted with the thread varnish *and* plaque/substratum interface (Benedict and Waite 1986b). Since it seems unlikely that one protein serves two such different functions, perhaps two proteins share one or more epitopes. Indeed, the sequences SYP*P*TY* (MeFP-1) and KSPPSY* (MeFP-2) may be just such an epitope. Only mono-specific antibodies will resolve such ambiguity. In vitro behavior of the proteins indicates MeFP-1 to be the more adhesive of the two adsorbing to any surface from solution at pH 6 or greater. The need for such a surface active-protein in the accessory gland is not obvious.

3.3 Catecholoxidase

Catecholoxidase (EC 1.14.18.1) is an enzyme long associated with melanin formation, quinone-tanning and the formation of byssus in mussels. Histochemical tests have been used to demonstrate catecholoxidase activity in the byssus as well as the foot of *M. edulis* and other mytilids (Smyth 1954; Gerzeli 1961; Pujol 1967; Banu et al. 1980), where it seems limited to mottled granules in the accessory gland (Bharathi and Ramalingam 1983). In the byssus, the enzyme catalyzes the following reaction:

$$2\text{Peptidyl-DOPA} + O_2 \rightarrow 2\text{Peptidyl-}o\text{-quinone} + 2H_2O$$

Some have suggested that catecholoxidase is also present in the phenol gland, but evidence for this is not compelling. Catecholoxidase activity was shown to be extractable from the threads of *M. edulis* and resides in a large protein (120 kDa) with an abundance of acidic amino acids, serine, and glycine following hydrolysis (Waite 1985). An enzyme with similar amino acid composition but lower molecular weight (37 kDa) was isolated from the threads of *G. demissa* (S.J. Samulewicz and J.H. Waite unpubl. results). Attempts to find the precursor enzyme in the mussel foot suggest that in both species catecholoxidase exists as a zymogen or latent enzyme prior to secretion. Activation can be effected in vitro by treatment with α-chymotrypsin which may or may not resemble the natural activator (Waite 1985). Since enzyme activity is latent in the accessory gland, histochemical detection as reported by Smyth (1954) and others should not have been possible unless inadvertant activation during routine fixation and/or dehydration had occurred.

The byssal catecholoxidase is not merely intriguing as a natural catalyst of protein cross-linking (Waite 1990b). Quinone-tanning catecholoxidases also need to be scrutinized with regard to their *structural* roles in the end product. The significance of this is apparent from the cytochemical studies of the formation of byssus (Smyth 1954; Pujol 1967; Vitellaro-Zucarrello 1981). According to these, latent catecholoxidase is present together with DOPA-containing proteins in mottled granules. Upon secretion, the enzyme is activated by an unknown mechanism (perhaps proteolytic) and acts upon DOPA residues in the protein. Although the mechanism by which two different proteins are stored and distributed in the same package is unknown, the function of co-packaging is clearly to improve mixing of the two. Incomplete mixing of catalyst and resin is widely recognized as leading to very inferior properties in synthetic thermoset, i.e., cross-linked materials (Cook 1991). Premixing is particularly important here because of the low diffusivity (high molecular weights) of enzyme and DOPA-containing protein, and also because the two components may actually be present in highly viscous or nonmiscible phases reacting only at their interface (Vitellaro-Zucarrello 1981). If, for whatever reason, the diffusibility of enzyme or protein-bound quinone is limited, then one would expect rather higher enzyme-substrate ratios than physiological models such as Michaelis-Menten recommend. Indeed our preliminary studies of catecholoxidase in the byssus of *G. demissa* suggest that two to three molecules of MeFP-1 protein may be present per enzyme (Samulewicz SJ and Waite JH, unpubl. results). At these concentrations, the enzyme must participate in the structural integrity of the material or run the risk of undermining its cohesiveness. Curiously, most of the enzyme is extractable from the cured byssus whilst the MeFP-1 is not.

4 The Process for Making Byssus

4.1 Cues and Preferences

There are probably many cues for adult mussels to make new byssal threads. Water velocity and agitation appear to be among the most effective in *Mytilus*. Up to 100% more threads can be made in agitated vis-à-vis still water (Young 1985). This is an effective adaptive compensation since it enables mussels to strengthen their holdfast promptly whenever turbulence (such as the onset of a storm) threatens dislodgement (Price 1980). Mussels take fastidious steps to prepare sites for the deposition of new byssal threads: the foot emerges from a gap in the two valves of the shell and, with tip-toeing motions, explores its surroundings for surfaces (Fig. 3). Mussels are not frivolous in their preferences for surfaces and

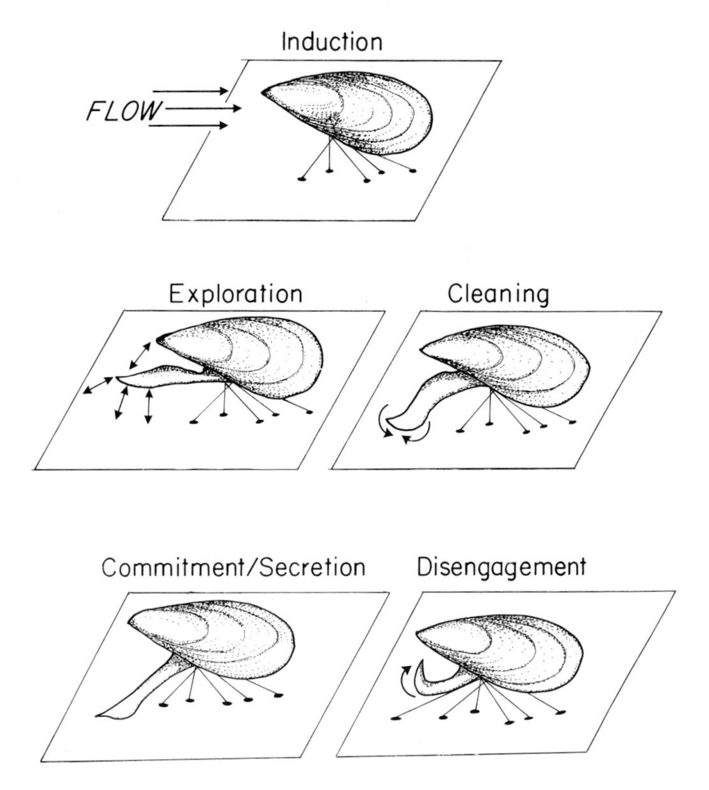

Fig. 3. Mussel (*Mytilus edulis*) foot behavior prior to the formation of a new thread and plaque. The various stages of behavior are described at length in the text

substrata. They definitely prefer slate to paraffin and even slate to glass (Young 1983a), although when not given the choice they can adhere to all three. They also prefer rocks to pebbles and certainly to gravel (Young 1983b). Having identified a suitable surface, it is intimately inspected and possibly scrubbed or coated by the tip of the foot. Scrubbing would serve to remove fouling films typically present on surfaces of substrata in natural waters (Characklis 1981). This author and others have observed the tip of the mussel foot squirming this way and that over the surface prior to thread formation as if it were sensing or clearing away debris (Price 1983; Waite 1987). Although the function of the squirming has not been established, film removal is absolutely necessary for strong attachment particularly if the film is cohesively weak. Thereafter the foot becomes quite motionless, even rigid. In this state it has become committed to the production of a new thread (Fig. 3). Duration of this stage can be from 1 to 5 min. Disturbance of the foot at this stage can have two abortive outcomes: (1) a rudimentum of the adhesive plaque if interrupted within the first minute (Maheo 1970); (2) a plaque *plus* a dangling thread if after 1 min. This suggests that the plaque is produced first, followed by the thread growing proximally from the plaque toward the stem.

4.2 Thread Formation

The process of thread and plaque assembly during the committed stage has yet to be examined in any detail. This is because the foot effectively conceals everything within the ventral groove. A conservatively fanciful attempt to reconstruct the process is depicted in Fig. 4. In thread formation, all the action for assembly occurs

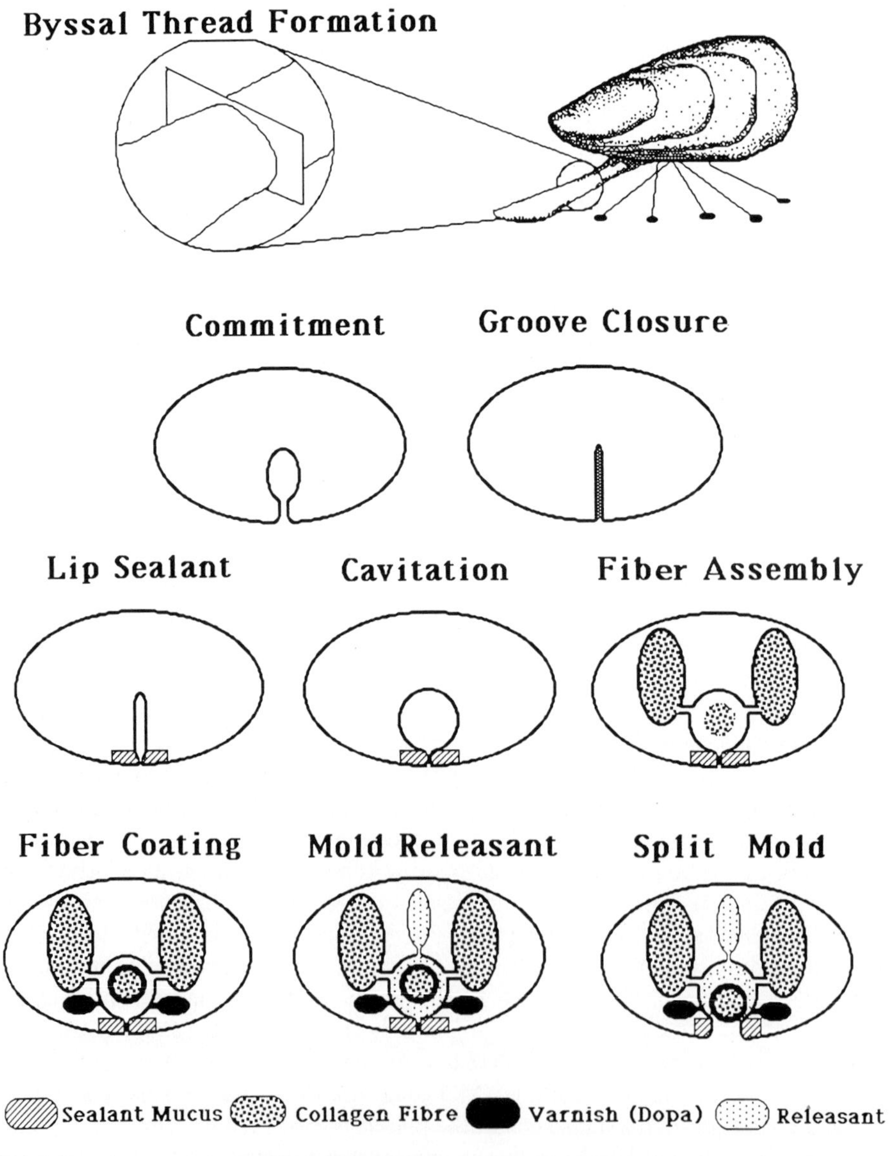

Fig. 4. Putative processing during thread formation as represented by a highly schematic cross-sectional view of the foot. See text for commentary

in the groove along the ventral surface of the foot. Initially the foot groove is concave and open to sea water. Subsequently, the groove walls mesh at their common surfaces, thus eliminating all bulk water. With the outer lips of the groove sealed, the inner walls draw away from one another, creating a space of negative pressure much like that produced by the pedal disk of limpets (Smith 1991). The negative pressure might be relieved by admission of a colloidal suspension of granules from the collagen gland. These enter by numerous cellular processes terminating in the groove. The granules are of several types including those with granular contents and banded filaments. These were described earlier (Vitellaro-Zucarrello 1980). The granules with banded filaments assemble or are massaged into highly oriented (anisotropic) fibers within the groove lining and are coated by a thin cuticular layer derived from the accessory or enzyme gland. Peristaltic contractions of the groove lining and the actions of cilia may contribute to the uniform distribution of the coating. The thread coating is quite sticky initially, therefore some provision must be made to prevent thread adhesion to the groove lining. A lubricating mucus probably functions as a releasant (Vitellaro-Zucarrello 1983). Finally the sealed lips of the ventral groove are opened, and the completed thread is released and disengaged.

4.3 Plaque Formation

The sequence of events in the distal depression where the adhesive plaque is made differs somewhat from that described above (Fig. 5). This process is more accessible than thread formation to direct observation because a transparent glass plate can be used as the substratum (Tamarin et al. 1976). If a mussel is tethered near the side of a glass aquarium, plaque formation can be followed through the glass by viewing events in the distal depression. The distal depression of the immobilized or committed foot is immediately evident by its relative absence of pigmentation. The mussel presses the tip of the foot against the surface (Fig. 5: extrusion of water) and apparently keeps it in place by suction similar to that used by the gastropod foot (Smith 1991). This can be achieved by (1) sealing the edges of the interface between the foot and the substratum with a mucosubstance and (2) creating a pocket of negative pressure that is consequent upon the muscular elevation of the ceiling of the distal depression (cavitation). This temporary adhesion is sufficient to sustain the weight of the mussel underwater (Williamson 1907). The negative pressure is probably relieved by the injection of plaque precursors through six or more pores in the ceiling (Seydel 1909; Tamarin et al. 1976). After covering a patch of the substratum (3–13 mm^2) with a layer of adhesive 2–10 μm thick (Tamarin et al. 1976; Benedict and Waite 1986a), the adhesive is spliced with fibrous material from the thread, and all is coated with varnish from the accessory gland (coating). Adhesion of the varnish to the foot is prevented by mucus. The sealant under the tip is broken prior to the lifting and disengagement of the foot from the new plaque.

The cellular and subcellular events involved in plaque formation are quite extraordinary. Flowing into the distal depression is a microemulsion consisting of a

Byssal Plaque Formation

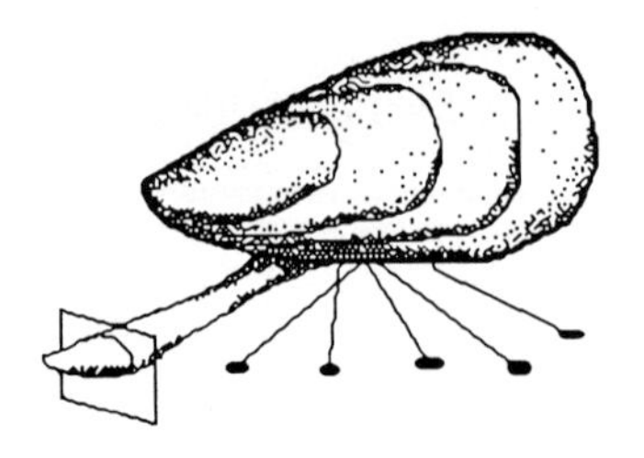

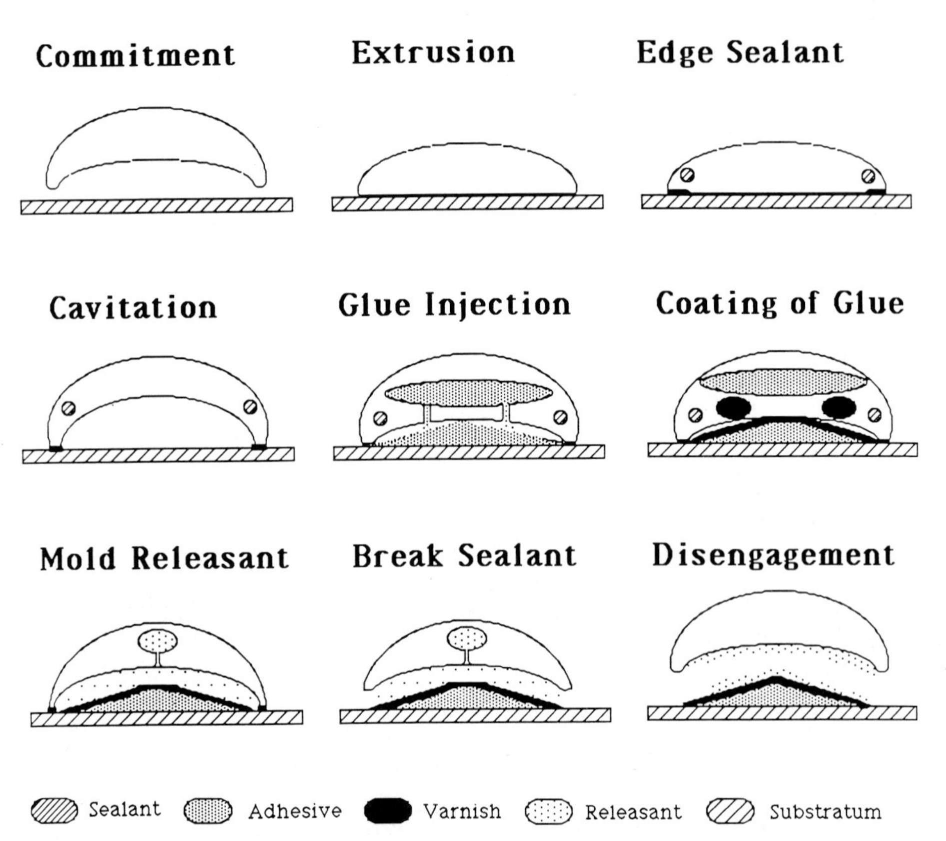

Fig. 5. Putative processing during plaque formation as represented by a highly schematic cross-sectional view of the foot tip. See text for commentary

fluid containing 1 μm diameter granules (Tamarin et al. 1976; Benedict and Waite 1986a). These are derived from the phenol gland and elsewhere and conducted to the depression by way of ciliated longitudinal ducts. Somewhere in the course of the delivery of this dispersion there occurs a phase inversion — a term coined by colloid chemists to describe a change in the continuous phase of oil : water mixtures. In the present case, the dispersion begins with water or some physiological fluid as the continuous phase and phenolic granules as the discontinuous phase. In normal nascent plaque, however, the granule contents become the continuous phase and the fluid the discontinuous. Sometimes the inversion is incomplete (Benedict and Waite 1986a). The mechanism of inversion in byssal plaques is unknown; however, it would be remiss not to mention that similar inversions can be induced by shear during the aqueous polymerization of styrene in micelles of a nonionic surfactant (Williams and Wrobleski 1988). Here, apparently,

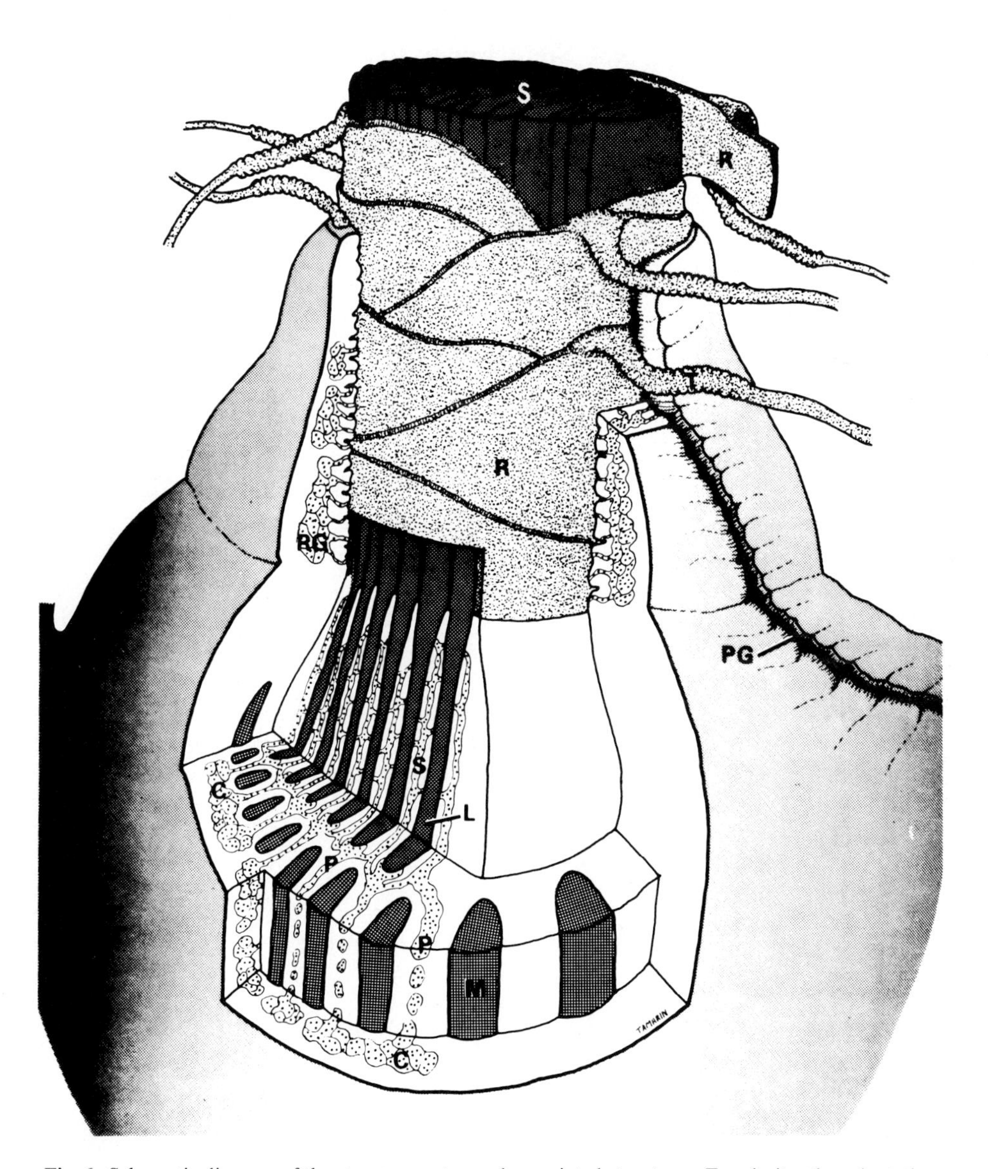

Fig. 6. Schematic diagram of the stem generator and associated structures. For clarity, the orientation plane of lamellae and septa has been rotated almost 90°. Normally, they lie in a plane parallel to the direction of the ventral pedal groove (*PG*). Byssus retractor muscles (*M*); byssus stem secretory cells (*C*); granule conducting processes (*P*); septa (*S*); lamellae (*L*); glands of thread attachment cuffs (*RG*); thread attachment cuffs (*R*); byssus threads (*T*); stem core (*S*). (Tamarin 1975)

polymerization and micelle collision occur simultaneously, and inversion takes place when the molecular weight and viscosity of growing polymer chains in two colliding micelles are sufficient to sustain contact despite the transfer of kinetic energy. The thorough cross-linking of proteins prevents the collapse of the cellular structure formed by phase inversion.

4.4 Stem Formation

The byssal threads are secured at the base of the foot by a peculiar structure known as the stem. The stem which has a segmented telescopic appearance is rooted in the byssal retractor muscles, and each thread is tethered to it by way of a collar or cuff. The cuffs provide the sheathing for the stem core, a dense bundle of lamellae that extends from the distal tip of the stem to the musculo-glandular zone deep within the base of the foot (Tamarin 1975). Growth of the stem is nothing less than an episodic eruption perhaps like the pulsed extrusion of toothpaste from a squeezed tube. There are a fixed number of lamellae in the core that grow larger with the passage of each successive new thread collar (Tamarin 1975). Growth of the lamellae in the musculo-glandular region may provide the force that pushes the structure upward. Collars are formed in the cervical crevice. Each collar is skewed in its girth peaking at the point of attachment to the thread (Fig. 6). The composition of the stem remains a matter of conjecture. Histologists would have it that two secretory cell types contribute to the lamellae. One of these (type I cells) contains granules with dimensions (1.2×0.7 μm), ellipsoid shapes, and contents (banded fibrous elements) that resemble those of the collagen gland. Type II cells have smaller (0.7×0.3 μm), denser granules. The collar is produced by type I cells. All external surfaces are coated by a varnish derived from a looped extension of the accessory gland (Fig. 2). Many mussels produce stems with scores even hundreds of attached threads. The advantage of the stem is that it can maintain the cohesive continuity of the byssus while undergoing continuous rejuvenation by the secretion of new threads to replace obsolescent old ones; moreover, it can be completely jettisoned leaving the mussel to start anew elsewhere.

5 Manufacturing Models

5.1 Themes of Industrial Materials Processing

Polymeric materials in industry are fabricated by as many different processes as there are materials. In the case of thermoplast polymers (materials that can be melted and reformed on cooling), all processes rely largely on the shaping of molten polymer by mechanical forces followed by the cooling of the polymer melt below its glass transition temperature. In thermosets (cross-linked materials that decompose rather than melt on heating), however, precursors such as prepolymer, catalyst, and cross-linker are mixed as liquids, and cross-linking is not initiated until some time after the material is spun or cast. Both thermosets and thermoplastic materials can be shaped by processes such as spinning, extrusion, and calendering, which lead to the formation of fibers, molds and sheets of polymer, respectively (Birley and Scott 1982). A number of natural processes such as silk formation by spiders and silkworms can be represented by analogous industrial machines (Iizuka 1983). Are natural materials thermoplasts, thermosets, or a combination of the two? This question is crucial as far as the precursors of processed materials are concerned. Precursors of thermosets must be stored separately or in nonreactive forms so as to prevent premature cross-linking. Whether thermoset or thermoplast is difficult to determine categorically with respect to structural polymers in Nature since rather few of these are thoroughly understood. On the face of it, the byssus like many natural materials would seem to be a composite of thermoplasts like uncross-linked collagen *and* thermosets like quinone-tanned DOPA-containing proteins.

Comparisons with industrial processes are valuable in that they foster a new perspective by provoking a functional dissection of natural processes. Let me attempt to demonstrate this using extrusion molding as a primitive prototype. The extruder consists of a cylindrical barrel with an aperture at one end and a motor-drive at the other (Fig. 7a). Attached to the motor by a shaft is a screw which runs the length of the cylindrical barrel. In extrusion molding, granular thermoplastic polymer is added to the hopper. Upon entering the heated chamber, the polymer melts and flows. The motor-drive turns the screw, which mixes and pushes the molten polymer toward the aperture, where a die gives shape to the extruded material. As the material cools it solidifies, and the shape assumes permanence. Although many of the mechanical details are irrelevant in a biological context, the following features are noteworthy: (1) polymeric precursors are granular (easy to store, pour); (2) processing is performed on molten polymer (more cohesive, easy to shape and mix); (3) processing often involves mechanical forces, and (4) polymeric materials need to solidify or harden before they become functional.

5.2 Extrusion and Calendering of Threads

Due to the complex microarchitecture of byssal threads, it is improbable that one process will suffice to produce all the parts. The ventral groove of the foot is often described as a die-like structure in which threads are cast, but in many respects extrusion molding followed by calendering and sizing (Birley and Scott 1982) more closely approximate thread formation (Fig. 7). Let us say we have *n* hoppers, each containing a special blend of collagen gland granules. At the proximal end of the gland the blend is 5 to 1 elastic protein to collagen, whereas at the distal end collagen predominates 5 to 1. In between, the gradient of proportions may be linear, as suggested by Gathercole and Keller (1975), or hyperbolic (Mascolo and Waite 1986). Experimental observation suggests that not all the hoppers feed their contents into the ventral groove at once. Those at the extreme distal end go first, followed progressively by their more proximal neighbors. As the

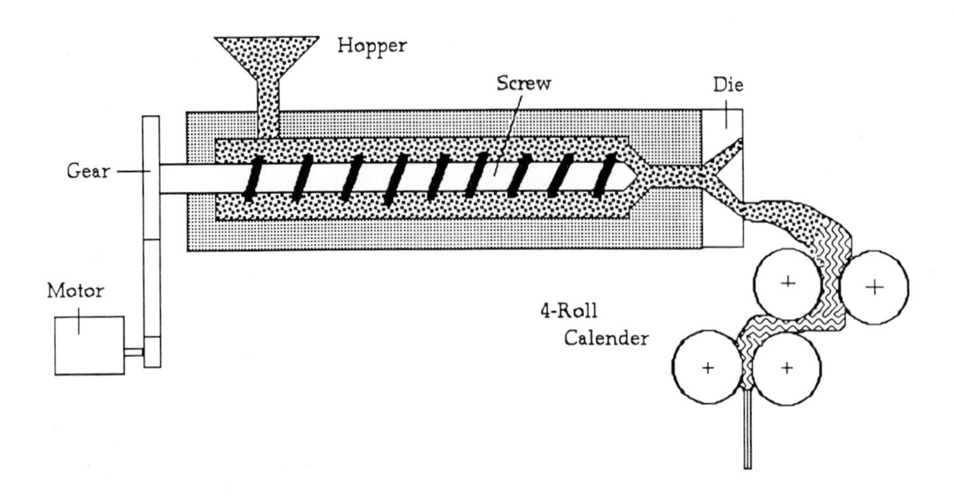

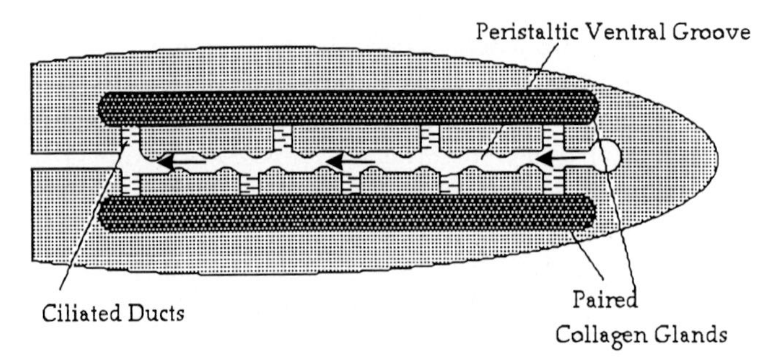

Fig. 7a, b. Manufacturing model of thread formation in Mytilus. **a** Schematic of extrusion molding and calendering. The extrudant is captured by a four-roll calender and rolled to a uniform thickness. **b** Thread-forming region of the foot. Paired collagen glands are the "hoppers" containing granular precursors that feed into the extrusion cylinder at many points. Granule contents coalesce or melt and are massaged into fibers by peristalsis. The *arrow* indicates the direction of thread growth. The multifeed connection between the glands and the ventral groove may be essential to introduce protein gradients into the thread

granules are fed into the groove, they coalesce into a mass that is extended and possibly made uniform by calendering.

Calendering is a process by which a molten polymer is squeezed into a uniform thickness (usually a sheet) through a succession of rotating rolls (Fig. 7). The rolls in the mussel would be the peristaltic movements in the walls of the ventral groove that push the molten blend of collagen and elastic protein along the groove. As Maheo (1970) suggests, the thread grows from the distal end towards the proximal, thus the prevailing direction of peristalsis should be towards the stem. Reliance on calendering to produce fibers is most unusual in Nature as well as industrial manufacturing both of which typically spin fibers to achieve constant thickness (Montcrieff 1975). The mechanism by which such extraordinarily uniform diameter and fibril orientation (Benedict and Waite 1986a) are achieved within the thread defies imagination. In fibers produced by spinning, polymer anisotropy and crystallization are typically induced by a post-spin process called cold-drawing (Montcrieff 1975). To be sure, much fiber orientation can result from the extrusion process per se (Matsuoka et al. 1991), but increased molecular alignment and strength typically result from cold-drawing which amounts to applying a moderate load to the newly spun or extruded material (Montcrieff 1975). Given the extensive byssal retractor muscle system, there is no reason why this could not be employed in cold-drawing a freshly secreted thread, but this remains conjecture. In view of the self-assembly that characterizes the in vitro fibrillogenesis of collagen (type I) (Veis 1982), there may be no need for cold-drawing of collagen fibers in mussels. In *Bombyx mori* silk formation, fibers are spun through a spinneret from a concentrated silk solution (15–30% fibroin) (Iizuka 1983). At a critical shear rate of 10^2 to $10^3\ s^{-1}$, the fibroin undergoes denaturation and coagulates forming microfibrils that merge into a fiber. The spinning speed ranges from 0.4 to 1.5 $cm\ s^{-1}$ and each fiber is lengthened to three times its original length by something resembling cold-drawing.

That aquatic midges can spin silk under water (this Vol. Case and Wieslander) in itself rules out any insuperable impediment posed by water to spinning. Why, then, should mussels have shunned the spinning of byssal threads? The answer to this may lie in the life history of the animal. The mussel foot has many responsibilities besides making byssal threads. As a juvenile before a single thread secures it to the substratum, the mussel, like the snail, has only its foot to hold it down. Thus threads must be made by the foot while it temporarily and simultaneously secures the mussel. This may essentially preempt spinning. Moreover, collagen may not lend itself to spinning. With the possible exception of a single silk (sawfly silk), there is no known spun silk with a composition like collagen (Rudall 1962).

5.3 Sizing of the Byssus

The deposition of a skin on the thread and plaque might be compared to a coating or sizing process. Sizing is a technique adopted from the textile industry involving the application of binding materials that have some "special significance in protecting or strengthening" the fabric (Collocott and Dobson 1974). Sizing materials can be applied by brushing, rolling, dipping, etc. The sizing liquor in this case comes from accessory gland granules which coalesce and cure on the surface of the plaque to form a dense varnish. Timing for the release of these granules is critical and should not be before the thread and plaque have been completely cast and spliced. Cilia on the ceiling of the distal depression (Tamarin et al. 1974) and in the ventral groove probably brush on the sizing. The skin is certainly a thermoset consisting as described above of catecholoxidase and MeFP-1 and perhaps others. The two are present together in the same accessory gland-derived granules in this way fixing their stoichiometry prior to metering and circumventing the need for mixing. The activation of the curing process may be proteolytic, but the nature and source of the protease is not known. Transmission electron micrographs of accessory gland granules and the skin of plaques and threads (Vitellaro-Zucarrello 1981) reveals a morphology with striking similarities to interpenetrating polymer networks (IPN) (Sperling 1981). IPNs are two-phased interpolymer blends in which one polymer is cross-linked in the immediate presence of the other. IPNs are particularly recommended when coatings and substrata have greatly different physical properties such as elastic moduli. Addition of the extra polymer and phase acts as to damp the incompatibilities between the two (Sperling 1981). A stiff quinone-tanned varnish over a more extensible collagenous core (strain 0.4) is likely to require such damping.

5.4 Injection Molding of Plaques

Byssal plaques resemble a foam with a skin in which the skin and foam are likely to be of different compositions (Tamarin et al. 1976; Benedict and Waite 1986a; Waite 1987). There are three known ways for making structural foams: (1) physical blowing, (2) chemical blowing, and (3) emulsive phase inversion (Shutov 1986; Williams and Wrobleski 1988). The first involves blowing an inert gas into molten or precured polymer; the second, mixing molten or precured polymer with a liquid chemical that decomposes upon heating into a gas, and the third has already been described above (Sect. 4.3). Since the plaques are not bouyant, it is unlikely that they arose by either of the first two methods. If we liken the distal depression to a mold and the plaque to a cast, then the industrial process coming closest to plaque formation would seem to be that of reaction injection molding (RIM) (Shutov 1986). RIM (Fig. 8) seems most appropriate since DOPA-containing proteins, which are the prepolymers of natural thermosets by virtue of their cross-linking, are prominent in the plaque. This process starts with two or possibly more liquid streams of prepolymers A, B, etc. which are carefully metered, mixed, and delivered to a mold. The flow rate of the two prepolymers must be carefully regulated to ensure correct stoichiometry. The mixture takes the form of the mold, reacts, and cures rapidly into a solid material. A helpful way of viewing the RIM process is to break it down into several unit operations (Shutov 1986):

Prepolymer A ↘
meter/mix → fill → cure → eject → postcure
Prepolymer B ↗

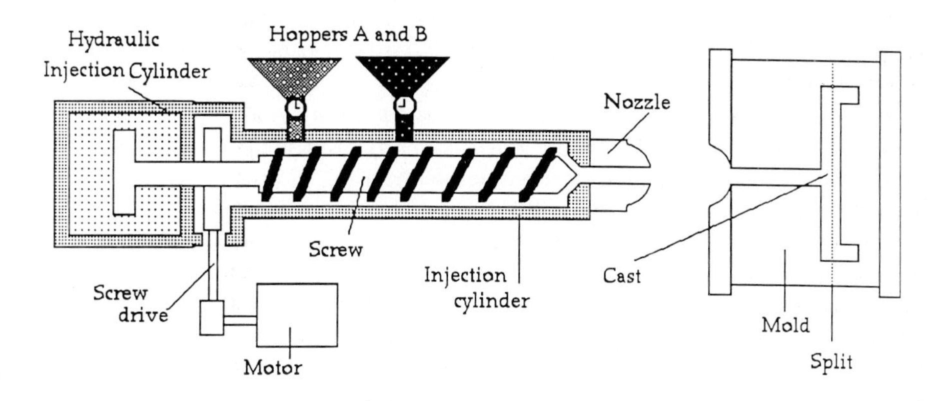

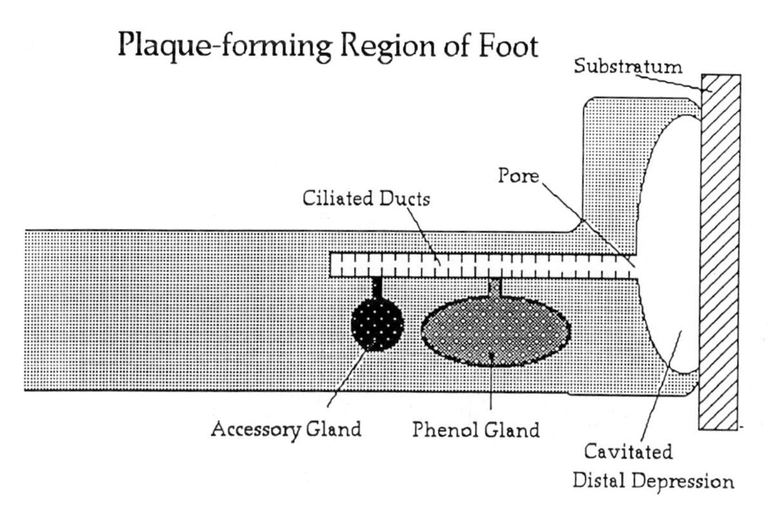

Fig. 8a, b. Manufacturing model of plaque formation in *Mytilus*. **a** Schematic of reaction injection molding. Granular precursors are metered into the injector from hoppers A and B, where they are thoroughly mixed by the motor-drive screw. When a sufficient amount of prepolymer mix has entered the injection cylinder, the hydraulic unit is engaged first to draw back the screw, and then to push out a preset quantity known as the shot volume. The shot is forced out through the nozzle which would be tightly coupled to the mold during cast formation. **b** Precursors are in the phenol and possibly accessory glands. Mixing and injection are driven by ciliated ducts and the cavitated distal depression. The nozzle and mold are represented by the pores and distal depression/substratum, respectively. The shot volume may be determined by the volume of the cavitated space

The mixing is typically mechanical and the cure and postcure generate or require heat in lieu of a catalyst to proceed. In our model of plaque formation, one or more hopper is metered into an injector. The contents of the hoppers are likely to be MeFP-2, MeFP-1, and catecholoxidase. Some stoichiometric mixture of these produces the bulk of the cast perhaps by an emulsive phase inversion that is shear-induced by the flow through the nozzle or the screw-drive, i.e., the transit through the pores or the beating cilia of the longitudinal ducts (Fig. 8B). If catecholoxidase is indeed included among the precursors of the foam, then cure of the cast involves catalytic oxidation of peptidyl-DOPA to peptidyl-*o*-quinone, the latter progressing spontaneously to quinone-derived cross-links. If not, then the mode of curing will involve another mechanism. The recent report (Okumura et al. 1989) that disulfide bond formation can lead to the enhanced stability of protein foams suggests that there may be a significant role for cystine-rich MeFP-2 in plaque foam formation. Cast ejection occurs when the foot disengages the plaque, and postcure, when the plaque tans to a golden-brown upon incubation in aerated sea water. Note that the precursor consists of granular packages of "molten protein" that coalesce when the packages are ruptured.

6 Future Challenges

Natural materials are being increasingly regarded by biologists *and* materials scientists as masterpieces of design, economy, and function. Whilst design and economy have been genetically honed by the relentless pressures of natural selection, functional performance relies, in addition, on a formidable task called proper assembly. How animals contrive to assemble exquisite materials such as byssus under the most challenging conditions defies the imagination. Up to the present, very little effort has been invested in understanding how animals make materials. The reasons for this, however, are quite understandable: the materials themselves are not as yet understood, and the available technologies for observing and dissecting assembly processes are inadequate.

In the case of the byssus, we must determine what it is made of before we can intelligently ask how it is made. All major byssal precursors must therefore be isolated and characterized, and specific probes must be made to detect them. Using these probes, precursors need to be localized in the byssus *and* in the foot. When this has been accomplished, we can then inquire when each precursor is introduced into the nascent byssus. This question has already been superficially addressed by the ingeniously simple discontinuous method of Maheo (1970), who observed that if an extended foot, preoccupied with the formation of a new thread, is gently tapped with a glass rod, it will abort its partially formed thread. By experimentally inducing thread abortion at progressively longer intervals beginning at the onset of commitment, it should be possible to reconstruct an agenda for the addition of various precursors. Finally we can return to the question of *how* the byssus is assembled. Except for plaque formation, the byssus is simply not accessible to direct observation during its assembly in the ventral groove. Fiber optics and video microscopy may some day be sufficiently miniaturized to spy on thread formation in situ, allowing investigators a dynamic glimpse of the biological versions of extrusion molding, calendering, and foam formation. But even if such technological innovation succeeds, there remains the major hurdle — the animal itself — whose behavior in the presence of humans is testy at best. Coaxing its cooperation will be the ultimate coup de grâce.

Why bother with all this? For some, including this author, the pure elegance of biological materials is sufficient inspiration. There are, however, practical reasons as well for such studies. Industry has long shown an intense interest in the manufacture of plastics with oriented polymers (Flam 1991), microcellular structures (Jackson et al. 1991), microengineered mechanical properties (Amato 1991) etc. Yet it continues to be bedeviled by the manufacturing costs of such superior materials. Ironically, materials of this kind are routinely and economically made in Nature, and the byssus is but one of these. While it is certainly unrealistic to expect that Nature's materials will have much relevance for industry's more "outlandish" materials applications (in space, at very high temperatures, and in very harsh chemical environments), the relevance of the *natural principles* governing materials design and manufacturing is not likely to be limited to Nature.

References

Adams E (1978) Invertebrate collagens. Science 202:591–597

Allen JA, Cook M, Jackson DJ, Preston S, Worth EM (1976) Observations on the rate of production and mechanical properties of the byssus threads of *Mytilus edulis* L. J Molliscan Stud 42:279–289

Amato I (1991) Heeding the call of the wild. Science 253:966–968

Bairati A, Vitellaro L (1973) The occurrence of filamentous banded elements as components of *Mytilus galloprovincialis* byssus. Experientia 29:593–594

Banu A, Shyamasundari K, Rao KH (1980) Histological and histochemical observations on the foot glands of some byssus bearing bivalves of the Waltair coast. Histochem J 12:553–563

Benedict CV, Waite JH (1986a) Composition and ultrastructure of the byssus of *Mytilus edulis*. J Morphol 189:261–270

Benedict CV, Waite JH (1986b) Location and analysis of byssal structural proteins. J Morphol 189:171–181

Bharathi N, Ramalingam K (1983) Electrophortic study of the enzyme phenoloxidase from the enzyme gland in the foot of *Perna viridis* L. J Exp Mar Biol Ecol 70:123–128

Birley AW, Scott MJ (1982) Plastics materials: properties and applications. Leonard Hill, Glasgow

Bowen JH, Mitchell PWD, Ohannessian TD (1974) Dental cement from marine sources. Tech Rep Franklin Inst Res Lab C2781:1–29

Brown CH (1950) A review of the methods available for the determination of forces stabilising proteins in animals. Q J Microsc Sci 91:331–339

Brown CH (1952) Some structural proteins of *Mytilus edulis*. Q J Microsc Sci 92:487–502

Carriker MR (1990) Functional significance of the pediveliger in bivalve development. In: Morton B (ed) The Bivalvia — proceedings of a memorial symposium in honour of Sir Charles Maurice Yonge. Hong Kong University Press, Hong Kong, pp 267–282

Characklis WG (1981) Fouling biofilm development: a process analysis. Biotechnol Bioeng 23:1923–1960

Collocott TC, Dobson AB (eds) (1974) Chambers dictionary of science and technology. Chambers Ltd, Edinburgh

Comyn J (1981) The relationship between joint durability and water diffusion. Dev Adhes 2:279–313

Cook WD (1991) Fracture and structure of highly cross-linked polymer composites. J Appl Polym Sci 42:1259–1269

Cox LR (1969) General features of Bivalvia. In: Moore RC (ed) Treatise on invertebrate paleontology, Part N, vol I/6. Geological Society of America Inc and University of Kansas Press, Manhattan KS, pp N2–N129

Day RW, Barkai A, Wickens PA (1991) Trapping of three drilling whelks by two species of mussel. J Exp Mar Biol Ecol 149:109–122

Denny MW (1988) Biology and mechanics of the wave-swept environment. Princeton University Press, Princeton, NJ
Eagland D (1988) Adhesion and adhesive performance. Endeavour 12:179–184
Filpula DR, Lee SM, Link RP, Strausberg SL, Strausberg RL (1990) Structural and functional repetition in a marine mussel adhesive protein. Biotechnol Prog 6:171–177
Flam F (1991) Plastics get oriented — and get new properties. Science 251:874–876
Fraenger W (1983) Hieronymus Bosch. GP Putnam's Sons, New York
Gathercole LJ, Keller A (1975) Light microscopic waveforms in collagenous tissues and their structural implications. In: Atkins EDT, Keller A (eds) Structure of fibrous biopolymers. Butterworths, London, pp 153–175
Gerzeli G (1961) Ricerche istomorfologiche e istochimiche sulla formazione del bisso in *Mytilus galloprovincialis*. Pubbl Stn Zool Napoli 32:88–103
Hamilton EI (1980) Concentration and distribution of uranium in *Mytilus edulis* and associated materials. Mar Ecol Prog Ser 2:61–73
Harger JRE (1970) The effect of wave impact on some aspects of the biology of sea mussels. Veliger 12:401–414
Harris VA (1990) Sessile animals of the seashore. Chapman and Hall, London
Iizuka E (1983) The physico-chemical properties of silk fibers and the fiber spinning process. Experientia 39:449–454
Jackson CL, Shaw MT, Aubert JH (1991) The linear elastic properties of microcellular foams. Polymer 32:221–225
Koide M, Lee DS, Goldberg ED (1982) Metal and transuranic records in mussel shells, byssal threads and tissues. Estuarine Coastal Shelf Sci 15:679–695
Lane DJW, Nott JA (1975) Study of the morphology, fine structure and histochemistry of the foot of the pediveliger of *Mytilus edulis* L. J Mar Biol Assoc UK 55:477–495
Lane DJW, Beaumont AR, Hunter JR (1985) Byssus drifting and drifting threads of the young post-larval mussel *Mytilus edulis*. Mar Biol 84:301–308
Lucas F, Shaw JTB, Smith SG (1955) The chemical constitution of some silk fibroins and its bearing on their physical properties. J Textile Inst T 46:440–452
Maheo R (1970) Étude de la pose et de l'activité de secretion du byssus de *Mytilus edulis* L. Cah Biol Mar 11:475–483
Mascolo JM, Waite JH (1986) Protein gradients in the byssal threads of some marine bivalve molluscs. J Exp Zool 240:1–7
Matsuoka T, Inoue Y, Takabatake J (1991) Fiber orientation and material properties analysis in injection molding. Kobunshi Ronbunshu 48:151–157 (in Japanese)
McMahon RF (1988) Respiratory response to periodic emergence in intertidal molluscs. Am Zool 244:97–114
Mercer EH (1952) Observations on the molecular structure of byssus fibers. Aust J Mar Freshwater Res 3:199–204
Montcrieff RW (1975) Man-made fibers. Newnes-Butterworths, London
Ockelmann KW (1983) Descriptions of mytilid species and definition of the Dacrydiinae N. Subfam. (Mytilacea-Bivalvia). Ophelia 22:81–123
Okumura K, Miyake Y, Taguchi H, Shimabayashi Y (1989) Enhanced stability of protein foam due to disulfide bond formation just after foaming. Agric Biol Chem 53:2029–2030
Pikkarainen J, Rantanen J, Vastamäki M, Kari K, Lampiaho K, Kulonen E (1968) On collagens of invertebrates with special reference to *Mytilus edulis*. Eur J Biochem 4:555–560
Price HA (1980) Seasonal variation in the strength of byssal attachment of the common mussel *Mytilus edulis*. J Mar Biol Assoc UK 60:1035–1037
Price HA (1983) Structure and formation of the byssus complex in *Mytilus* (Mollusca, Bivalvia). J Moluscan Stud 49:9–17
Pujol JP (1967) Le complex byssogene des mollusques bivalves: Histochimie comparee des secretions chez *Mytilus edulis* et *Pinna nobilis*. Bull Soc Linn Normandie 10:308–332
Rudall KM (1955) Distribution of collagen and chitin. Symp Soc Exp Biol 9:49–79
Rudall KM (1962) Silk and other cocoon proteins. In: Florkin M, Mason HS (eds) Comparative biochemistry, vol IV. Academic Press, New York, p 397
Rzepecki LM, Chin SS, Waite JH, Lavin MF (1991) Molecular diversity of marine glues: polyphenolic proteins from five mussel species. Mar Mol Biol Biotechnol 1:68–79
Seydel E (1909) Untersuchungen über den Byssusapparat der Lamellibanchiaten. Zool Jahrb 27:465–582
Shutov FA (1986) Integral structural polymer foams. Springer, Berlin Heidelberg New York
Smeathers JE, Vincent JFV (1979) Mechanical properties of mussel byssus threads. J Molluscan Stud 45:219–230

Smith AM (1991) The role of suction in the adhesion of limpets. J Exp Biol 161:151–169
Smyth JD (1954) A technique for the histochemical demonstration of polyphenoloxidase and its application to egg-shell formation in helminths and byssus formation in *Mytilus*. Q J Microsc Sci 95:139–152
Southgate T, Myers AA (1985) Mussel fouling on the Celtic Sea Kinsale Field gas platforms. Estuarine Coastal Shelf Sci 20:651–659
Sperling LH (1981) Interpenetrating polymer networks and related materials. Plenum, New York
Stanley SM (1972) Functional morphology and evolution of byssally attached bivalve mollusks. J Paleontol 46:165–212
Strausberg RL, Link RP (1990) Protein-based medical adhesives. Trends Biotechnol 8:53–57
Tamarin A (1975) An ultrastructural study of byssus stem formation in *Mytilus californianus*. J Morphol 145:151–178
Tamarin A, Keller PJ (1972) An ultrastructural study of the byssal thread forming system in *Mytilus*. J Ultrastruct Res 40:401–416
Tamarin A, Lewis P, Askey J (1974) Specialized cilia of the byssus attachment plaque forming region in *Mytilus californianus*. J Morphol 142:321–328
Tamarin A, Lewis P, Askey J (1976) The structure and formation of the byssal attachment plaque-forming region in *Mytilus californianus*. J Morphol 149:199–220
Turner RD, Rosewater J (1958) The family Pinnidae in the western Atlantic. Johnsonia 3:285–325
Udhayakumar M, Karande AA (1989) Byssal threads of *Mytilopsis sallei* (Recluz) and their adhesive strength. Proc Indian Acad Sci 98:65–76
Van Winkle W (1969) Mechanical properties of byssal threads. NSF-RPCT Rep Part II:1–14
Veis A (1982) Collagen fibrillogenesis. Connect Tissue Res 10:11–24
Vincent JFV (1990) Structural biomaterials, revised edn. Princeton University Press, Princeton NJ
Vitellaro-Zucarrello L (1980) The collagen gland of *Mytilus*: an ultrastructural and cytochemical study on secretory granules. J Ultrastruct Res 73:135–147
Vitellaro-Zucarrello L (1981) Ultrastructural and cytochemical study on the enzyme gland of the foot of a mollusc. Tissue Cell 13:701–713
Vitellaro-Zucarello L (1983) Ultrastructural and cytochemical study of a mucous gland of the foot of *Mytilus galloprovincialis*. Basic Appl Histochem 27:103–115
Vitellaro-Zucarrello L, De Biasi S, Blum I (1983a) Histochemical and ultrastructural study on the innervation of the byssus glands of *Mytilus galloprovincialis*. Cell Tissue Res 233:403–413
Vitellaro-Zucarrello L, De Biasi S, Bairati A (1983b) The ultrastructure of the byssal apparatus of a mussel. V. Localization of the collagenic and elastic components in the threads. Tissue Cell 15:547–554
Waite JH (1983) Adhesion in byssally attached bivalves. Biol Rev 58:209–231
Waite JH (1985) Catecholoxidase in the byssus of the common mussel. J Mar Biol Assoc UK 65:359–371
Waite JH (1986) Mussel glue from *Mytilus californianus* Conrad: a comparative study. J Comp Physiol B 156:491–496
Waite JH (1987) Nature's underwater adhesive specialist. Int J Adhesion Adhes 7:9–14
Waite JH (1990a) Marine adhesive proteins: natural composite thermosets. Int J Biol Macromol 12:139–144
Waite JH (1990b) Phylogeny and chemical diversity of quinone-tanned glues and varnishes. Comp Biochem Physiol 97B:19–29
Waite JH, Tanzer ML (1983) Polyphenolic protein of *Mytilus edulis*: novel adhesive containing L-dopa and hydroxyproline. Science 212:1038–1040
Waite JH, Housley TJ, Tanzer ML (1985) Peptide repeats in a mussel glue protein: variations on a theme. Biochemistry 24:5010–5014
Waite JH, Hansen DC, Little KT (1989) The glue protein of ribbed mussels (*Geukensia demissa*): a natural adhesive with some features of collagen. J Comp Physiol B159:517–525
Ward JE (1985) Functional morphology and histochemistry of the attachment thread of the ectoparasite *Boonea impressa*. Bull Am Malacol Union 3:97 (Abstr)
Witman JD, Suchanek TH (1984) Mussels in flow: drag and dislodgement by epizoans. Mar Ecol Prog Ser 16:259–268
Williams JM, Wrobleski DA (1988) Spatial distribution of the phases of water in oil emulsions. Langmuir 4:656–662
Williamson HC (1907) The spawning, growth and movement of the mussel (*Mytilus edulis*), horse-mussel (*Modiolus modiolus*, L), and the spoutfish (*Solen siliqua*, L). Annu Rep Fish Board Scot 25 (part III):221–254
Yonge CM (1955) Adaptation to rock boring in *Botula* and *Lithophaga* with a discussion on the evolution of this habit. Q J Microsc Sci 96:383–410

Yonge CM (1962) The primitive significance of the byssus in the bivalvia and its effects in evolution. J Mar Biol Assoc UK 42:113–125

Yonge CM (1979) Cementation in bivalves. In: van der Spoel S, van Bruggen AC, Lever J (eds) Pathways in malacology. Junk Publ, the Hague, p 83

Yonge CM, Thompson TE (1976) Living marine molluscs. Collins, London, p 167

Young GA (1985) Byssus thread formation by the mussel *Mytilus edulis*: effects of environmental factors. Mar Ecol Prog Ser 24:261–271

Young GA (1983a) Response to and selection between firm substrata by *Mytilus edulis*. J Mar Biol Assoc UK 63:653–659

Young GA (1983b) The effect of sediment type upon the position and depth at which byssal attachment occurs in *Mytilus edulis*. J Mar Biol Assoc UK 63:641–651

Young GA, Crisp DJ (1982) Marine animals and adhesion. In: Allen KW (ed) Adhesion 6. Applied Science Publishers, Barking, England, pp 19–39

Chapter 3

Reflections on the Structure of Mussel Adhesive Proteins

Richard A. Laursen[1]

1 Adhesion in Nature

Adhesion is a ubiquitous phenomenon in Nature: it is what holds things together in the living world. Examples of bioadhesion include the polysaccharide gels produced by bacteria to immobilize themselves on rocks in a rushing mountain stream (Savage and Fletcher 1985), and the receptor-mediated binding of the Arg-Gly-Asp (RGD) domains of fibrin, laminin, and other proteins involved in cell-cell adhesion (Ruoslahti and Pierschbacher 1987). These can be referred to as examples of microscopic bioadhesion, since they involve the immobilization or joining together of microscopic entities. There are, however, many examples of macroscopic bioadhesion, whereby large species of plants, animals, and insects anchor themselves to surfaces for some advantage.

One of the most striking examples of this sort of adhesion is seen in the sessile organisms that inhabit the often turbulent intertidal zones of the seacoast. In this environment, submerged half the time and exposed to air and extremes of heat and cold the other half, practically the only large permanent dwellers are certain kelps and seaweeds, barnacles and mussels, which have devised mechanisms for attaching themselves, in a more or less permanent manner, to rocks, pilings, and other surfaces. Barnacles and mussels accomplish this feat through the use of proteinaceous adhesives, which they secrete in liquid form and then cause to set into an insoluble high polymer, in the process, gluing themselves to the surface. It is remarkable that they are able to do this under water.

2 Mussel Adhesion

Macroscopic bioadhesion as practiced by mussels, especially the common blue mussel (*Mytilus edulis*), is best understood. As described in more detail elsewhere in this volume (Waite, this Vol.), these mussels attach themselves to surfaces by means of a byssus (or several of them), a collagenous thread which is attached to the animal at one end and to the surface at the other. The byssus is formed by the foot in a ventral groove that extends from the base, where the byssus attaches to the

[1] Department of Chemistry, Boston University, Boston, MA 02215, USA

Results and Problems in Cell Differentiation 19
Biopolymers
Case, S.T. (Ed.)

animal, to the tip. The distal end of the byssus is flattened into a round disk, which is glued to the surface. The byssal thread and disk are formed by secretion of collagen from accessory glands in the foot into the groove. Simultaneously, the phenol gland, located in the tip of the foot secretes the adhesive protein, which not only adheres the disk to the surface, but coats the byssus, acting as a sort of varnish to protect the byssus (Waite 1990). When the mussel wishes to make a byssal thread, it extends the foot and presses the distal end to the surface, where it apparently accomplishes some cleaning of the surface by a scrubbing action (Tamarin et al. 1976). The various proteins are secreted, and within minutes, the foot is retracted from the already insoluble thread.

During the past decade, certain mussel adhesive proteins have been isolated and at least partially characterized. In what follows, the structures and properties of these novel proteins will be discussed with the view of trying to understand their structures and how they are processed or assembled into a functional biopolymer.

3 The Structure of Mussel Adhesive Proteins

3.1 Primary Structure

3.1.1 Peptide Sequence Studies

Characterization of the mussel adhesive proteins began with the pioneering work of Waite (Waite and Tanzer 1981), who isolated, by acid extraction of the phenol glands from the feet of *M. edulis* and subsequent chromatography, a highly basic protein having an apparent molecular weight of 130 000. Amino acid analysis of the protein showed a predominance of tyrosine, lysine, proline, serine, threonine, as well as the hydroxylated amino acids, 3- and 4-hydroxyproline, and dihydroxyphenylalanine (DOPA). The presence of DOPA was of particular interest, because it is rarely found in proteins, and yet accounts for about 10% of the amino acids in the adhesive protein.

Subsequent studies by Waite and coworkers (Waite 1983a; Waite et al. 1985) suggested that the protein was composed largely of repeats, with variations, of the decapeptide, Ala-Lys-Pro-Ser-Tyr-Hyp-Thr-DOPA-Lys, and the hexapeptide Ala-Lys-Pro-Thr-DOPA-Lys, where Hyp is 3- or 4-hydroxyproline. Waite (1986) also examined the adhesive protein of the Pacific blue mussel, *Mytilus californianus*, and found that it, too, consists mostly of decapeptide repeats very similar to those found in *M. edulis*, except for the presence of significantly higher amounts of isoleucine in position 3 of the repeat decapeptide. More recently, Waite et al. (1989) isolated tryptic peptides from the adhesive protein of the ribbed mussel, *Geukensia demissa* and found several with the sequences Gly-DOPA-Lys and X-Gly-DOPA-Y-Z-Gly-DOPA/Tyr-Lys, where X was Ala in octapeptides or Gln-Thr in nonapeptides, Y was frequently hydrophobic, and Z was often Pro or Hyp. Amino acid analysis of this protein showed the presence of large amounts of Gly and Gln, as well as smaller amounts of hydroxylysine, all of which are absent in the *Mytilus* proteins, and a relatively low abundance of Pro. Both types of proteins show high percentages of Tyr/DOPA and Lys, however.

3.1.2 DNA Sequence Studies

The size and likelihood that the adhesive proteins consist largely of tandem peptide repeats precludes direct sequencing, and prompted us and investigators at Genex Corporation to initiate DNA sequencing studies on the adhesive protein genes of *M. edulis*. Figure 1 shows the amino acid sequence for the *M. edulis*

```
M E G I K L N L C L L C I F T F D V L G F S N G
N I Y N A H V S S Y A G A S A G A Y K K
L P N A Y P Y G/T K P E P V Y K P V K T
S Y S A P Y K P P T Y Q P L K K K V D Y
R P T K S Y P P T Y G S K T N Y L P L A
K K L S S Y K P I K T T Y N   (94)

A K T N Y P P V Y K   1 (104)
P K M T Y P P T Y K
P K P S Y P P T Y K
S K P         T Y K
P K I T Y P P T Y K
A K P S Y P P T Y K
P K K T Y P P T Y K
P K L T Y P P T Y K
P K P S Y P P T Y K
S K P         T Y K  10 (186)
P K I T Y P P T Y K
A K P S Y P P T Y K
P K K T Y P P T Y K
P K L T Y P P T Y K
P K P S Y P P T Y K
P K P S Y P P S Y K
T K K T Y P P T Y K
P K L T Y P P T Y K
P K P S Y P P S Y K
P K K T Y P P T Y K  20 (286)
P K L T Y P P T Y K
A K P S Y P P T Y K
A K P S Y P P T Y K
A K P S Y P P T Y K
A K P         T Y K
A K P T Y P S T Y K
A K P T Y P P T Y K
A K P S Y P P T Y K
A K P S Y P P T Y K
A K L        .T Y K  30 (378)
A K P S Y P P T Y K
A K P S Y P P T Y K
A K P S Y P P T Y K
A K P S Y P P T Y K
A K P S Y P P T Y K
V K P         T Y K
A K P T Y P S T Y K
A K P S Y P P T Y K
A K P S Y P P T Y K
A K P S Y P P T Y K  40 (474)
A K P S Y P P T Y K
A K P T Y P S T Y K
A K P         T Y K

A K P S Y P P T Y K
A K P S Y P P T Y K
A K P S Y P P T Y K
A K P         T Y K
V K P T Y P S T Y K
A K P S Y P P T Y K
A K P S Y P P T Y K  50 (566)
A K P S Y P P T Y K
A K P S Y P P T Y K
A K P S Y P P T Y K
A K P S Y P P T Y K
A K P         T Y K
A K P T Y P S T Y K
A K P S Y P P T Y K
A K P S Y P P T Y K
A K P         T Y K
A K P         T Y K  60 (654)
A K P T Y P S T Y K
A K P S Y P P T Y K
A K P         A Y K
A K P T Y P S T Y K
A K P T Y P S T Y K
A K P S Y P P T Y K
P K I S Y P P T Y K
A K P S Y P S T Y K
A K S S Y P P T Y K
A K P         T Y K  70 (746)
A K P T Y P S T Y K
A K P         T Y K
A K P T Y P P T Y K
A K P S Y P P T Y K
P M P S Y P P T Y K
S K S S Y P S S Y K
P K K T Y P P T Y K
P K L T Y P P T Y K
P K P S Y P A S Y K
P K I T Y P S T Y K  80 (842)
L K P S Y P P T Y K
S K T S Y P P T Y N
K K I S Y P S S Y K
A K T S Y P P A Y K  84
P T N R Y *             (897)
```

Fig. 1. Amino acid sequence of *M. edulis* adhesive protein deduced from cDNA sequence (Ou 1990). The *underlined portion* is the signal peptide. *Numbers to the right* indicate the tandem repeats; *numbers in parentheses* refer to positions in the amino acid sequence. The *slash* (/) indicates the location of a presumed intron, based on the genomic DNA sequence (see Fig. 2); the location of a stop codon in the DNA sequence is indicated by an *asterisk* (*)

protein translated from the cDNA sequence (Ou 1990). The precursor protein contains 897 amino acids, including a signal sequence of 24 residues. Cleavage of the signal peptide during processing yields the mature protein with the amino-terminal sequence Asn-Ile-Tyr-Asn-. . . , which was confirmed by direct sequencing of the protein. This protein contains 84 tandem peptide repeats, of which 71 are decapeptides and 13 are hexapeptides. A relatively short nonrepeat domain containing 94 residues is found at the amino-terminus. Figure 2 shows the sequence deduced by the Genex group (Filpula et al. 1990) from the genomic DNA sequence. This sequence, which contains 86 repeats, is incomplete, lacking the first 52 residues seen in the cDNA sequence, due apparently to the presence of an intron between the codons for Gly-52 and Thr-53 (cf. Figure 1). It is remarkable that the

```
. . intron. . . T K H E P V Y K P V K T
S Y S A P Y K P P T Y Q P L K K K V D Y
R P T K S Y P P T Y G S K T N Y L P L A
K K L S S Y K P I K T T Y N

A K T N Y P P V Y K   1
P K M T Y P P T Y K
P K P S Y P P T Y K
S K P         T Y K
P K I T Y P P T Y K
A K P S Y P S S Y K
P K K T Y P P T Y K
P K L T Y P P T Y K
P K P S Y P P T Y K
P K P S Y P P S Y K  10
T K K T Y P S S Y K
A K P S Y P P T Y K
A K P S Y P P T Y K
A K P S Y P P T Y K
A K P         T Y K
A K P T Y P S T Y K
A K P S Y P P T Y K
A K P         T Y K
A K P S Y P P T Y K
A K P S Y P P T Y K  20
A K P S Y P P T Y K
A K P         T Y K
A K P         T Y K
A K P         T Y K
A K P S Y P P T Y K
A K P S Y P P T Y K
A K P S Y P P T Y K
A K P S Y P P T Y K
A K P S Y P P T Y K
A K P S Y P P T Y K  30
A K P         T Y K
A K P T Y P S T Y K
A K P S Y P P T Y K
A K P S Y P P T Y K
A K P         T Y K
A K P T Y P S T Y K
A K P S Y P P T Y K
A K P S Y P P T Y K
A K P         T Y K
A K P T Y P S T Y K  40
A K P S Y P P T Y K
A K P S Y P P T Y K
A K P         T Y K
A K P T Y P S T Y K
A K P S Y A S T Y K
A K P S Y P P T Y K
S K S S Y P S S Y K
P K K T Y P P T Y K
P K L T Y K P T Y K
P K P S Y P P S Y K  50
P K T T Y P P T Y K
P K I S Y P P T Y K
A K P S Y P P T Y K
A K P S Y P P T Y K
A K P S Y P P T Y K
A K P S Y P P T Y K
A K P         T Y K
A K P T Y P S T Y K
A K P S Y P P T Y K
A K P S Y P P T Y K  60
A K P S Y P P T Y K
A K P T Y P S T Y K
A K P S Y P P T Y K
P K I S Y P P T Y K
A K P S Y P P T Y K
A K P S Y P P T Y K
A K P         T Y K
A K P T N P S T Y K
A K P S Y P P T Y K
A K P S Y P P T Y K  70
A K P S Y P P T Y K
A K P       P T Y K
A K P T Y P S T Y K
A K P         T Y K
A K P T Y P P T Y K
A K P S Y P P T Y K
P K P S Y P P T Y K
S K S I Y P S S Y K
P K K T Y P P T Y K
P K L T Y P P T Y K  80
P K P S Y P P S Y K
P K I T Y P S T Y K
L K P S Y P P T Y K
S K T S Y P P T Y N
K K I S Y P S S Y K
A K T S Y P P A Y K  86
P T N R Y *
```

Fig. 2. Amino acid sequence of *M. edulis* adhesive protein deduced from the genomic DNA sequence (Filpula et al. 1990). Tandem repeats are numbered on the *right*. Regions which are identical in sequence to the cDNA sequence (see Fig. 1) are *underlined*

entire coding region for the sequence depicted in Fig. 2 consists of a single giant exon, i.e., there are no introns within the tandem repeat region.

The calculated molecular weights for the *M. edulis* adhesive protein are 101 845 for the cDNA-derived sequence, and 100 412 for the genomic fragment. Assuming approximately 70% hydroxylation of the Tyr and Pro residues, it can be estimated that the molecular weight of the mature *M. edulis* adhesive protein is about 106 000, which is substantially lower than the 130 000 reported by Waite and Tanzer (1981). It is possible that the high apparent molecular weight reflects a highly extended structure, resulting in retarded migration during polyacrylamide gel electrophoresis.

Sequencing of a small cDNA clone from *M. californianus* (Fig. 3) revealed a close similarity with the *M. edulis* protein, except for the appearance of Arg in position 1 of the decapeptide repeat and a higher frequency of Ile and Thr in position 3.

Although the adhesive proteins from the two *Mytilus* species are clearly related, that from *G. demissa* seems to be a member of a different class. cDNA sequencing (Shen 1991; Fig. 4) showed that the *G. demissa* protein differs not only in amino

```
                Y K
R K I S Y P P T Y K
T K P S Y P A T Y K
R K I S T P P T Y K
T K P S Y P A S Y K
R K T S Y P S T Y K
R K T S Y P P T Y K
P K I S Y P S T Y K
T K P S Y P P T Y K
P K P S Y
```

Fig. 3. Amino acid sequence of a *M. californianus* adhesive protein fragment deduced from cDNA (M. Connors, unpubl. results)

```
          TGYSADYK      AKP       SPYVPGYK
GKP       SSYDPGYK        GQQKQ   TGYLPGYK
  GQQKQ   TGYDTGYK        GQQKQ   TAYDPGYK
  GQQKQ   TAYDPGYK        GQQKQ   TAYNPGYK
     GGVKKTGYSADYK           GGVQKTGNSAGYK
GKP       SSNVPGYK      AKP       SPYVPGYK
  GQQKQ   TGYVPGYK        GQQKQ   TGYVPGYK
  GQQKQ   TGYDKGYK        GQQKQ   TGYDLGYK
     GGVQKSGYSAGYK        GQQKQ   TAYDPGYK
AKP       SPYVPGYK           GGVQKTGYSADYK
  GQQKK   TGHVPGYK      GKP       SSNVPGYK
     AGVQKTGYSAGYK        GQQKQ   TGYDPGYK
AKP       SPYVPGYK        GQQKQ   TGYDKGYK
  GQQKQ   TGYLPGYK        AQQKQ   TGYDKGYK
  GQQKQ   TAYDPGYK           GGVQKTGYSAGSK
  GQQKK   NGYVPGYK      GKP       SSYDPGNK
     GGVQKTGYSAGYK        GQQKQ   TGYDTGYK
AKP       SPYVPGYK           GVAQKSGYTAGYK
  GQQKQ   TGYLPGYK      GKP       NLFDAGYK
  GQQKQ   TAYDPGYK        GQQKQ   TGYDPLYK
  GQQKQ   TAYNPGYK           GGVQKTGYSIGYK
     GGVQKTGNSAGYK        GQTPY (COOH)
```

Fig. 4. Partial amino acid sequence of *G. demissa* adhesive protein deduced from cDNA (clone λGd6.10) sequence (Shen 1991). A smaller clone (not shown) had a different repeat pattern, but was identical at the carboxyl terminus (*underlined*)

acid composition, being rich in Gly and Gln, and relatively poor in Pro, but also in the repeat structure, which consists of variants of a tripeptide and two pentapeptides that alternate with an octapeptide. The consensus peptides are:

Gly/Ala-Lys-Pro

Gly-Gln-Gln-Lys-Gln

Gly-Gly-Val-Gln-Lys

Thr/Ser-X-Tyr-X-X-Gly-Tyr-Lys.

Despite these differences, there are similarities between the *Mytilus* and *Geukensia* proteins, in that they both contain about 20% each of Lys and Tyr and that these residues are similarly located in the peptide chain:

xKxxYxxxYK (*Mytilus* and *Geukensia*)

KxxxYxxxYK (*Geukensia*).

3.1.3 Motifs

Forming a coherent picture of a possible common secondary or tertiary structure for the adhesive proteins has become increasingly problematic as more primary structure data have appeared. Rzepecki et al. (1991) have recently reported peptide sequence data on five new species of adhesive protein. Taking into account these data (Figure 5), about the only common motif that can be discerned is the sequence, – –K– – –Y– – –Y– –K– –, where, typically, there are two to four intervening residues between the Tyr and Lys residues. Usually these residues are the strong β-turn formers Pro/Hyp and Gly or the hydroxylated amino acids, Thr and Ser. In all of the proteins, a high proportion of the tyrosines appear as the hydroxylated derivative, DOPA. The irreducible common feature shared by these proteins seems to be the presence of roughly equal proportions (about 20% each) of

Sequence	Species
G Q Q K Q T G Y X P G Y K G K P S P Y X P G Y K	*Geukensia demissa* and *Brachidontes exustus*
Y P A K P S S Y G T G Y K	*Septifer bifurcans*
A K P S - Y P P T Y K	*Mytilus edulis, M. californianus*
X H K P - - Y - T G Y K	*Mytella guyanesis*
X Y Y P K,G Y G A K	*Trichomya hirsuta*
S S Y Y P X G Y G K	*Modiolus modiolus squamosus*
G D C G N - Y G G G Y G G G C	*Schistosoma mansoni* (eggshell)
· · K · · · Y · · · Y · · K	Consensus sequence

Fig. 5. Sequence motifs in mussel adhesive proteins. Only some representative sequences are included here; other variants have also been described (Rzepecki et al. 1991). A sequence motif from *S. mansoni* eggshell protein (Rodriques et al. 1989) is included for comparison. In this case cysteines apparently replace lysine in forming cross-links

lysine and tyrosine (or DOPA), which is understandable given the suspected role of these amino acids in forming cross-links (see Sect. 3.1.6).

3.1.4 Microheterogeneity

Even in the early stages of DNA sequencing, it became evident that some sort of heterogeneity exists in the adhesive proteins. Comparison of the sequences for *M. edulis* shown in Figs. 1 and 2 gives an indication of the differences. Furthermore, in our own studies (Ou 1990), we isolated several small clones that only partially overlapped with the sequence shown in Fig. 1; and the Genex group (Strausberg et al. 1989) sequenced a cDNA clone, 14-1, that only partially overlapped their genomic sequence (Fig. 2). In addition, we sequenced another *G. demissa* clone, slightly smaller than that shown in Fig. 4, that was identical in the carboxyl terminal 67 amino acids, but differed in sequence and repeat pattern in the amino terminal region.

There are several possible explanations for such microheterogeneity: rearrangement during cloning, multiple genes, alternative exon splicing mechanisms, population variation and others. Recombination during cloning seems unlikely, since in our studies (Ou 1990) we used *rec*$^-$ host cells. Furthermore, variation was seen in both the repeat patterns and in individual amino acids within repeats; and repeat domains were always intact, indicating that slippage or rearrangement of blocks of sequence had not occurred. In their studies of the genomic sequence, Filpula et al. (1990) found no evidence for more than one genomic sequence. Alternative exon splicing seems unlikely, at least for the *M. edulis* protein, since the entire tandem repeat domain is located within a single exon (Filpula et al. 1990). Population variation cannot be ruled out at this point, since the mussels used in these studies were collected at locations ranging from Massachusetts to Delaware.

Heterogeneity is also seen at the protein and RNA level. We, and Waite (Waite 1983a; Waite et al. 1989) have noted that both the *M. edulis* and *G. demissa* adhesive proteins run as closely spaced doublets on acid gel electrophoresis, but not as a broad smear, as might be expected for a large population of proteins with similar molecular weights. Actually the lack of greater heterogeneity is a little surprising, considering that hydroxylation of the Tyr and Pro residues can increase the molecular weight by nearly 5000. This implies a similar degree of modification of all protein molecules by the time the protein is isolated. Shen (1991) noted in studies of the *G. demissa* sequence the presence of two RNA bands that hybridize with probes complementary to the adhesive protein gene sequence. Whether these RNAs encode proteins of different sequence is not known, however.

Whatever the reasons for the heterogeneity, it is not without precedent for repetitive proteins. For example, Bobek et al. (1988), find some variation in the eggshell protein of *Schistosoma. mansoni*, though less than that seen for the adhesive proteins. The nature of the variation seen suggests several things about the relationship of structure to function for the adhesive proteins: (1) the overall repeat domain structure is essential, since repeat sequences are never interrupted, (2) some variation of sequence is permitted within repeat domains, particularly at certain positions, (3) the pattern of repeats (e.g., of hexa- and decapeptides in the

M. edulis protein) is not critical, and (4) the presence and location f the Tyr and Lys residues is critical, since they are almost completely invariant.

3.1.5 Similarities with Other Proteins

The mussel adhesive proteins are unusual in two respects, the first being their high content of the amino acid DOPA, which is rarely seen in proteins, and the second being their tandem repeat structures. It is therefore of interest to see whether other proteins share these properties. Waite (1990) has catalogued known DOPA-containing proteins, only a few of which have been sequenced. All of these appear to be proteins with adhesive or other structure maintaining properties, as found in egg cases, mollusk shells, byssus, etc. In these cases, the DOPA is presumed to participate in the quinone-tanning process (see below) that imparts stability to the structures.

A search of the protein sequence data base reveals a few proteins that display some sequence similarity with the mussel adhesive proteins, but no convincing homology. The major eggshell protein of the fluke, *S. mansoni*, is extremely rich in Gly (44%) and Tyr/DOPA (11%) (Bobek et al. 1986; Rodriques et al. 1989), and in this respect resembles the *G. demissa* adhesive protein (Fig. 4). The eggshell protein contains repeats of 16–18 residues and differs from the mussel proteins in containing Cys rather than Lys in many of the repeat domains. It is interesting, however, that the spacing of the Tyr and Cys residues is similar to the Tyr and Lys spacings in the mussel proteins (see Fig. 5 and discussion in Sect. 3.2.4.3). Other proteins that have high alignment scores with the *G. demissa* protein are collagen, which consists largely of tripeptide repeats (Gly-X-X/Pro) and glutenin, the seed storage protein of wheat endosperm. One of the glutenin proteins consists largely of two repeats: Gly-Gln-Gln-Pro-Gly-Gln and Gln-Gln-Pro-Gly-Gln-Gly-Gln-Gln-Gly-Tyr-Pro-Thr-Ser (Sugiyama et al. 1985). With both collagen and glutenin, the high apparent similarity with the adhesive protein seems to be due to high content of Gln and/or Gly, rather than true homology. It is noteworthy, however, that both collagen and glutenin have adhesive properties. Degraded forms of collagen extracted from hides have long been used as glues (see Sect. 4.2.2), and glutenin gives wheat flour some of its stickiness, as in bread dough.

Several proteins show similarities with *M. edulis* adhesive protein, although there is no convincing homology. The similarity seems to be due to the fact that all of these proteins are rich in Pro, Ser, Thr, Tyr, and/or Lys, and consist largely of tandem repeats. These include the cell wall proteins, soybean SbPRP1 (Pro-Pro-Val-Tyr-Lys) (Hong et al. 1987) and carrot extensin (variations of Ser-Pro-Pro-Pro-Pro-Thr-Pro-Val-Tyr-Lys-Tyr-Lys) (Chen and Varner 1985), and the sgs-3 glue protein from *Drosophila melanogaster* (Pro-Thr-Thr-Thr-Lys) (Garfinkel et al. 1983) the latter of which has adhesive properties.

3.1.6 Cross-Links

For proteins as intrinsically water-soluble as those under discussion to function as adhesives in a wet environment, they must be converted to a water-insoluble form, for example, by cross-linking. Based on the presence of high levels of DOPA, it has been postulated that the adhesive proteins undergo cross-linkage by the process known as quinone-tanning, whereby the catechol moiety is further oxidized to an ortho-quinone, which then reacts with a lysine amino group (Fig. 6). To date, the structure of the cross-link has not been established, and other mechanisms of cross-linking are possible (for a review of quinone-tanning, see Waite 1990). However, the presence of roughly equimolar amounts of lysine in all of the adhesive proteins is consistent with the Michael addition mechanism shown in Fig. 6.

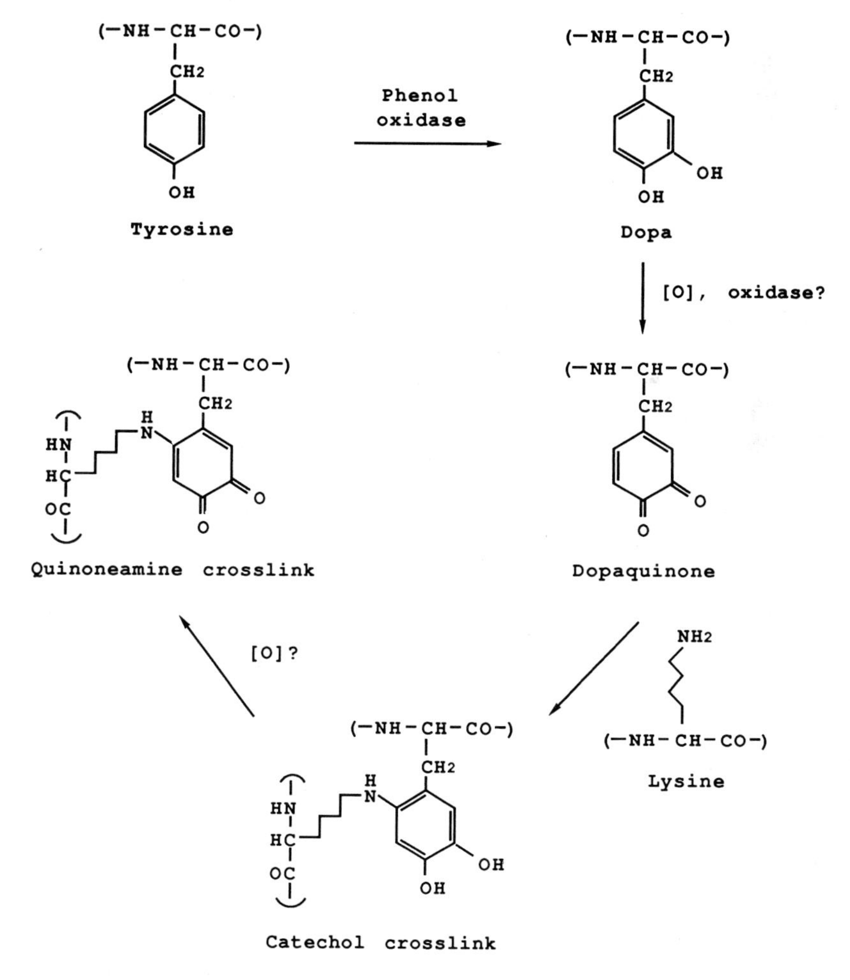

Fig. 6. Postulated mechanism for cross-link formation in mussel adhesive proteins involving Michael addition of lysine to dopaquinone. Other cross-link structures are possible. (Waite 1990)

3.2 Secondary Structure

3.2.1 Do the Adhesive Proteins Have Secondary Structure?

It has been argued (Waite 1983b, Williams et al. 1989) that the adhesive proteins may possess an open, unordered conformation in solution, rather than the ordered or polycrystalline structures found in certain synthetic polymers and natural structural proteins such as collagen, silk, keratin, etc. An unordered conformation could be an advantage for adhesives, since it would maximize the opportunities binding of the polymer side chains to surfaces. On the other hand, there are several arguments a priori in favor of some kind of regular structure in the functional state.

First, it would be an unusual, perhaps unique, protein that did *not* have a definite conformation. If there were no general structural requirement for function (adhesion to surfaces, and cohesion due to crosslinking of polymer chains) what would be the evolutionary pressure to maintain so precisely the repeat units and the arrangement of Tyr and Lys residues within repeats? In regard to this point, von Heijne and Nishikawa (1991) have postulated that chloroplast transit peptides, which, like the adhesive proteins, are rich in Ser, Thr, Gln, and Asn, and are "perfect random coils", i.e., they are devoid of any regular secondary structure. However, these peptides, as well as random coil domains seen in proteins of known three-dimensional structure, are *not* found in tandem repeat domains. Rather they occur as peptides of varying length, with turn-promoting residues dispersed irregularly throughout the sequence.

Second, Waite and coworkers (Waite et al. 1985; Marumo and Waite 1986; Williams et al. 1989) have noted that there is differential reactivity of Tyr and Pro residues with hydroxylating enzymes in the *M. edulis* protein and a recombinant analog. In particular, Tyr-9 is hydroxylated up to about nine times faster than Tyr-5, and Pro-6 and Pro-7 are hydroxylated predominantly at the 3- and 4-positions, respectively. In *G. demissa*, it is the tyrosine corresponding to Tyr-5 in *M. edulis* that is more rapidly hydroxylated (Waite et al. 1989). These differences imply that the local environments of the Tyr and Pro residues are different, as they would be if the repeat domain had a discrete conformation.

Finally, there is the analogy with collagen, a protein rich in proline and glycine, which itself has been used (in the degraded form, gelatin) as an adhesive. Although gelatin in solution is essentially random coil, it forms the characteristic, if imperfectly aligned (Engel and Prockop 1991) collagen triple helix on gelation (Veis 1964), and it only becomes a good adhesive after reaching the gel stage (Hubbard 1977).

In summary, an inspection of the sequence data in Figs. 1–4 shows very clearly that there is ordering of the repeat domains along the polypeptide chain of the adhesive proteins. In addition, the conservation of sequence and the hydroxylation data argue for a definite folded structure within the repeat units. It follows then that these units should be arranged in some sort of higher order array, since each repeat unit is in more or less the same environment. The question then is, "What is that higher order structure?"

3.2.2 Physicochemical Studies

There have been few studies of adhesive protein secondary structure using physical chemical methods, no doubt in part because of the difficulties of obtaining large amounts of the protein. Williams et al. (1989) analyzed both native *M. edulis* protein and a non-hydroxylated recombinant protein consisting of 20 repeats of the peptide, AKPSYPPTYK, by circular dichroism (CD) spectrometry under a variety of conditions. Their results suggested the presence of 70–80% of "random" structure with perhaps about 18% of β-turn, which was unchanged by the presence of denaturants such as 6 M guanidinium chloride. These conclusions were supported by the Chou and Fasman (1978) analysis, which predicted a similar amount of β-turn, but no β-sheet or α-helix.

However, the conclusions from the CD studies cannot be regarded as definitive for the situation in vivo, becuase they were done with very dilute (0.03–0.1%) solutions and under conditions that probably do not duplicate well those (which are essentially unknown) that exist during secretion and processing of the protein by the mussel. In particular, the cured adhesive plaques appear to contain from about 30–100% solids (Waite 1986; Waite et al. 1989), presumably much of which is adhesive protein, so it seems likely that the protein performs its function at much higher concentration and under conditions that might alter its conformation. Furthermore, Chou and Fasman (1978) analysis is not very useful for deducing the conformations of structural proteins with repeat structures, since these tend to have unusual conformations (e.g., the triple polyproline helix of collagen and the β-spiral of elastin). The prediction of β-turns probably has validity, though one might predict a priori that any protein that contains nearly 30% of glycine or proline is likely to contain turns.

Recently, Trumbore (1991) examined the structure of *M. edulis* adhesive protein using small angle X-ray diffraction methods, and concluded that the protein had structure of a prolate ellipsoid 20 Å in diameter and 284 Å long, and that it was not aggregated, at least under the experimental conditions used. On the basis of molecular modeling and energy minimization studies, he further concluded that the consensus decapeptide repeat unit was most likely a 3_{10} helix, although, since a fully extended 3_{10} would have a length of 1200 Å, it would have to be folded to fit the diffraction data model. This is an interesting proposal, but it is difficult to see what would provide the stabilizing force for such a structure. Long stretches of 3_{10} helix are rarely seen, though the 3_{10} helix frequently appears as the last turn of an α-helix (Richardson 1981). Two or more consecutive residues in the 3_{10} conformation can make good tight turns, however, which may be significant in the structures proposed below.

3.2.3 Antibody Cross-Reactivity

There has been no comprehensive study of the antigenic properties of the mussel adhesive proteins, but there is some immunochemical evidence for structural similarities between the *G. demissa* and *M. edulis* proteins. Shen (1991) raised polyclonal antibodies against the *G. demissa* protein which were used for immunoscreening cDNA clones. Since the expression system used did not contain tyrosine and proline hydroxylases, a positive reaction means that the antibodies recognized the nonhydroxylated form of the protein as well as the hydroxylated. Furthermore, these antibodies also cross-reacted with the *M. edulis* protein, an observation confirmed by Waite et al. (1989). No cross-reactivity was seen with the *Modiolus modiolus* adhesive protein, however (Connors M, unpubl. observation). It is possible that this cross-reactivity is due to some common conformational epitope or simply to a sequence epitope, such as Tyr-Lys, which is common to the two classes of protein. The latter might explain why cross-reactivity was not seen with the *Modiolus* protein, which appears to lack this sequence (Rzepecki et al. 1991; see Fig. 5).

3.2.4 Models

3.2.4.1 Fibrous Structural Proteins

Structural proteins are widespread in Nature, but relatively little is known about their structures, compared with soluble globular proteins such as enzymes. As a class, these proteins tend to be large, insoluble, noncrystalline (and hence not amenable to X-ray crystallography), and they tend not to have easily assayable functions, making structure-function studies difficult. These proteins have been designed for many properties, including high tensile strength (α-keratin, silk, collagen), elasticity (elastin), and performance as binding agents (bone collagen, shell matrix proteins) (Dickerson and Geis 1969).

In trying to arrive at possible structures for the adhesive proteins, it is necessary to consider what functions they need to perform. As will be discussed in the following section, the protein must bind to surfaces and achieve cohesive strength. Therefore functional groups must be available for surface-surface interactions and for interchain cross-linking. It is probably desirable that the cured (cross-linked) protein have some elasticity, since the adhesive plaque between the surface and the byssal disk, is subjected to constant tugs and pulls as the mussel is buffeted by the action of the waves. A rigid structure would tend to break, while a flexible one would give. The model should also take into account the high net positive charge (+187 in the case of the *M. edulis* protein), which needs to be dispersed to minimize repulsive interactions.

The best understood structural proteins, from a secondary structure point of view, are the fibrous proteins, silk, α-keratin, and collagen (cf. Dickerson and Geis 1969), because their physical states make them amenable to study by fiber X-ray diffraction methods. α-Keratin, as found in wool, consists of three α-helixes arranged in a left-handed coil. This "protofibril" is then cross-linked through disulfide bonds and bundled together with other fibrils and embedded in a matrix protein to form a hair. α-Keratins are highly extensible, due to stretching of the α-helixes and to the restorative effect of the disulfide bonds. The high-sulfur α-keratin matrix protein (Ellenen and Dopheide 1972) is perhaps more analogous to the mussel adhesive proteins, since it contains tandem repeats and acts as a sort of glue to bind the fibrils together. Its secondary structure is not known, however.

The crystalline portions of silkworm silk are composed of repeats of the peptide, Gly-Ser-Gly-Ala-Gly-Ala, arranged as layers of β-sheet. Silk has a very high tensile strength because the resistance to tension is borne by covalent bonds. It is not very extensible, however, and what extensibility it does have is due mainly to noncrystalline regions arising from bulky side chains. Neither silk nor α-keratin seems like a very good model for the mussel adhesive proteins, since it is unlikely these proteins contain α-helix or β-sheet.

On the other hand, there are a number of analogies between the adhesive proteins and collagen, one being the high content of proline/hydroxyproline and glycine. Collagen consists mainly of repeats of the sequence, Gly-X-Y, where Y is usually Pro or Hypro. Collagen chains form left-handed polyproline helixes, three of which coil together to form a right-handed triple helix, which is stabilized by interchain hydrogen bonds. There are many types of collagen, ranging from the strong, rigid, highly helical fibrous types I, II, and III (and others) to the nonfibrillar, networked type IV, which occurs in basement membranes (Mayne and Burgeson 1987). Furthermore, during byssus formation, both the collagen filaments,which make up the stem and disk of the byssal thread, and the adhesive protein, which coats and embeds the filaments in the adhesive plaque (Waite 1983b), are laid down simultaneously. Thus the adhesive protein may play a role similar to type IV anchoring collagen.

The elastomeric protein, elastin, contains many repeats of the peptide, Val-Pro-Gly-Val-Gly, which are believed to fold into a helical array of type II β-turns called a β-spiral (Urry 1990). The spiral is stabilized by hydrophobic interactions between turns of the spiral. Stretching the spiral exposes the hydrophobic groups to water, which is energetically unfavorable, and thus restoration to the normal β-spiral conformation is hydration-mediated. This type of structure is probably not applicable to the adhesive proteins, however, since they contain few hydrophobic groups.

3.2.4.2 Ice Nucleation Protein

The bacterial ice-nucleation proteins are not structural proteins in the usual sense, but they do have structural similarities to the adhesive proteins. The ice nucleation protein from *Pseudomonas syringae* contains 122 imperfect repeats of the octapeptide, Ala-Gly-Tyr-Gly-Ser-Thr-Leu-Thr, but apparently no cross-links. The secondary structure of this protein is unknown, but Mizuno (1989) has proposed, based on minimum energy calculations, that the octapeptides fold into discrete multi-β-turn domains, which are arranged into a helix or spiral containing six domains per turn. Stabilization for this structure is provided, in part, by the hydrophobic side chains of leucine, which project into the core of the helix.

3.2.4.3 Eggshell Proteins

As discussed above, the eggshells of the parasite, *S. mansoni*, contain glycine-rich proteins which are believed to undergo cross-linkage by quinone tanning. Rodriques et al. (1989) have proposed a β-sheet model for the eggshell protein in which several 16–18-residue repeat domains form β-hairpins which lie in a simple meandering plane, with the glycines in the center of the sheet and the tyrosines and cysteines at the turns. In this model, Tyr (or DOPA) Cys are located at the edges of the sheet where they could cross-link with other residues in other sheets. Another *S. mansoni* protein contains 80 repeats of the sequence, GYDKY, and Middaugh et al. (1990) have concluded, based on studies with a synthetic model, that this protein may exist as a left-handed α-helix, which, if true, would be the first known example of such a structure.

3.2.4.4 Mussel Adhesive Protein: a Hypothetical Model

Despite the sequence diversity seen with the mussel adhesive proteins (see Fig. 5), one might expect that they have similar secondary or tertiary structures, as is observed with many classes of globular proteins (Creighton 1983). For example, the myoglobins and hemoglobins vary up to about 80% in amino acid sequence, but their folded structures are practically identical with respect to their function of binding heme.

Although it is difficult to propose an exact structure for the mussel adhesive proteins at this time, some sort of helical structure seems most attractive. The regularity of the repeat domains seen in both the *Mytilus* and *Geukensia* proteins suggests that they do have definite folded structures, probably consisting of multiple β-turns, as postulated for elastin (Urry 1990) and the ice nucleation protein (Mizuno 1989). Given a linear array of essentially identical subunits, and assuming that they fold into some sort of regular higher order structure, it follows that their environments within that structure will be identical, in order to achieve minimum energy. As pointed out by Mizuno (1989), the structure that best satisfies these conditions is a helix.

A coiled coil type of helix could accomodate several types of domain (e.g., three types in the case of the *G. demissa* protein; see Fig. 4), permit dispersal of charge due to the lysine ammonium groups, allow for maximum exposure of other side chains involved in cross-linking and binding to surfaces, and show some elasticity. A rigid, rod-like structure is not envisioned, but rather one where the helix is interrupted periodically by bends. For example, in the *M. edulis* protein, hexapeptides are dispersed irregularly among the decapeptide repeats. These may act as defects and induce breaks in the helical structure, in analogy with type IV collagen (Glanville 1987). As can be seen by comparing the sequences in Figs. 1 and 2, the patterns of the hexapeptide insertions are completely different. It may be unimportant for function where the defects are as long as the helix is broken up periodically.

Thus the model taking shape is that of a helix with periodic, but irregularly spaced kinks in it — like a screen door spring that has come undone and been slammed in the door too many times. Such a structure, after cross-linking, would give rise to a flexible, mesh-like matrix that would serve to embed the byssal collagen fibrils and bind them to surfaces, in analogy with the collagen fibril "anchoring plaques" composed of type IV and VII collagen (Keene et al. 1987). A β-sheet type of structure, such as proposed (Rodriques et al. 1989) for the *S. mansoni* eggshell protein, cannot be discounted entirely, but a flattened structure such as this would seem to make more sense for the construction of a sac-like structure, such as an egg case, than an embedding material.

4 Assembly of Monomeric Proteins to Form a Functional Adhesive

4.1 General Considerations

Given the lack of evidence (cf. Williams et al. 1989) that the mussel adhesive proteins have any structure at all, how could a highly hydrated, polycationic polymer coalesce to form an ordered, relatively dense, insoluble adhesive plaque? How is the net positive charge reduced, which would seem necessary not only to prevent repulsion between polymer chains, but also because a neutral amino group is needed for nucleophilic attack on dopaquinone (Fig. 6) in the cross-linking reaction. And how can a dilute protein solution become concentrated enough, in a totally aqueous environment, to achieve sufficient cohesive strength? These are all problems that the mussels have solved, and that man will also have to solve, if he is to make practical application of these materials.

4.2 Characteristics of Adhesives

4.2.1 Adhesion and Cohesion

Adhesion, particularly as it relates to the mussel adhesive proteins, has been reviewed by Waite (1983b, 1990) and will not be discussed in detail here. Adhesion is the joining of two dissimilar materials, in this case, the adhesive protein with a surface (rock, shell, etc.) or with collagen fibrils. Cohesion is the joining together of similar materials, e.g., the cross-linking of two polymer chains. The strength of the adhesive depends on both of these properties.

It is not surprising that the mussel adhesive proteins adhere well to natural surfaces, though the detailed mechanism(s) of binding are still mostly speculative. These proteins are studded with hydroxyl and amide groups capable of forming hydrogen bonds with polar surfaces; catechol groups which form extremely tight complexes with metal ions, metal oxides and minerals (Hansen and Waite 1991); and possibly quinone-amines, resulting from cross-linkage (see Fig. 6), which reportedly have a great avidity for metals and other surfaces (Nithianandam and Erhan 1991). Cohesive strength presumably arises from the formation of cross-links, but it also depends on having a sufficiently high mass of adhesive material. As will be discussed below, the adhesive plaques contain 30–100% solids. How the mussel is able to form these essentially solid masses is a question that needs to be addressed.

4.2.2 Collagen-Based Glues

Collagen, or rather degraded forms of it extracted from hides, has been used as an adhesive for centuries (Hubbard 1977). When dry, hide glues have enormous cohesive and adhesive strength, with tensile strengths of up to 10000 pounds per square inch. However, unlike the mussel adhesive proteins, hide glues are not cross-linked and are water-soluble. The adhesive strength of hide glue is demonstrated by its use in a technique called "glass chipping", wherein glue is applied to the surface of glass and allowed to dry. On drying, it shrinks and pulls up chips of glass, leaving a decorative surface (Swift 1989), demonstrating that the protein-glass bond is stronger than the bonds in the glass. Hide glues only achieve their high adhesive strength when dry, usually by evaporation of water, a process not possible for mussel glues, which cure under water. Furthermore, secondary structure formation seems to play a part in the curing of hide and other protein glues and of polysaccharide glues. Cooling of a warm solution of hide glue results in formation of a gel (gelatin). The gel has some adhesive strength, but drying is required to achieve full strength (Hubbard 1977). Gelatin shows an X-ray diffraction pattern similar to that native collagen, indicating the presence of triple helical structure (Veis 1964). It should be noted, however, that the triple helix structure seen in gelatin is probably degenerate. The correct folding of a collagen triple helix in vivo is governed by carboxyl-terminal globular domains that are cleaved after secretion of collagen protofibrils. Processed collagen chains cannot refold except to form short triple helical regions (Engel and Prockop 1991).

4.3 Attainment of a Condensed State

4.3.1 Charge Reduction

It would seem essential that the high positive charge due to lysine in the mussel adhesive proteins be reduced in order for the polymer chains to get close enough to cross-link. There are at least two possible mechanisms for reducing these charges. One is by cross-link formation, which reduces the pK_a of the amino group of lysine from about 10.4 to that of an arylamine (ca. 5) in the lysine-dopaquinone adduct. Of course there has to be some mechanism for removing protons, since deprotonation of the lysine N^{ε}-amino group must precede attack of dopaquinone. The second is by introduction of anions, which break up the hydration shell around the ammonium groups. For example, it is well-known that ion exchangers such as diethylaminoethyl (DEAE) Sephadex shrink 20- to 40-fold on changing from very low to high ionic strength buffers. A polyanion, such as an acidic mucopolysaccharide, should be particularly effective, since it could act in a cooperative manner and might even help proper folding of the protein. The observation of Tamarin et al. (1976) of a "mucosubstance" (of unknown composition) that is secreted during byssus plaque formation is consistent with this idea. A possibly related effect is the demonstration by Kubota et al. (1983) that sodium dodecyl sulfate can cause an ordering of certain Ala-Lys copolymers.

4.3.2 Aggregation

Hide and many other glues achieve maximum cohesive strength by evaporation of solvent, whereupon the polymer chains can interact strongly with each other. Since mussel glues cure under water, evaporation of water is impossible. How, then, can water be eliminated to permit strong interchain interactions? Tamarin et al. (1976) and Waite (1983b, 1986) have remarked on the sponge-like structure of the cured adhesive protein on byssal disks. Essentially, the plaque consists of voids and dense, proteinaceous material, the latter occupying approximately 30, 70, and 100% of space in the *M. edulis*, *M. californianus*, and *G. demissa* proteins, respectively (Waite 1986; Waite et al. 1989). Waite (1986) has discussed this structure in terms of a foam, but a foam is a two-phase system of a gas and a liquid, and it is difficult to see how foaming would lead to a net concentration of the polymer phase. The answer to the question above may lie in considerations of the gel structure of the adhesive.

Virtually all adhesives and other synthetic polymers go through a gel phase during curing (Flory 1953). There is a spectrum of types of gel structure, ranging from the microreticular gel, which has a uniform matrix of polymer chains (like a loose ball of spaghetti), to the macroreticular (macroporous) gel, described by Kun and Kunin (1964), in which the polymer chains have coalesced to form regions of high density and others of almost zero density, e.g., as in a sponge. Rüchel et al. (1978) have shown that the pore size and macrostructure of polyacrylamide gels can vary widely, depending on the relative concentrations of monomer and cross-linker. In particular, high degrees of cross-linkage give large pore gels that look strikingly similar to the adhesive plaques described by Waite (1986). A biological analogy for this type of process is the retraction or shrinking and strengthening of fibrin clots (gels) which occurs on cross-linking (Doolittle 1975).

5 Processing of Adhesive Proteins by the Mussel

Tamarin et al. (1976) have described the formation of the byssal attachment plaque at the macroscopic level (e.g., at the level seen in the electron microscope), but little is known at the biochemical level. Hydroxylation of the protein occurs within the cells of the phenol gland, which can be assumed because the hydroxylated protein can be isolated by extraction of the phenol gland. But where does oxidation to dopaquinone occur? Is the polypeptide chain further processed after secretion? Is cross-linking enzyme-catalyzed? Does the adhesive protein cross-link with lysine groups in collagen? What is the concentration of the secreted protein? How does the mussel deal with the high positive charge of the protein?

All of these questions await answers, but perhaps one can gain insights by considering the biosynthesis of collagen (Byers 1990), which is co-secreted (by a different gland) along with the adhesive protein during byssus formation. Procollagen, which contains nonhelical globular domains at the amino- and carboxyl-termini of the chain, is hydroxylated and assembled into the triple helix in the rough endoplasmic reticulum. Hydroxylation of proline enhances the stability of the triple helix. After transport through the Golgi and secretion, the globular ends are excised, leaving triple-helical tropocollagen, which then forms the characteristic fibrils. The nonhelical domain is essential for proper folding to the triple helix (Engel and Prockop 1991). In the case of the *M. edulis* adhesive protein, we know that the protein that can be isolated has lost the signal peptide, but that it still contains the nonrepeat domain at the amino terminus (Ou 1990). Does this region perform an analogous role for folding of the mussel adhesive protein? Is it present in the cured adhesive?

6 Conclusions

As primary sequence data accumulate, the structural picture of the mussel adhesive proteins appears increasingly complex. However, it is possible, as seen with other classes of protein, that these different sequences represent different ways of achieving a more or less common secondary or tertiary structure. It is proposed here that the mussel adhesive proteins have helical structures (yet to be defined) that place the lysine and DOPA residues on the surface where they can easily form cross-links. It is further suggested that the helix contains defects or kinks to give it flexibility, and that it forms a mesh-like matrix that can entrap collagen fibrils and interact with surfaces. Effective elimination of water from the matrix is achieved by a high degree of cross-linking, resulting in a macroporous gel structure. The adhesive proteins are associated with, and have some structural similarities with collagen, and it therefore may be useful to use collagen as a model for planning future studies of the structure and function of the adhesive proteins.

Acknowledgment. Some of the work described here was supported by a grant from the Office of Naval Research.

References

Bobek L, Rekosh DW, Van KH, LoVerde PT (1986) Characterization of female-specific cDNA derived from a developmentally regulated mRNA in the human blood fluke *Schistosoma mansoni*. Proc Natl Acad Sci USA 83:5544–5548

Bobek LA, Rekosh DM, LoVerde PT (1988) Small gene family encoding an eggshell (chorion) protein of the human parasite *Schistosoma mansoni*. Mol Cell Biol 8:3008–3016

Byers PH (1990) Folding of collagen molecules containing mutant chains. In: Gierasch LM, King J (eds) Protein folding: deciphering the second half of the genetic code. American Association for the Advancement of Science, Washington DC, pp 241–247

Chen J, Varner JE (1985) An extracellular matrix protein in plants: characterization of a genomic clone for carrot extension. EMBO J 4:2145–2151

Chou PY, Fasman GD (1978) Prediction of the secondary structure of proteins from their amino acid sequence. Adv Enzymol 47:45–148

Creighton TE (1983) Proteins: structures and molecular properties. Freeman, New York, pp 252–264

Dickerson RF, Geis I (1969) The structure and action of proteins. Benjamin, Menlo Park, pp 33–43

Doolittle RF (1975) Fibrinogen and fibrin. In: Putnam FW (ed) The plasma proteins, vol II, 2nd edn. Academic Press, New York, pp 109–161

Ellenen TC, Dopheide TA (1972) The sequence of SCMK-B2B, a high-sulfur protein from wool keratin. J Biol Chem 247:3900–3909
Engel J, Prockop DJ (1991) The zipper-like folding of collagen triple helices and the effects of mutations that disrupt the zipper. Annu Rev Biophys Biophys Chem 20:137–152
Filpula DR, Lee S-M, Link RP, Strausberg SL, Strausberg RL (1990) Structural and functional repetition in a marine mussel adhesive protein. Biotechnol Prog 6:171–177
Flory PJ (1953) Principles of polymer chemistry. Cornell University Press, Ithaca
Garfinkel MD, Pruitt RE, Meyerwitz EM (1983) DNA sequences, gene regulation and molecular protein evolution in the *Drosophila* 68C glue gene cluster. J Mol Biol 168:765–789
Glanville RW (1987) Type IV collagen. In: Mayne R, Burgeson RE (eds) Structure and function of collagen types. Academic Press, New York, pp 43–79
Hansen DC, Waite JH (1991) Marine metal chelating proteins. In: Sikes CS, Wheeler AP (eds) Surface reactive peptides and polymers: discovery and commercialization. ACS Symp Ser 444, American Chemical Society, Washington DC, pp 256–262
Hong JC, Nagao RT, Key JL (1987) Characterization and sequence analysis of a developmentally regulated putative cell wall protein gene isolated from soybean. J Biol Chem 262:8367–8376
Hubbard JR (1977) Animal glues. In: Skeist I (ed) Handbook of adhesives (2nd edn). Reinhold, New York, pp 139–151
Keene DR, Sakai LY, Lunstrum GP, Morris NP, Burgeson RE (1987) Type VII collagen forms an extended network of anchoring fibrils. J Cell Biol 104:611–621
Kubota S, Ikeda K, Yang JT (1983) Conformation of sequential and random copolypeptides of lysine and alanine in sodium dodecyl sulfate solution. Biopolymers 22:2219–2236
Kun KA, Kunin R (1964) Pore structure of some macroreticular ion exchange resins. J Polym Sci Polym Lett 2:587–591
Marumo K, Waite JH (1986) Prolyl 4-hydroxylase in the foot of the marine mussel *Mytilus edulis* L.: purification and characterization. Biochem Biophys Acta 872:98–103
Mayne R, Burgeson RE (1987) Structure and function of collagen types. Academic Press, New York
Middaugh CR, Thomson JA, Ji H, Mach H, Johnson BA, Cordingley JS (1990) Evidence for a left handed alpha-helix in an egg shell protein from *Schistosoma mansoni*. Biophys J 57:419a
Mizuno H (1989) Prediction of the conformation of ice-nucleation protein by conformational energy calculation. Proteins: Struct Funct Genet 5:47–65
Nithianandam VS, Erhan S (1991) Quinone-amine polymers: 10. use of calcium hypochlorite in the syntheses of polyamine-quinone (PAQ) polymers. Polymer 32:1146–1149
Ou J-J (1990) Cloning and sequencing of complementary DNA encoding the polyphenolic adhesive protein from *Mytilus edulis*. Thesis, Boston University
Richardson JS (1981) Protein anatomy. Adv Prot Chem 34:167–399
Rodriques V, Chaudri M, Knight M, Meadows H, Chambers AE, Taylor WR, Kelly C, Simpson AJG (1989) Predicted structure of a major *Schistosoma mansoni* eggshell protein. Mol Biochem Parasitol 32:7–14
Rüchel R, Steere RL, Erbe EF (1978) Transmission-electron microscopic observations of freeze etched polyacrylamide gels. J Chromatogr 166:563–575
Ruoslahti E, Pierschbacher MD (1987) New perspectives in cell adhesion: RGD and integrins. Science 238:491–497
Rzepecki LM, Chin S-S, Waite JH, Lavin MF (1991) Molecular diversity of marine glues: polyphenolic proteins from five mussel species. Mol Mar Biol Biotechnol 1:78–88
Savage DC, Fletcher M (1985) Bacterial adhesion: mechanisms and physiological significance. Plenum, New York
Shen X-T (1991) Cloning and sequencing of complementary DNA encoding the adhesive protein from *Geukensia demissa*. Thesis, Boston University
Strausberg RL, Anderson DM, Filpula D, Finkelman M, Link R, McCandliss R, Orndorf SA, Strausberg SL, Wei T (1989) Development of a microbial system for production of mussel adhesive protein. In: Hemingway RW, Conner AH (eds) Adhesives from renewable resources. ACS Symp Ser 385, American Chemical Society, Washington DC, pp 453–464
Sugiyama T, Rafalski A, Peterson D, Söll D (1985) A wheat HMW glutenin subunit gene reveals a highly repeated structure. Nucleic Acids Res 13:8729–8737
Swift (1989) Glass chipping with animal glue. Bulletin 902, Swift Adhesives, Downers Grove, IL
Tamarin A, Lewis P, Askey J (1976) The structure and formation of the byssus attachment plaque in *Mytilus edulis*. J Morphol 149:199–222
Trumbore MW (1991) Investigations of the three-dimensional conformation of the adhesive polyphenolic protein of *Mytilus edulis* using small-angle scattering and molecular modeling. Thesis, University of Connecticut

Urry DW (1990) Protein folding and assembly: an hydration-mediated free energy driving force. In: Gierasch LM, King J (eds) Protein folding: deciphering the second half of the genetic code. American Association for the Advancement of Science, Washington DC, pp 63–71

Veis A (1964) The macromolecular chemistry of gelatin, Academic Press, New York, pp 270–275

von Heijne G, Nishikawa K (1991) Chloroplast transit peptides: the perfect random coil? FEBS Lett 278:1–3

Waite JH (1983a) Evidence for a repeating 3,4-dihydroxyphenylalanine- and hydroxyproline-containing decapeptide in the adhesive protein of the mussel, *Mytilus edulis* L. J. Biol Chem 258:2911–2915

Waite JH (1983b) Adhesion in byssally attached bivalves. Biol Rev 58:209–231

Waite JH (1986) Mussel glue from *Mytilus californianus* Conrad: a comparative study. J Comp Physiol B 156:491–496

Waite JH (1990) The phylogeny and chemical diversity of quinone-tanned glues and varnishes. Comp Biochem Physiol 97B:19–29

Waite JH, Tanzer ML (1981) Polyphenolic substance of *Mytilus edulis*: a novel adhesive containing L-DOPA and hydroxyproline. Science 212:1308–1040

Waite JH, Housley TJ, Tanzer ML (1985) Peptide repeats in a mussel glue protein: theme and variations. Biochemistry 24:5010–5014

Waite JH, Hansen DC, Little KT (1989) The glue protein of ribbed mussel (*Geukensia demissa*): a natural adhesive with some features of collagen. J Comp Physiol B 159:517–525

Williams T, Marumo K, Waite JH, Henkens RW (1989) Mussel glue protein has an open conformation. Arch Biochem Biophys 269:415–422

Chapter 4

Composition and Design of *Fasciola hepatica* Eggshells

Allison C. Rice-Ficht[1]

1 Introduction

1.1 Life Cycle of *Fasciola hepatica*

Fasciola hepatica is a parasitic flatworm of cosmopolitan distribution which infects a wide range of mammalian hosts including man and domestic livestock. First described by Linnaeus in 1758, the adult form is a large, leaf-shaped worm (3.0 × 1.3 cm) which inhabits the bile ducts of its mammalian hosts. Fascioliasis is a serious human health problem particularly in the Caribbean, South America, and parts of Northern Europe, and a threat to livestock worldwide; its transmission is dependent upon both a snail intermediate host and the presence of aquatic vegetation upon which the infectious form (metacercaria) encysts. Ingestion of vegetation carrying the cyst is the mechanism by which the mammalian host acquires infection.

Once ingested, the metacercarial form is swept through the digestive tract to the duodenum where it hatches within 2 h and penetrates the peritoneal cavity. Subsequently, the juvenile worm migrates to the liver capsule within a 4–6-day period and begins an extensive foray (6–7 weeks) through the liver parenchyma. This migration is detrimental to the host, causing anemia, necrosis, and fibrotic thickening of the liver tissue. At 7 weeks most juveniles have entered an adult stage and become resident in the bile ducts and may remain for up to 12 years. The bile ducts become progressively thickened and even calcified by the infection. It is this form of the worm which produces large quantities of eggs (up to 2500/day) and is the focus of the study of eggshell protein production. *F. hepatica*, like most trematodes, is hermaphroditic and capable of self-fertilization although multiple-worm infections are more robust and cross-fertilization is predominant.

Eggs produced by the adults are delivered to the duodenum in bile and pass out of the host in feces as a "nonembryonated" form. The eggs are reported to require at least 8 days to develop with longer periods (up to 23 weeks) required in cooler climates. The eggs may lie dormant for some months under field conditions especially at lower temperatures (under 10°C). Once hatched, the free-swimming

[1] Department of Medical Biochemistry and Genetics, Texas A & M University, College Station, Texas 77843-1114, USA

Results and Problems in Cell Differentiation 19
Biopolymers
Case, S.T. (Ed.)

miracidia or larvae may infect a snail intermediate host of the genus *Lymnaea*. After snail penetration the miracidium develops into a sporocyst which undergoes asexual reproduction and after 4–7 weeks releases a number of cercariae from the snail host; within minutes the cercariae come to rest on vegetation near the water surface, shed their tails, and begin to secrete a substance which forms a hard, sticky coating on the organism. The resulting metacercaria is infectious and may remain so for up to 1 year under laboratory conditions; viability appears to depend upon the presence of a moist environment with desiccated organisms surviving only 1–3 months. A comprehensive review of the life cycle is offered by Schmidt and Roberts (1985).

1.2 The Role of Shell Polymers and Sclerotization in Development

F. hepatica employs unique biopolymers throughout the life cycle to form protective materials for the developing embryo and cercaria. One such polymer, the *F. hepatica* eggshell, is composed primarily of proteins which have undergone covalent cross-linking. The polymer consists of a spectrum of protein shell precursors which undergo post-translational modification of tyrosine to form dihydroxyphenylalanine (DOPA) residues in the protein backbone (Waite and Rice-Ficht 1987, 1989). The presence of these residues permits subsequent enzymatic oxidation and cross-linking of the proteins to produce a hard, porous shell. The use of DOPA-derived cross-linking is widespread, especially among insects (Eisner et al. 1966; Walker and Menzer 1969) while DOPA-mediated cross-linking prevails in marine invertebrates (Waite 1990). DOPA-containing proteins, which also have distinct adhesive properties (Waite 1990), may have additional importance in formation of the quinone-tanned capsule of the metacercaria which mediates adherence of the organism to aquatic vegetation (Dixon and Mercer 1967).

2 Eggshell Morphogenesis

The vivid process of *F. hepatica* eggshell formation, which is quite accessible at the level of light microscopy, has captured the attention of parasitologists and biochemists alike since the late 1800s (Smyth and Clegg 1959, review). The origin of the shell material has been debated to be either the (1) the vitelline cells of the vitellaria in the worm periphery or (2) Mehlis' gland, the site of shell assembly (Fig. 1). Although it was originally suggested that Mehlis' gland contributed the bulk of the shell material, it is now widely accepted that the shell protein originates with the vitelline glands based on direct histochemical evidence in *F. hepatica* and through analogy with related systems (Smyth 1954; Burton 1963, 1967). The eggshell precursor proteins, produced in the vitellaria, are manufactured in highly

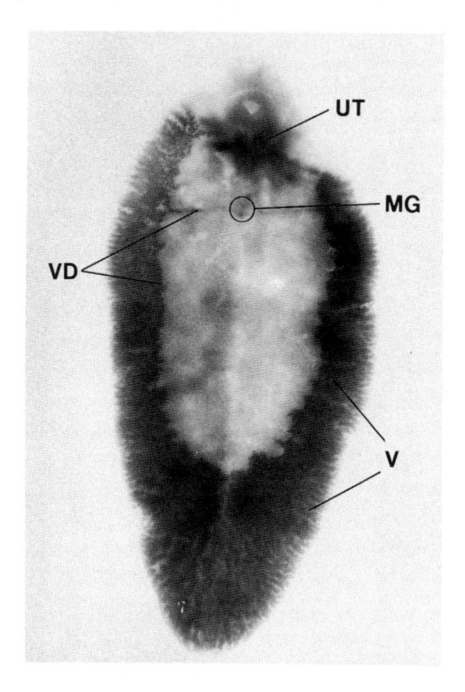

Fig. 1. Dorsal view of a mature liver fluke (*F. hepatica*) stained for DOPA with Arnow's reagent (nitrite-molybdate). The extensive vitellaria (*V*) consist of clusters of vitelline follicles which stain intensely red for DOPA as do the vitelline ducts (*VD*) and eggs located in proximal portions of the uterus (*UT*). Mehlis' gland (*MG*), the initial site of shell formation, is positioned between the vitelline ducts and the uterus

specialized vitelline cells which "stockpile" the protein precursors and enzymes needed for shell formation as well as yolk and glycogen for the developing embryo. At the level of light microscopy, mature vitelline cells are observed to contain large numbers of translucent globules lining the cytoplasmic membrane (Fig. 2). The large, highly visible globules each consist of a membrane-bounded compartment enclosing both the protein precursors of the shell and enzymes destined to carry out enzymatic oxidation and cross-linking of these proteins. Histochemically these globules are basophilic (Smyth 1951) and polyphenolic (Stephenson 1947; Smyth and Clegg 1959) in character. At higher resolution the contents of a single globule are not at all homogeneous, consisting of droplets of basophilic material which stain distinctly with malachite green or bromophenol blue to yield a mulberry pattern (Smyth 1954; Johri and Smyth 1956). The mixing of the two phases does not appear to occur until release of protein from the globules during shell formation and may be important in compartmentation of enzyme and substrate. Other much smaller granules not involved in shell formation are distributed throughout the cytoplasm; these granules are carbohydrate-rich based on isotopic labelling studies with ^{3}H-labeled glucose and ^{3}H-labeled galactose and serve to nourish the developing embryo (Hanna 1976).

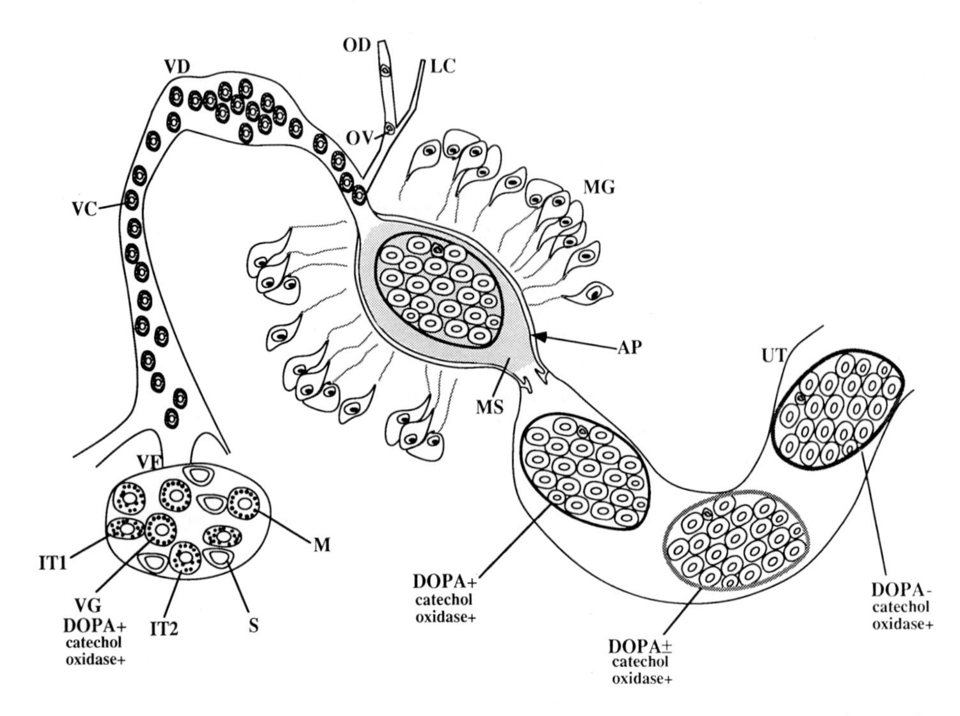

Fig. 2. Eggshell formation in the female reproductive tract of *F. hepatica*. Eggshell formation from the vitelline gland or follicle (*lower left*) through Mehlis' gland, the biochemical "black box" of eggshell formation (*center*), to the uterus where eggshells reach maturity (*lower right*). *AP* Alkaline phosphatase; DOPA dihydroxyphenylalanine; *IT1* intermediate type 1 vitelline cell; *IT2* intermediate type 2 vitelline cell; *LC* Laurer's canal; *M* mature cell; *MG* Mehlis' gland; *MS* Mehlis' secretions; *OD* oviduct; *OV* ovum; *S* immature vitelline cell or stem cell; *UT* uterus; *VC* vitelline cell; *VD* vitelline duct; *VF* vitelline follicle; *VG* vitelline gland. The portion of the uterus passing through Mehlis' gland is often referred to as the ootype

The development of the vitelline cell within the vitelline follicle has been divided by Threadgold (1982) into four stages (Fig. 2). These include (1) stem cells (S) located peripherally in the follicle with exiguous GER (granular endoplasmic reticulum), and completely lacking eggshell globules, (2) intermediate type 1 cells (It1) in the follicle periphery with extensive GER and minimal numbers of globules, (3) intermediate type 2 cells (It2) in a medullary position in the follicle with extensive GER and many large shell globules and (4) mature cells (M) in the medulla of the follicle with degenerating GER, many large shell globules lining the cell periphery and glycogen granules abundant in the cytoplasm. The highest rate of protein production within the cells is evident in early vitelline cell development as determined by pulse labeling experiments with ^{3}H-labeled tyrosine (Hanna 1976); eggshell protein precursors are produced rapidly in the early vitelline cells (morphologically It1, It2) which accumulate protein globules at the cell margin. Protein production is virtually complete before the cells leave the vitelline gland (Hanna 1976; Threadgold 1982). Glycogen granules are synthesized later in development following degeneration of the GER but prior to release of the vitelline cell to the vitelline duct.

The production and release of globules by the vitelline cells appears formally equivalent to regulated secretory pathways present in higher eukaryotes (Burgess and Kelly 1987), while the overall process of vitelline cell production and eggshell formation may be considered a type of holocrine secretion. At high resolution the It2 vitelline cells reveal small electron-dense sperules in proximity with a highly developed rough endoplasmic reticulum (Bjorkman and Thorsell 1963; Irwin and Threadgold 1972). The sperules appear to fuse, giving rise to the much larger biphasic globules visible at lower magnifications. It is not clear whether the polyphenolic protein and the polyphenol oxidase are secreted together in the same "spherules" or if they are secreted in separate sperules which later fuse to form the larger biphasic globules. Morphological studies of *F. hepatica* reveal the presence of rough endoplasmic reticulum and a Golgi apparatus, hallmarks of secretory pathways in vertebrate organisms (Irwin and Threadgold 1970, 1972; Hanna 1976; Threadgold 1982). Accordingly, the calcium ionophore monensin perturbs the secretory process in *F. hepatica* much as it affects secretion in higher eukaryotes (Ledger and Tanzer 1984; Skuce and Fairweather 1988).

Mature vitelline cells exit the vitelline gland follicles and migrate into collecting ducts which direct them to the ootype within the lumen of Mehlis' gland; it is here that a chemical signal is postulated to trigger release of the vitelline globule contents via regulated secretion. Approximately 30 vitelline cells and one ovum are partitioned in the ootype during initial stages of egg formation. Mehlis' gland then secretes a "mucus"-like substance (Rao 1959; Burton 1967; Irwin and Threadgold 1972; Bogitsh 1985) into the ootype enveloping the vitelline cell/egg mass. Mehlis' gland is documented to consist of at least two types of secretory cells — an S1 cell producing a filamentous type of secretory body with a dense core and an S2 cell producing dense secretory bodies with a "packed fibrous appearance" (Irwin and Threadgold 1972). The S1 secretion has been localized in the ootype lumen and is seen to form a layer over the entire ootype epithelium (Irwin and Threadgold 1972); histochemically the S1 secretion is lipoprotein in nature and also forms a layer or membrane around the vitelline cells and ovum, collectively (Clegg 1965). The S2 secretory bodies contain an acid mucopolysaccharide based on positive staining with periodic acid Schiff's (PAS) and Alcian blue (Irwin and Threadgold 1972) although the localization of this secretion in the ootype is in question.

The vitelline cells are engulfed in a secretion originating possibly in the vitelline follicles (Irwin and Threadgold 1972) which forms a discrete interface with the S1 lipoprotein; it is upon this interface or membrane that eggshell protein deposition proceeds. Small quantities of S1 secretion are also observed to cross the interface and intermix with vitelline globules. As eggshell protein globules are released they appear to (1) spread on the layer of S1 secretion lining the ootype and (2) coalesce forming larger globules also coated with S1 secretion which migrate and fuse with the developing shell layer. It has been suggested that the coating of S1 secretion or lipoprotein facilitates globule fusion with the growing shell layer (Gönnert 1962; Burton 1967). The fully formed eggshell retains its outer coating of lipoprotein as it traverses the uterus. It has also been suggested that the lipoprotein secretion may give rise to an inner or vitelline membrane lining the finished shell (Clegg 1965; Wilson 1967; Wharton 1983).

The polyphenol oxidase or catechol oxidase component of the vitelline globule (Smyth 1954) is activated at this time by an unknown mechanism; oxidation and cross-linking or sclerotization of the shell begins here, continues as the egg passes into the uterus and is virtually complete before the egg reaches the genital pore (Waite and Rice-Ficht 1989). Since autotanning of the vitellaria does not occur even in sectioned worms, Smyth (1954) has postulated that a blocking mechanism is at work. Activation of a proenzyme by processing or introduction of an oxygen-carrying cofactor by Mehlis' gland are alternate possibilities. The eggshell residing in the ootype is initially very electron dense but becomes mottled in appearance as it matures in the uterus (Irwin and Threadgold 1972); the change in appearance coincides with oxidation of the eggshell precursors (Waite and Rice-Ficht 1989).

The finished eggshell consists of three layers including (1) an inner lipoprotein layer (vitelline membrane), (2) a middle proteinaceous layer formed from the contents of vitelline cell globules, and (3) an outer layer composed of lipoprotein (Clegg 1965; Irwin and Threadgold 1972). Wilson has investigated the fine structure of fully cured shells and finds them homogeneous in appearance, lacking discrete layers. Staining with phosphotungstate reveals randomly oriented fibrils and "occasional dense globules". The lack of detail is in contrast to the ultrastructure observed by Neill et al. (1988) in the *Schistosoma mansoni* egg where three distinct layers and cribiform pores are noted. The use of newly formed *F. hepatica* eggs and the freeze-substitution technique employed by Neill (1988) could possibly reveal more detail in the *F. hepatica* eggshell.

Permeabilities of the fully tanned eggshell have been assessed by Rowan (1962), who concluded that the shell itself is permeable to a series of straight chain alcohols under a molecular weight of 150 Da. Wilson (1967) likewise concluded that a number of inorganic and small organic molecules will penetrate the shell; he further tested the permeability of the vitelline membrane (inner membrane of the shell) but results were equivocal. Presumably the osmotic barrier needed in a freshwater environment would be maintained by a membrane within the shell rather than the porous, inert shell itself.

3 Biochemical Characteristics of *F. hepatica* Eggshell (Chorion) Proteins and Precursors

3.1 Chemical Stability of the Eggshell Is Based on Quinone Tanning

F. hepatica produces a variety of eggshell precursors which are protein and phenolic in nature and contain the unusual post-translational modification of tyrosine to DOPA (Waite and Rice-Ficht 1987). The catecholic proteins are ultimately cross-linked or quinone-tanned (Pryor 1940), imparting extraordinary stability to the eggshell. Quinone-tanned biological material of this type is reported to be refractory to proteases, acids, bases, and many organic solvents (Brown 1975).

Catecholic polymers of this type are widespread in nature and perform diverse functions. In bacterial systems catecholic peptides sequester ferric iron from the environment (Ong et al. 1979; Raymond and Carrano 1979); in marine invertebrates the catecholic polymers are proteins modified by the presence of DOPA residues which function as underwater adhesives and sealants (Waite et al. 1985; Waite 1986). The synthetic counterparts of these polymers have found widespread application in industry as semi-conductors (Jaegfeldt et al. 1983; Lau and Miller 1983), metal chelators (Pecoraro et al. 1981), electrocatalysts (Degrand 1985), and adhesives (Pizzi 1985). Additional interest in DOPA compounds has been generated by the recent finding that DOPA derivatives may act as redox cofactors in the active site in amine oxidases (Janes et al. 1990).

Documentation of *F. hepatica* eggshell production has historically included a quinone-tanning process which involves cross-linking of proteins containing catecholic groups (Smyth and Clegg 1959). However, the nature of the substrates involved in the cross-link and the structure of the cross-link itself have been the subject of heated debate (Ramalingam 1973) and novel investigative approaches. The presence of phenolic compounds has been established in *F. hepatica* and related helminths by a variety of histochemical techniques including ferric chloride, Fast Red Salt B, argentaffin, chromaffin, sodium iodate, and ammonium molybdate (Smyth 1954; Johri and Smyth 1956; Bell and Smyth 1958). A reagent specific for dihydroxyphenols or 1,2 benzenediols was not employed to study shell formation until Bogitsh (1985) employed an osmiophilic tetraazotized salt specific for DOPA, dopamine, catechol, and pyrogallol, but unreactive with tyrosine or phenylalanine. A second reagent developed by Arnow (1937; nitrate-molybdate), which is quite specific for mono-substituted 1,2 benzenediols (Waite and Tanzer 1981; Waite and Benedict 1984) has been employed to document the presence of diphenolic residues throughout the *F. hepatica* vitellaria (Waite and Rice-Ficht 1989; Fig. 1). The reaction, which produces a bright red chromophore, indicates that the initial formation of 1,2 benzenediols occurs in the vitelline follicles collectively referred to as the vitellaria. Vitelline cells maintain intense staining properties in the vitelline duct which they impart to the eggshell upon passage through the Mehlis' gland region. The reactivity of eggshells with Arnow's reagent dissipates as the shells pass through the uterus; eggs in proximity to the genital pore are refractory to staining. Characterization of the diphenolic component as DOPA has been achieved through direct sequencing of protein precursors of the eggshell (Waite and Rice-Ficht 1987).

3.2 Protein Complement of the Eggshell; Implications of Post-Translational Modification

Three DOPA-containing proteins, constituting approximately 7% ($\pm$ 1.5%) of the total protein of the worm have been purified to homogeneity from the vitellaria of *F. hepatica*. DOPA is the single type of catecholic group found in the *F. hepatica*

Table 1. Major amino acid composition of *F. hepatica* eggshell precursor proteins

Protein	Mol. wt. (kDa)	Amino acids (%)		%Tyr to DOPA[a]
vpA	70	Asx	21	≥30
		Gly	11	
		DOPA	4	
		Tyr	9	
		Arg	8	
vpB	31	Asx	14	60
		Gly	16	
		DOPA	10	
		Lys	13	
vpC	17	Gly	41	100
		DOPA	20	
		His	20	

[a] Represents average percent conversion of tyrosine residues to DOPA.

proteins of 70 kDa (vitelline protein A, vpA), 31 kDa (vpB), and 17 kDa (vpC; Waite and Rice-Ficht 1987; 1989). VpB is the major component of the shell by weight and is the best-characterized protein of the group (Waite and Rice-Ficht 1987, 1992; Rice-Ficht et al. 1992). The major amino acids represented in each protein are outlined in Table 1 and indicate several common features among the eggshell precursors.

Among the amino acids listed in Table 1, glycine has proven to be a major component of all precursors. This is not unexpected, since many other well-characterized structural biopolymers including collagen (Eyre 1980) and elastin (Urry 1983) are also glycine-rich. From a materials chemist's viewpoint, polyglycine is an ideal polymer for physical resilience since it is identical to the formulation for nylon 2. vpA and vpB also contain a high concentration of asparagine/aspartate; in vpB, for which the DNA sequence is available, aspartate predominates (21 residues compared to 16 asparagine) and is dispersed throughout the protein. Both proteins also exhibit a high number of basic residues with arginine predominant in vpA (8%) and lysine predominant in vpB (13%). The lysine of vpB is likewise distributed throughout the protein and shows no apparent clustering; since lysines are implicated as nucleophiles in the process of sclerotization (as discussed below), the even distribution of these residues may be an extremely important feature of eggshell precursor primary sequence. Compositionally, vpC is the most unusual protein of the shell in that 81% of its residues are glycine, DOPA, or histidine (Waite and Rice-Ficht 1989). Peptide sequence of the protein reveals that a major portion is composed of a $(\text{Gly-X})_n$ where X is serine, DOPA, or histidine. Its Gly-His repeat motif is quite similar to that of high molecular weight kininogen I (Kitamura et al. 1983), the implications of which are discussed below.

All three precursors also contain substantial quantities of DOPA as components of the protein primary sequence; this was first demonstrated in the *F. hepatica* vpB protein through sequencing of tryptic peptides (Waite and Rice-Ficht 1987). At the same time the gene sequence of vpB indicated the presence of tyrosine codons at positions corresponding to all DOPA residues identified in the protein. This finding implied post-translational modification of tyrosine residues as the basis for the covalent association of DOPA with these proteins (Rice-Ficht et al. 1992; Waite and Rice-Ficht 1992). Furthermore, fractionation of DOPA-containing proteins through reversed phase high pressure liquid chromatography (HPLC) indicates an inverse relationship between the number of tyrosine and DOPA residues in individual fractions, further supporting a precursor-product relationship between the two amino acids. This evidence, coupled with DOPA-specific histological staining, suggests that a tyrosyl hydroxylase present in early vitelline cell ontogeny is responsible for hydroxylation of tyrosine residues in nascent eggshell precursors (Fig. 3).

In examining the DOPA and tyrosine content in all three eggshell precursors, it is apparent that the degree of post-translational modification varies from 30% conversion of tyrosine to DOPA in vpA to 100% in vpC. Unlike other DOPA-

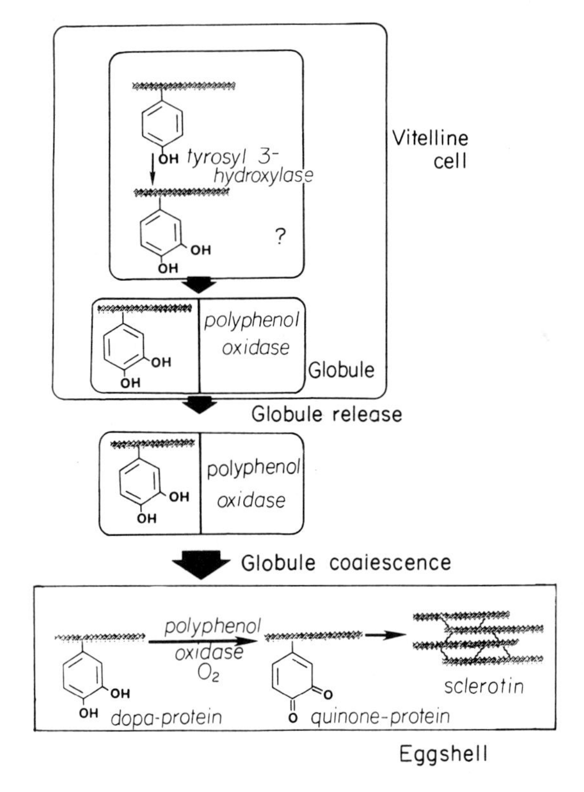

Fig. 3. Model illustrating the origin and fate of DOPA in the eggshell precursor protein of *F. hepatica*. This amino acid may be formed by the action of a putative tyrosine 3-hydroxylase on a protoeggshell precursor protein. The cellular compartment of this hydroxylation is unknown. The DOPA-containing eggshell precursor is then stockpiled along with polyphenol oxidase in the biphasic vitelline globules. Vitelline globules are released from vitelline cells and coalesce during eggshell formation, during which polyphenol oxidase acts on the DOPA residues, converting them into cross-linking *o*-quinones. (Waite and Rice-Ficht 1987)

containing proteins with highly conserved peptide repeats and clearly defined target sequences for modification (Waite 1990), *F. hepatica* vitelline proteins have a highly degenerate target sequence and the basis for hydroxylation of specific tyrosines is not entirely clear. The concensus target site for hydroxylation may be described as follows

$$[X_1]\text{-Tyr*-}[X_2]\text{-Tyr*-}[X_3]\text{-Lys/Arg,}$$

where X_1 represents 0–3 amino acids, X_2 represents two to three amino acids, one of which is Asp or Gly, X_3 represents one or two amino acids, often Gly, Asp, or Ala, and Tyr* indicates that the residue may occur as Tyr or DOPA. The requirement for this sequence is not in itself sufficient to permit modification since, as indicated, some sites carrying the consensus flank completely unmodified tyrosines (Waite and Rice-Ficht 1992). The data of Table 1 implies a relationship between surface to volume ratios of proteins and degree of hydroxylation. Could simply the exposure of tyrosine residues on the surface of a vitelline protein molecule govern the sites of modification? Surface probability algorithms (Emini et al. 1985) have been applied to translations of the vpB gene sequence; but surface presentation predicted by this algorithm and known protein modification sites do not favorably compare. A study of the purified tyrosyl hydroxylase may be required to determine the basis for site-specific modification.

The presence of DOPA residues in these proteins implicates them as substrates in the quinone-tanning process. In considering the histochemical data discussed earlier, the initiation of DOPA-protein oxidation to DOPA-quinone occurs only after deposition of the shell in the ootype (Figs. 2, 3); the highly reactive nature of *o*-quinones suggests that formation at earlier stages would be counterproductive. Nucleophilic addition to the quinone then produces protein cross-linking and shell hardening. Direct demonstration that DOPA residues of the shell precursors are consumed (1) during maturation of the eggshell in the *F. hepatica* uterus and (2) during treatment in vitro of purified precursors with mushroom tyrosinase supports this model (Waite and Rice-Ficht 1987). Accompanying the oxidative loss of DOPA residues in vivo and in vitro is a concomitant decrease in the quantitative recovery of lysine residues (120 mol/1000 to 86 mol/1000). If one considers that removal of DOPA residues from amino acid composition via oxidation and cross-linking increases the mol% of other amino acids present in the mixture, the decrease in lysine content is even more suggestive. Lysine is therefore strongly implicated as the major nucleophile in the cross-linking reaction (Waite and Rice-Ficht 1987).

Studies on DOPA oxidation with model compounds such as N-acetyl-DOPAethyl ester suggest that the *o*-quinone spontaneously rearranges to a quinone methide, and to N-acetyl-α,β-dehydroDOPAethyl ester (Rzepecki et al. 1991). This implies that sclerotization may disrupt the peptide backbone with the formation of each cross-link; if correct, the outcome is profound, since it indicates that individual peptides are "processed" during tanning. This may encumber future attempts to extract vitelline proteins from mature shells and to analyze protein-protein interactions in mature shells. Another intriguing prospect in protein cross-linking is the possible use by *F. hepatica* of metals to mediate the process. It is well known that *o*-diphenols, quinones, and semi-quinones are metal chelators of high affinity (Pierpoint and Buchanan 1981); an example of this avidity is the observation by Kretchmar and Raymond (1988) that *o*-diphenols can scavenge Fe(III) from transferrin, an iron-binding blood protein. If metals present were chelated by more than one DOPA residue, then another type of cross-link would effectively be formed. Metal cross-linking of eggshell proteins has been observed in *Gryllus mitratus* (oriental garden cricket) and employs cross-linking of protein carboxyl groups with calcium (Kawasaki et al. 1971).

Enzymatic oxidation of DOPA residues appears to be mediated by a polyphenol oxidase or catechol oxidase whose presence has been documented in the vitelline globule along with its polyphenolic substrates (Smyth 1954). Although its mode of activation is unknown, the activity of the enzyme itself has been partially characterized by Mansour (1958), who found that the enzyme could be assayed in spite of extreme insolubility at neutral pH. The enzyme is active upon a variety of substrates including dopamine, L-epineprine, tyrosine ethyl ester, catechol, and DOPA. The enzyme is inhibited by diethyldithiocarbamate and stimulated by the addition of $CuSO_4$, indicating that copper is likely to be a prosthetic group of the enzyme. Phenolases with similar properties have reportedly been isolated from other helminths including *F. gigantica* (Nellaiappan and Ramalingam 1980; Thangaraj et al. 1982), *S. mansoni* (Seed et al. 1978; Seed and Bennett 1980), *S. japonicum* (Wang et al. 1986) and *Parapleurus sauridae* (Nellaiappan and Ramalingam 1980).

It is an intriguing observation that trematodes and other invertebrates have apparently separated their tyrosyl hydroxylase and catechol oxidase activities into discrete enzymes (Waite 1990), whereas fungi and plants do not separate the two activities in their polyphenol oxidases. Why would a trematode opt for two enzymes when one enzyme could carry out the entire job of tyrosine conversion to dopaquinone? There are currently two theories about the function of phenolics in sclerotized structures. The first, the so-called dehydration theory, states that *o*-diphenols such as DOPA contribute to sclerotization by extensive hydrogen-bonding to the peptide backbone (Vincent and Hillerton 1979). This has the effect of dehydrating the protein (since phenolics are superior to water in hydrogen-bonding) and also confers resistance to proteases. In such a representation, tyrosyl hydroxylase plays a special role because its product, the *o*-diphenolic moiety of DOPA, is without peer in hydrogen-bonding. This model, however, lacks a distinct role for catechol oxidase. The second theory (cross-linking theory) postulates that *o*-diphenols are intermediates of *o*-quinones, the desired end products for cross-linking. Catechol oxidase is vital in this scheme since it catalyzes the formation of *o*-quinones from peptidyl-DOPA. If there are single enzymes capable of converting tyrosine to quinone, why invoke the DOPA intermediate or two enzymes? In well-characterized polyphenol oxidases, tyrosine hydroxylation occurs at a rate that is 10–30 times slower than DOPA oxidation (Lerch and Ettlinger 1972; Vanni and Gastaldi 1990). Perhaps the formation of quinones in trematode eggshells cannot afford to be held hostage to the rate-limiting hydroxylation of tyrosines. Consequently, this step was reassigned to the less stressful milieu of the vitelline cells. In any case, physical separation of the two enzyme activities was necessitated perhaps to ensure that hydroxylation is complete before oxidation and cross-linking begins.

3.3 Microheterogeneity of Shell Precursors

Initial investigations of the vpB protein suggested that it is a single polypeptide of molecular weight 31 kDa (Waite and Rice-Ficht 1987). Subsequent characterization of vpC (Waite and Rice-Ficht 1989) and further analysis of vpB (Waite and Rice-Ficht 1992; Rice-Ficht et al. 1992) revealed that microheterogeneity exists among individual classes of the eggshell precursors. In the case of vpC at least four electrophoretic variants ranging from 16 to 18.5 kDa could be detected on Tris-borate/sodium dodecyl sulfate polyacrylamide (SDS-PAGE) gels; all four variants are present in six independently isolated worms and are not the result of allelic variants of the gene. Re-examination of vpB reveals as many as 21 electrophoretic variants based on isoelectric focusing. The heterogeneity probably does not reflect differences in hydroxylation, since even the highest pH achieved in the isoelectrofocusing gradient is well below the pK of either phenolic hydroxyl group. This suggests that the differences may derive instead from variation in primary sequence. Supporting evidence for the heterogeneity is seen in the isolation of two different cDNAs representing primary transcripts of vpB and the presence of at least six genomic copies of the vpB gene (Rice-Ficht et al. 1992). Some heterogeneity must also exist at the level of protein tyrosyl hydroxylation, since reversed phase HPLC of vpB reveals a linear DOPA/Tyr gradient starting in the leading edge and ending in the trailing edge of the eluting peak (Waite and Rice-Ficht 1987). vpB tryptic peptide sequences suggest that nearly all the tyrosines in the protein are fair game for hydroxylation although some are clearly more efficiently converted than others (Waite and Rice-Ficht 1992). The function of such a shotgun approach to post-translational modification is not yet clear. Perhaps tyrosyl hydroxylase modification originally served to modify tyrosines in the uncomplicated [-Gly-Tyr-Gly-] motifs of vpC (Waite and Rice-Ficht 1989), in which it does a most effective job (Table 1). As other precursor proteins developed participatory roles in eggshell formation, these were also modified, but adventitiously so, according to the resemblance of their sequence to vpC. Indeed, most of the tyrosines modified to DOPA in vpB have at least one flanking glycine (Waite and Rice-Ficht 1992). Overall, microheterogeneity within size classes of vitelline proteins exists both at the level of primary gene sequence and at the level of tyrosine to DOPA conversion.

3.4 Shell Assembly

Based on histochemical data outlined previously as well as the biochemical analysis of eggshell proteins, both a biological (Fig. 2) and biochemical (Fig. 3) model of shell assembly are proposed. In these models all three shell precursor proteins are translated in the extensive GER of the early vitelline cells and stockpiled in globules prior to the time that the cells leave the follicle. Nascent protein precursors are modified by protein-specific tyrosyl hydroxylase to produce DOPA residues in the protein backbone. The shell precursors and polyphenol oxidase are transported to the ootype packed away in vitelline cell globules. As vitelline cells congregate in the ootype along with one fertilized ovum, the ootype becomes distended causing extrusion of the S2 secretory bodies of Mehlis' gland into the ootype. The S2 secretions form a lipoprotein membrane around the vitelline cell/egg mass which is followed by release of globular material from the vitelline cells through regulated secretion. As the globule contents spread on the lipoprotein membrane, catechol oxidase becomes activated and oxidation and cross-linking ensues. DOPA residues are abundant in newly formed shells and shells of the proximal uterus. As the shells pass through the uterus, DOPA residues are consumed in the cross-linking reactions and the shells are completely "cured" prior to extrusion. Cross-linking also appears to coincide with a "shrinkage" of the shell by approximately 35% as the shell proceeds through the uterus (Waite JH and Rice-Ficht AC, unpubl. observ.).

Some clues to the role of individual proteins in shell formation have presented themselves. vpB is the major shell component by weight and is present in molar quantities approximately 30 times that of vpA or vpC. Of the three proteins studied, only one, vpC, has substantial homology with any published protein sequence; vpC shows strong homology with the His-rich domain of high molecular weight kininogen I whose function is to bind to negatively charged surfaces and accelerate binding of other blood-clotting factors to that surface (Kitamura et al. 1983). Based on this homology, we postulate that vpC is the first of the shell proteins to associate with the lipoprotein layer, facilitating binding and dispersion of vpB and vpA during shell formation.

Histochemical studies of *F. hepatica* have revealed the presence of large quantities of alkaline phosphatase activity exclusively in the ootype epithelium (Rice-Ficht, unpubl. observ.). In many systems, such as that of bone remodeling and bone deposition, alkaline phosphatase is an integral part of the mineralization process (Anderson 1989); in these cases the enzyme serves to remove phosphate from many different types of macromolecules making it available for the formation of calcium phosphate crystals. Although the presence of some quantities of metals has been detected in the vitellaria of *S. mansoni* through electron probe microanalysis (Shaw and Erasmus 1984), there is no evidence for extensive mineralization either in *F. hepatica* or in the related turbellarian *Bdelloura candida* (Huggins LG, unpubl. results). We favor the theory that the enzyme is present to produce phosphate needed as a buffering system in the ootype. The presence of an alkaline environment in the ootype would support the contention (Cordingley JS, pers. commun.) that a slightly alkaline environment would have a dramatic effect on histidine-rich eggshell precursors. A slight alteration of the pH could lead to a shift in equilibrium toward deprotonation of the imidazole ring of histidine residues in dispersed eggshell proteins. Proteins with large quantities of histidine such as vpC or the histidine-rich protein described by Johnson et al. (1987), would experience a large shift in the net charge and perhaps drive globule coalescence.

4 Comparative Strategies in Eggshell Production

4.1 Gene Structure

Isolation of cDNAs encoding vpB protein has been achieved using a combination of antibody selection and hybridization with degenerate oligonucleotides (Rice-Ficht et al. 1991). The proteins encoded, approximately 30.9 kDa in size, show amino acid compositions in good agreement with that of the purified protein. Sequencing of cDNAs has revealed the presence of at least two distinct mRNAs encoding vitelline protein B1 (vpB1) and vitelline protein B2 (vpB2) which differ significantly in nucleotide and amino acid sequence but are quite similar in amino acid composition (Table 2). A third member of the vpB protein family which is very closely related in nucleotide and amino acid sequence to vpB1 has been described by Zurita et al. (1987, 1989). In fact, only a few point changes in the nucleotide sequence relative to vpB1 are detectable; the changes do prove to be significant since the introduction of a single nucleotide into the female genital complex gene leads to early termination of translation relative to vpB1 and production of a 20 kDa protein. And, although there are no apparent repeating motifs in either the nucleotide or amino acid sequence of vpB1 or vpB2, a rigorous statistical analysis of protein sequence has revealed a highly degenerate hexapeptide motif with glycine conserved at the fifth position (Rzepecki LM, pers. comm.).

Table 2. Amino acid composition generated from gene sequence of various chorion or eggshell proteins (mol%)

Amino acid	vpB1 cDNA[a]	vpB2 cDNA[a]	s15-1[b]	pSMf[c]	ESG1[d]
ASP	9.0	8.7	1.0	5.6	5.2
ASN	6.7	4.7	4.1	5.1	6.1
THR	1.6	1.6	1.0	3.4	0.9
SER	5.5	5.1	8.2	6.2	5.7
GLU	6.3	6.7	5.2	0.6	0.9
GLN	2.0	0.8	3.1	0.6	0.0
PRO	1.2	1.2	7.2	1.7	2.8
GLY	15.0	15.4	21.6	44.1	50.5
ALA	7.1	7.1	15.5	2.8	1.9
CYS/2	0	0	0	5.6	2.8
VAL	0.8	0.4	5.2	1.1	0.9
MET	2.4	2.4	0	0.6	0.5
ILE	0.4	0	2.1	0.6	0.9
LEU	3.6	3.6	2.1	1.7	1.9
TYR	13.4	13.8	14.4	10.7	10.8
PHE	3.6	3.6	0	2.3	0.9
HIS	3.3	5.5	1.0	1.7	0.5
LYS	11.5	12.7	1.0	5.6	5.7
ARG	6.3	6.7	7.2	0.0	0.5
TRP	0	0	0	0.0	0.5

[a] from Waite and Rice-Ficht (1991). Values represent mature vpB exclusive of signal sequence.
[b] from Wong et al. (1985); *D. melanogaster* chorion protein.
[c] from Bobek et al. (1988); *S. mansoni* chorion protein.
[d] from Henkle et al. (1990); *S. japonicum* chorion protein.

A comprehensive study of protein and nucleotide sequence databases has not yielded any additional significant sequence homologies with the vpB class of proteins. However, a comparison of amino acid composition between vpBs and other glycine-rich chorion proteins of related helminths and insects has indicated similarities based on amino acid content. Table 2 illustrates a comparison of amino acid composition of several insect and helminth chorion proteins. A high glycine content is noted in all proteins listed although the schistosome proteins described thus far are two- to three-fold as glycine-rich as those observed in *F. hepatica* or *Drosophila melanogaster*. Peptide sequencing in *Fasciola* suggests that among the vpB variants there may be additional family members of a higher glycine content yet to be sequenced (Waite JH and Rice-Ficht AC, unpubl. results). Tyrosine residues are abundant and roughly equivalent on a mol% basis from protein to protein. A high number of charged amino acids are also evident in the shell proteins of *D. melanogaster* (Wong et al. 1985), *S. mansoni* (Bobek et al. 1986), *S. haematobium* (Bobek et al. 1989), *S. japonicum* (Henkle et al. 1990), and *F. hepatica* (Waite and Rice-Ficht 1987); although *F. hepatica* contains two to three times the number of charged residues present in the other proteins. It is interesting to note that in each protein listed, the number of negatively charged residues approximates the number of positively charged ones. With few exceptions, charged residues are dispersed throughout the molecules with no clustering evident. vpB proteins are also distinctive in their lack of any detectable tryptophan; low to nondetectable amounts of Trp is a common feature of most chorion proteins. vpB contains no cysteine residues, indicating at least that disulfides are not involved in the cross-linking of that protein in the shell matrix. In contrast, the major glycine-rich protein of *S. mansoni* (Bobek et al. 1986), *S. japonicum* (Henkle et al. 1990) and *S. haematobium* (Bobek et al. 1989) contain two or more pairs of cysteines spaced four residues apart which are likely required for disulfide formation and the maintenance of secondary structure. Although schistosome eggshells cannot be solubilized with reducing agents alone, the assistance of disulfides in cross-link formation has not been ruled out.

A common sequence feature of glycine-rich chorion proteins is the presence of a Gly-Tyr sequence motif which recurs manyfold in each of the proteins. In *F. hepatica* and in *S. mansoni* (Waite JH, unpubl. result), this represents a target of post-translational modification by a tyrosine hydroxylase. In other systems where tyrosine hydroxylation does not occur, such as *D. melanogaster*, the sequence may be maintained for purposes of secondary structure. Alternatively, since strong covalent interactions between *Drosophila* spp. shell proteins exist in the form of di and tri-tyrosine cross-links (Petri et al. 1976), Gly-Tyr sequence motifs may be recognition sites for peroxidase cross-linking, although the site specificity of cross-links in *Drosophila* has not been reported. In *B. mori* it is clear that no covalent cross-linking other than disulfide formation takes place, since thiol urea will completely solubilize eggshells (Goldsmith and Kafatos 1984). A strong bias toward high glycine and tyrosine content is evident in glycine-rich chorion proteins of most genera regardless of cross-linking mechanism (Margaritis 1985).

Large-scale sequence comparisons among species are useful in understanding eggshell function but are unavailable as yet from other *Fasciola* species. Inspection of individual members of a protein or gene family such as vpB1 and vpB2 reveals great variation in sequence but a nearly identical amino acid composition. A comparison of the two cDNAs encoding vpB1 and vpB2 shows that the N-terminal 30% as well as the C-terminal 30% of the sequences are close to identity. The central region of 40% is widely divergent at the amino acid sequence level (33.3%) but overall composition is maintained (Table 2). Sequence heterogeneity among members of eggshell gene families is the norm rather than the exception. All genes listed in Table 2 represent one member of a multigene family whose members are divergent to varying degrees. It has been suggested that this allows organisms to adapt and to survive in different ecological niches or microenvironments (Goldsmith and Kafatos 1984).

In making sequence comparisons of the *Schistosoma* glycine-rich proteins from different species several features of gene structure are evident. First, there is substantial divergence between the glycine-rich chorion proteins of different species although sequence alignments suggest that they may have arisen from a common ancestor (Henkle et al. 1990). Secondly, in spite of divergence the amino acid compositions are similar. Finally, in contrast to the vpB family of *F. hepatica*, the individual members of a protein family show little sequence diversity with only a few nucleotide point changes detected (Bobek et al. 1986, 1988, 1989; Henkle et al. 1990).

While vpB demonstrates an amino acid composition similar to the glycine-rich chorion proteins of *Schistosoma* spp. and *Drosophila*, vpC is quite distinct in sequence and composition. The protein is composed primarily of a $(Gly\text{-}X)_n$ repeat motif in which X is Ser, DOPA, or His (Waite and Rice-Ficht 1989). The majority of histidine occurs as tandem Gly-His repeats found in a pepsin-resistant fragment of the protein. The N-terminus of the protein contains the sequence His-His-Trp-Asp-Gly-DOPA-Gly-DOPA-Gly with His also representing two N-terminal residues. The Gly-His repeat motif is the region which corresponds to a His-Gly repeating domain of high molecular weight kininogen I (Kitamura et al. 1983); this region of kininogen is capable of binding to kaolin and other negatively charged surfaces. It is argued that this binding property is involved in "kininogen-accelerated" surface-induced activation of factor XII and the triggering of coagulation (Sugo et al. 1980). By analogy vpC may be responsible for binding to the lipoprotein layer secreted by Mehlis' gland and accelerating binding of other vitelline proteins to that surface.

With respect to other helminth proteins, the high histidine content suggests that it may perform a function similar to that postulated for the C-terminal domain of the 47 kDa protein described by Johnson et al. (1987) which has a C-terminal region rich in histidine including a tract of up to nine histidine residues. A large domain flanking this area on the N-terminal side is composed of a distinct repeating pentameric amino acid sequence which includes tyrosine and glycine. With the exception of vpC, the 47-kDa protein is unique among helminth shell proteins in its extensive format of internally repetitive peptides.

Among helminths in general the eggshell is composed of a complex of proteins each represented by a multigene family. In *F. hepatica*, eggshell construction is achieved using three Gly/DOPA-rich proteins with individually distinct amino acid compositions: (1) a 70-kDa protein rich in Asp/Asn/Arg, (2) a 31-kDa protein (the major shell component) rich in Lys and (3) a 17-kDa His-rich protein. In *Schistosoma* spp. the shell precursor complement appears to include (1) a class of extremely glycine-rich (mol% ~50), tyrosine-rich proteins of approximately 20 kDa, (2) a 47-kDa protein with an extensively repeated Tyr/Gly-rich pentapeptide and a His-rich C-terminal domain, and (3) a protein whose transcript has been localized to the vitelline cell cytoplasm but whose transcript reading frame is uncertain (Reis et al. 1989). The minimal requirements for fashioning an eggshell in these systems appears to be (1) overall high glycine content, (2) Tyr/Gly sequence motifs, (3) abundance of charged residues, and (4) a His-rich domain.

4.2 Expression of Chorion Proteins

Expression of eggshell protein in *F. hepatica* appears to be both stage- and tissue-specific. During development, only the adult form of the worm contains active vitellaria and engages in egg-laying by histological inspection. The exact timing of transcription and translation of eggshell proteins during development has not been assessed. While eggshell protein transcription/translation is induced by worm pairing in *Schistosoma* spp. (Erasmus 1973; Bobek et al. 1986), induction of shell proteins in *F. hepatica*, a hermaphroditic worm, is induced through less well-defined signals.

However, a study of the *tissue-specific* transcription/translation of DOPA-containing proteins in *F. hepatica* has revealed much about shell protein synthesis. Histological localization studies of protein and transcripts have revealed that (1) DOPA proteins are detectable exclusively in the vitelline cells and in the eggshell itself (Waite and Rice-Ficht 1989), (2) the vpB protein is synthesized only in the vitelline cells and primarily accumulates during vitelline cell development in the follicle (Rice-Ficht et al. 1992) and (3) vpB transcripts are most abundant in the It1, It2, and M subclasses of vitelline cells, diminishing substantially as cells move to the vitelline duct (Rice-Ficht et al. 1992). These data, taken together with pulse labeling studies by Hanna (1976), point to synthesis and stockpiling of eggshell precursor protein in the early vitelline cell, a process which is a fait accompli prior to migration to the vitelline duct.

In focusing on transcription of the most abundant eggshell protein, vpB, hybridization of vpB cDNA to total mRNA from the worm reveals a continuous spectrum of transcript sizes ranging from 900–1400 bp and representing a number of different primary transcripts. Isolation of discrete cDNAs representing each extreme of this range (vpB1 — 900 bp; vpB2 — 1350 bp) supports the interpretation of this result. Coupled with the detection of a number of genomic copies of vpB, these data suggest that a substantial level of heterogeneity in the vpB protein pool is generated at the nucleic acid level. Additional heterogeneity is acquired at the level of post-translational modification. RNA splicing does not appear to play a significant role in eggshell production in helminths since genomic copies of *S. mansoni* (Bobek et al. 1988), *S. haematobium* (Bobek et al. 1989), and *F. hepatica* eggshell proteins (Rice-Ficht AC, unpubl. data) are unpunctuated by introns. This is in contrast to extensive and alternative processing of eggshell primary transcripts (*dec-1*) in *D. melanogaster* (Waring et al. 1990). Likewise the amplification of chorion genes such as that observed in *D. melanogaster* (Spradling and Mahowald 1980) has not been detected in helminths (Bobek et al. 1986). These differences may be attributable to (1) the continual maintenance of numerous gene copies of eggshell genes in helminths and (2) a different basic strategy for shell production than that employed by insects.

In helminths, cells dedicated to shell protein production develop in vitelline follicles and stockpile protein precursors of the shell. Vitelline cells later migrate to Mehlis' gland, where, contained by a lipoprotein membrane, the cells release protein and the shells form from the inside. In *D. melanogaster*, where gene amplification is observed, follicular epithelial cells surround the oocyte and are compelled to synthesize and secrete large quantities of shell protein over a short time frame (Margaritis 1985); proteins are applied to the oocyte in a temporally regulated fashion forming the shell from the outside. The follicular epithelia amplify chorion gene copies through saltatory or "onion-skin" replication producing a large number of chorion gene copies. A corresponding increase in chorion-specific transcript and protein production ensues, supporting synthesis of the required quantities of protein. The eggshells of insects also differ from helminths in that they are very complex, multilaminate structures, produced through carefully regulated temporal secretion of a number of different types of chorion proteins (comprehensively reviewed by Goldsmith and Kafatos 1984; Margaritis 1985). Helminth systems, in contrast, rely on a much more restricted set of shell proteins which are simultaneously released to spontaneously spread and form the shell layer or layers. The implications of the helminth process are fascinating, since much of the blueprint for structure and assembly must lie within the primary sequence of the shell precursors; future studies on the surface properties of these proteins, as well the abilities of the proteins to self-assemble, may reveal novel polymerization chemistries and processes.

Acknowledgments. I wish to extend special thanks to Dr. Herb Waite for extensive discussions, unlimited intellectual support, repeated readings of the manuscript, and for introducing me to the field of biopolymers. I also wish to thank Dr. Thomas Ficht for numerous helpful discussions and many critical readings of the manuscript, and Ms. Charlene Richardson for reading of the manuscript and technical support in putting it together.

References

Anderson HC (1989) Mechanism of mineral formation in bone. Lab Invest 60:320–326

Arnow LE (1937) Colorimetric determination of the components of 3,4 dihydroxyphenylalanine-tyrosine mixtures. J Biol Chem 118:531–537

Bell EJ, Smyth JD (1958) Cytological and histochemical criteria for evaluating development of trematodes and pseudophyllidean cestodes in vivo and in vitro. Parasitology 48:131–148

Bjorkman J, Thorsell W (1963) On the fine morphology of the formation of eggshell globules in the vitelline gland of the liver fluke (*Fasciola hepatica*, L.). Exp Cell Res 32:153–156

Bobek LA, Rekosh DM, van Keulen H, LoVerde PT (1986) Characterization of a female-specific cDNA derived from a developmentally regulated mRNA in the human blood fluke *Schistosoma mansoni*. Proc Natl Acad Sci USA 83:5544–5548

Bobek LA, Rekosh DM, LoVerde PT (1988) Small gene family encoding an eggshell (chorion) protein of the human parasite *Schistosoma mansoni*. Mol Cell Biol 8:3008–3016

Bobek LA, LoVerde PT, Rekosh DM (1989) *Schistosoma haematobium*: analysis of eggshell protein genes and their expression. Exp Parasitol 68:17–30

Bogitsh BJ (1985) Ultrastructural localization of polyphenols in the vitelline cells of *Haematoloechus medioplexus* with an osmiophilic tetrazotized salt. Trans Am Microsc Soc 104:95–99

Brown CH (1975) Structural materials in animals. Pitman, London, pp 162–182

Burgess TL, Kelly RB (1987) Constitutive and regulated secretion of proteins. Annu Rev Cell Biol 3:243–293

Burton PR (1963) A histochemical study of the vitelline cells, egg capsules and Mehlis' gland in the froglung-fluke, *Haematoloechus medioplexus*. J Exp Zool 154:247–257

Burton PR (1967) Fine structure of the reproductive system of a frog lung-fluke. 1. Mehlis' gland and associated ducts. J Parasitol 53:540–555

Clegg JA (1965) Secretion of lipoprotein by Mehlis' gland in *Fasciola hepatica*. Ann NY Acad Sci 118:969–986

Degrand C (1985) Electron transfer in quinonoid-modified electrodes: mediation and catalytic applications. Ann Chim 75:1–18

Dixon KE, Mercer EH (1967) The formation of the cyst wall in the metacercaria of *Fasciola hepatica*. L Z Zellforsch 77:345–360

Eisner T, Shepherd J, Happ GM (1966) Tanning of grasshopper eggs by an exocrine secretion. Science 152:95–97

Emini E, Hughes JV, Perlow DS, Boger J (1985) Induction of hepatitis A virus neutralizing antibody by a virus-specific synthetic peptide. J Virol 55:836–839

Erasmus DA (1973) A comparative study of the reproductive system of mature, immature and unisexual female *Schistosoma mansoni*. Parasitology 67:165–183

Eyre DR (1980) Collagen: molecular diversity in the body's protein scaffold. Science 207:1315–1322

Goldsmith MR, Kafatos FC (1984) Developmentally regulated genes in silkmoths. Ann Rev Genet 18:443–487

Gönnert R (1962) Histologische Untersuchungen über den Feinbau der Eibildungsstatte (Oogenotyp) von *Fasciola hepatica*. Z Parasitenkd 21:475–492

Hanna REB (1976) *Fasciola hepatica*: a light and electron autoradiographic study of shell-protein and glycogen synthesis by vitelline follicles in tissue slices. Exp Parasitol 39:18–28

Henkle KJ, Cook GA, Foster LA, Engman DM, Bobek LA, Cain GD, Donelson JE (1990) The gene family encoding eggshell proteins of *Schistosoma japonicum*. Mol Biochem Parasitol 42:69–82

Irwin SWB, Threadgold LT (1970) Electron microscope studies on *Fasciola hepatica*. VIII. The development of the vitelline cells. Exp Parasitol 28:399–411

Irwin SWB, Threadgold LT (1972) Electron microscope studies of *Fasciola hepatica*. X. Egg formation. Exp Parasitol 31:321–331

Jaegfeldt H, Kuwana T, Johansson G (1983) Electrochemical stability of catechols with a pyrene side-chain strongly adsorbed on graphite electrodes for catalytic oxidation of dihydronicotinamide adenine dinucleotide. J Am Chem Soc 105:1805–1814

Janes SM, Mu D, Wemmer D, Smith AJ, Kaur S, Maltby D, Burlingame AM, Klimman JP (1990) A new redox cofactor in eukaryotic enzymes: 6-hydroxydopa at the active site of bovine serum amine oxidase. Science 248:981–987

Johnson KS, Taylor DW, Cordingley JS (1987) Possible eggshell protein gene from *Schistosoma mansoni*. Mol Biochem Parasitol 22:89–100

Johri LN, Smyth JD (1956) A histochemical approach to the study of helminth morphology. Parasitology 46:107–116

Kawasaki H, Sato H, Suzuki M (1971) Structural proteins in the egg-shell of the oriental garden cricket, *Gryllus mitratus*. Biochem J 125:495–505

Kitamura N, Takagaki Y, Furuto S, Tanaka T, Nawa H, Nakanishi S (1983) A single gene for bovine high molecular weight and low molecular weight kininogens. Nature 305:545–549

Kretchmar SA, Raymond KN (1988) Effects of ionic strength on iron removal from monoferric transferrins. Inorg Chem 27:1435–1441

Lau ANK, Miller LL (1983) Electrochemical behavior of a dopamine polymer — dopamine release as a primitive analog of a synapse. J Am Chem Soc 105:5271–5277

Ledger PW, Tanzer ML (1984) Monensin — a perturbant of cellular physiology. Trends Biochem Sci 9:313–314

Lerch K, Ettlinger L (1972) Purification and characterization of a tyrosinase from *Streptomyces glaucescens*. Eur J Biochem 31:427–43

Mansour TE (1958) Effect of serotonin on phenol oxidase from the liver fluke *Fasciola hepatica* and other sources. Biochem Biophys Acta 30:492–500

Margaritis LH (1985) Structure and physiology of the eggshell. In: Kerkut GA, Gilbert LI (eds) Comprehensive insect physiology, biochemistry and pharmacology. vol 1. Pergamon Press, Elmsford, NY pp 153–280

Neill PJG, Smith JH, Doughty BL, Kemp WM (1988) The ultrastructure of the *Schistosoma mansoni* egg. Am J Trop Med Hyg 39:52–65

Nellaiappan K, Ramalingam K (1980) Specificity of the enzyme phenol oxidase and possible metabolic pathway of sclerotin in *Parapleurus sauridae*. J Parasitol 66:217–219

Nellaiappan K, Devasundari AF, Dhandayuthapani S (1989) Properties of phenol oxidase in *Fasciola gigantica*. Parasitology 99:403–407

Ong SA, Peterson T, Neilands JB (1979) Agrobactin, a siderophore from *Agrobacterium tumefaciens*. J Biol Chem 254:1860–1865

Pecoraro VL, Weitl FL, Raymond KN (1981) Ferric iron-specific sequestering agents. 7. Synthesis, iron exchange kinetics and stability constants of N-substituted sulphonated chatecholamide analogs of enterobactin. J Am Chem Soc 103:5133–5140

Petri WH, Wyman AR, Kafatos FC (1976) Specific protein synthesis in cellular differentiation. III. The eggshell proteins in *Drosophila melanogaster* and their program of synthesis. Dev Biol 49:185–199

Pierpoint CG, Buchanan RM (1981) Transition metal complexes of orthobenzoquinone and catecholate ligands. Coord Chem Rev 38:45–87

Pizzi A (1985) Condensed tannins for adhesives. Polym Mater Sci Eng 52:251–255

Pryor MGM (1940) On the hardening of the ootheca of *Blatta orientalis*. Proc R Soc Lond B 128:378–393

Rainsford KD (1972) The chemistry of eggshell formation in *Fasciola hepatica*. Comp Biochem Physiol 43B:983–989

Ramalingam K (1973) The chemical nature of the egg-shell of helminths. 1. Absence of quinone tanning in the egg-shell of the liver fluke *Fasciola hepatica*. Int J Parasitol 3:67–75

Rao KH (1959) Observations on the Mehlis' gland complex in the liver fluke *Fasciola hepatica*. J Parasitol 45:347–351

Raymond KN, Carrano CJ (1979) Coordination chemistry and microbial iron transport. Acc Chem Res 12:173–190

Reis MG, Kuhns J, Blanton R, Davis AH (1989) Localization and pattern of expression of a female specific mRNA in *Schistosoma mansoni*. Mol Biochem Parasitol 32:113–120

Rice-Ficht AC, Dusek KA, Kochevar GJ, Waite JH (1992) Eggshell precursor proteins of *Fasciola hepatica*: I. Structure and expression of vitelline protein B. Mol Biochem Parasitol 54:129–142

Rowan WB (1962) Permeability of the egg shell of *Fasciola hepatica*. J Parasitol 48:499

Rzepecki LM, Nagafuchi T, Waite JH (1991) alpha, beta-Dehydro-3,4-dihydroxyphenylalanine derivatives: potential sclerotization intermediates in natural composite materials. Arch Biochem Biophys 285:17–26

Schmidt GD, Roberts LS (1985) Subclass digenea: form, function, biology, and classification In: Schmidt GD, Roberts LS (eds) Foundations in parasitology. Times Mirror/Mosby, St. Louis, Missouri, pp 244–279

Seed JL, Bennett JL (1980) *Schistosoma mansoni*: phenol oxidase's role in eggshell formation. Exp Parasitol 49:430–441

Seed JL, Boff M, Bennett JL (1978) Phenol oxidase activity: induction in female schistosomes by in vitro incubation. J Parasitol 64:283–289

Shaw MK, Erasmus DA (1984) *Schistosoma mansoni*: electron probe microanalysis of the elemental composition of the tegument and subtegumental tissues of adult worms. Exp Parasitol 58:163–181

Skuce PJ, Fairweather I (1988) *Fasciola hepatica*: perturbation of secretory activity in the vitelline cells by the sodium ionophore monensin. Exp Parasit 65:20–30

Smyth JD (1951) Egg-shell formation in trematodes and cestodes as demonstrated by the methyl or malachite green techniques. Nature 168:322–323

Smyth JD (1954) A technique for the histochemical demonstration of polyphenol oxidase and its application to egg-shell formation in helminths and byssus formation in *Mytilus*. Q J Micros Sci 95:139–152

Smyth JD, Clegg J (1959) Egg-shell formation in trematodes and cestodes. Exp Parasitol 8:286–323

Spradling AC, Mahowald AP (1980) Amplification of genes for chorion proteins during oogenesis in *Drosophila melanogaster*. Proc Natl Acad Sci USA 77:1096–1100

Stephenson W (1947) Physiological and histochemical observations on the adult liver fluke *Fasciola hepatica*. Parasitology 38:128–139

Sugo T, Ikari N, Kato H, Iwanaga S, Fujii S (1980) Functional sites of high molecular weight kininogen as a cofactor in kaolin-mediated activation of factor XII (Hageman factor). Biochemistry 19:3215–3220

Thangaraj T, Nellaiappan K, Ramalingam K (1982) Activation of prophenol oxidase in the liver fluke *Fasciola gigantica* Cobbold. Parasitology 85:577–581

Threadgold LT (1982) *Fasciola hepatica*: Stereological analysis of vitelline cell development. Exp Parasitol 54:352–365

Urry DW (1983) What is elastin; what is not? Ultrastruct Pathol 4:227–251

Vanni C, Gastaldi D (1990) Kinetic investigations on the double enzymatic activity of the tyrosinase mushroom. Ann Chem 80:35–60

Vincent JFV, Hillerton E (1979) The tanning of insect cuticle — a critical review and revised mechanism. J Insect Physiol 25:653–658

Waite JH (1986) Mussel glue from *Mytilus californianus*: a comparative study. J Comp Physiol B 156:491–496

Waite JH (1990) The phylogeny and chemical diversity of quinone-tanned glues and varnishes. Comp Biochem Physiol B 97:19–29

Waite JH, Benedict CV (1984) Assay of dihydroxyphenylalanine (DOPA) in invertebrate structural proteins. Methods Enzymol 107:397–415

Waite JH, Rice-Ficht AC (1987) Presclerotized eggshell protein from the liver fluke *Fasciola hepatica*. Biochemistry 26:7819–7825

Waite JH, Rice-Ficht AC (1989) A histidine-rich protein from the vitellaria of the liver fluke *Fasciola hepatica*. Biochemistry 28:6104–6110

Waite JH, Rice-Ficht AC (1992) Eggshell precursor proteins of *Fasciola hepatica*: II. Microheterogeneity in vitelline protein B. (Mol Biochem Parasitol 54:143–152

Waite JH, Tanzer ML (1981) Polyphenolic substance of *Mytilus edulis*: novel adhesive containing L-DOPA and hydroxyproline. Science 212:1038–1040

Waite JH, Housley TJ, Tanzer ML (1985) Peptide repeats in mussel glue protein: theme and variations. Biochemistry 24:5010–5015

Walker WF, Menzer RE (1969) Chorionic melanization in the eggs of *Aedes aegypti*. Ann Entomol Soc Am 62:7–11

Wang FL, Su YF, Yang GM, Wang XZ, Qiu ZY, Zhou XK, Hu ZQ (1986) Isoenzymes of phenol oxidase in adult female *Schistosoma japonicum*. Mol Biochem Parasitol 18:69–72

Waring GL, Hawley RJ, Schoenfeld T (1990) Multiple proteins are produced from the *dec-1* eggshell gene in *Drosophila* by alternative RNA splicing and proteolytic cleavage events. Dev Biol 42:1–12

Wharton DA (1983) The production and functional morphology of helminth eggshells. Parasitology 86:85–97

Wilson RA (1967) The structure and permeability of the shell and vitelline membrane of the egg of *Fasciola hepatica*. Parasitology 57:47–58

Wong Y-C, Pustell J, Spoerel N, Kafatos FC (1985) Coding and potential regulatory sequences in *Drosophila melanogaster*. Chromosoma 92:124–135

Zurita M, Bieber D, Ringold G, Mansour TE (1987) Cloning and characterization of a female genital complex cDNA from the liver fluke *Fasciola hepatica*. Proc Natl Acad Sci USA 84:2340–2344

Zurita M, Bieber D, Mansour TE (1989) Identification, expression and in situ hybridization of an eggshell protein gene from *Fasciola hepatica*. Mol Biochem Parasitol 37:11–18

Chapter 5

The Cell and Molecular Biology of Eggshell Formation in *Schistosoma mansoni*

Kim E. Wells and John S. Cordingley[1]

1 Introduction. An Overview of Eggshell Formation in Schistosomes

Figure 1 is a micrograph of a newly formed egg within a female *Schistosoma mansoni*. Schistosome eggs are unusual amongst invertebrate eggs comprising many cells in addition to the fertilized ovum (Ov in Fig. 1). These so-called vitelline cells serve as a food source during embryogenesis within the egg in addition to being responsible for eggshell production. In Fig.1, there are vitelline cells on both the inside and outside of the eggshell. This egg was formed prior to perfusion of the mouse. The vitelline cells outside the eggshell probably leaked into the ootype during or after perfusion because the anesthetic used on the mouse host appears to relax the sphincter at the end of the vitelline duct allowing vitelline cells to leak into the ootype after eggshell formation.

At the time of eggshell formation, the vitelline cells secrete, or more accurately, they exocytose the eggshell precursors which fuse together and cross-link over the surface of the cell mass to form the eggshell. This simple statement describes a highly evolved, rapid, and effective mechanism for eggshell formation which has been rather glossed over in the past. When something is rapid and efficient there is a tendency to take the details for granted. Schistosome eggshells and their formation have suffered this indignity for many decades. Our ignorance of the details of this process is almost complete. Recent data discussed in this chapter are beginning to reveal the real complexity at the molecular and cellular level.

Perhaps the most exciting aspect of this process is what it can teach us in terms of material science. The schistosome eggshell is a remarkable molded-composite-material capsule. It is synthesized reproducibly and rapidly from components that are mixed together without cross-linking, stored for several days, and then molded and cross-linked in less than 5 min to form a precisely shaped capsule. The capsule is made out of materials with precisely tailored properties of porosity and strength, since the egg must take up nutrients and split consistently at the time of hatching.

This chapter will review our current state of understanding of the components of the eggshell and examine its mode of formation in some detail to focus on the implicit events necessary for this process to occur.

[1] University of Wyoming, Dept. of Molecular Biology, Box 3944 University Station, Laramie. WY 82071, U.S.A

Results and Problems in Cell Differentiation 19
Biopolymers
Case, S.T. (Ed.)

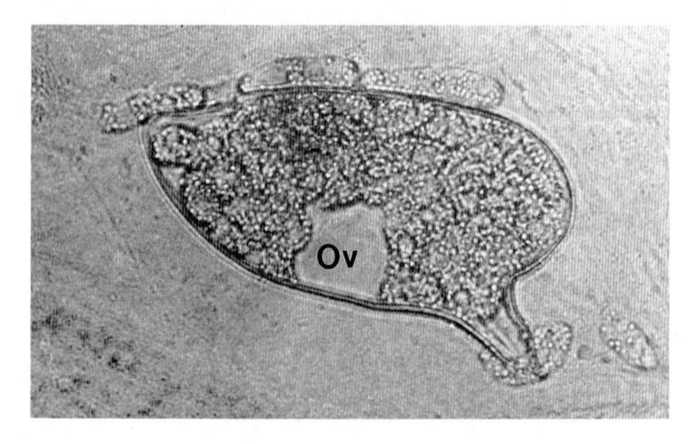

Fig. 1. A photomicrograph of a newly formed egg in the ootype of a female schistosome. The single fertilized ovum is indicated with the letters *Ov*. The remainder of the space inside the eggshell is filled with vitelline cells which appear granular due to the large numbers of refractile granules they contain

2 A Survey of Recent New Data

2.1 Female-Specific Genes and Proteins Are Not Yolk Proteins

There are several major proteins being synthesized by female *S. mansoni* which are not present in males. Early in our involvement with this problem we tacitly assumed that these major proteins were "yolk proteins". This stemmed from our knowledge of the yolk proteins in *Drosophila* and also reflected the naming of the "vitellaria" and the "vitelline cells" in schistosomes, which implied that they were yolk components and primarily food for the embryo. This has turned out to be only part of the story. The "vitelline cells" probably do serve as food for the developing embryo but not, it appears, by synthesizing any special proteins. In schistosomes, the major female-specific proteins appear to be exclusively eggshell components. When compared with *Drosophila*, this is not, in retrospect, surprising. *Drosophila* yolk proteins are synthesized by the fat bodies and transported via the hemolymph to the forming eggs. Since schistosomes have entirely different body plans, no hemolymph or fat body equivalents, and a very different way of making eggs, none of our subsequent findings seem surprising. Schistosomes are not insects. They are not even very close relatives (Field et al. 1988). Extrapolating from insects is only warranted with caution and any similarities probably reflect highly conserved aspects of the two groups' biology and therefore in some ways are their less interesting features. Within the platyhelminthes themselves, there are a variety of strategies for eggshell production, a point we will return to later in Section 5.

2.2 Glycine-Rich and Tyrosine-Rich Eggshell Proteins

Two classes of female-specific proteins have been characterized by way of their respective genes. Several groups have cloned and characterized the genes encoding one group of these proteins (Kunz et al. 1987; Simpson et al. 1987; Bobek et al. 1988; Bobek et al. 1991) which are glycine-rich and probably make up 90–95% of the weight of the eggshell (glycine-rich proteins, GRP). The gene encoding a distinct tyrosine-rich protein (TRP) was cloned by us in the early 1980s (Johnson et al. 1987) and we are finally beginning to make sense out of it.

The published sequences of the GRPs do not immediately suggest anything very obvious about the proteins' structural or functional properties. Published attempts to predict and model GRP structure are interesting without being very convincing or illuminating (Rodrigues et al. 1989). Current methods of predicting protein secondary structure are based upon the crystal structures of soluble enzymes (Chou and Fasman, 1974) and are consequently not very reliable guides to the structures of radically different kinds of structural proteins such as those in the schistosome eggshell. No direct biophysical measurements have been performed on the GRP to our knowledge.

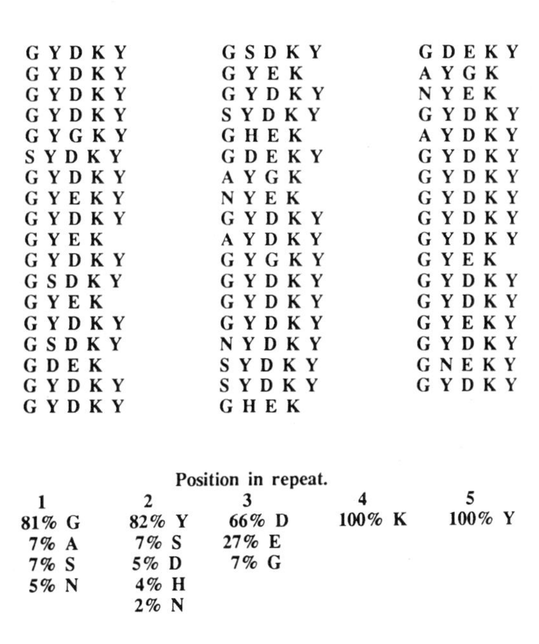

Fig. 2. A compilation of all of the repeats currently sequenced from the single copy of the TRP gene in our laboratory strain of *S. mansoni*. The repeats occur in the order presented with no sequences between adjacent repeats. The cumulative amino acid usage at each of the five positions are listed below the sequences. The "fifth" residue is not always present, but when it is, it is always tyrosine. This involves a certain amount of subjective division of the repeats in one or two places, but we think inspection of the sequence will support our decisions

However, the TRP is a quite different matter. The majority of the length of the protein is composed of simple repeating sequences with the consensus sequence Gly-Tyr-Asp-Lys-Tyr. Figure 2 is a compilation of all of the repeats from the sequence of the single genomic copy of this gene in our laboratory strain of *S. mansoni*. As summarized in Fig. 2, the conservation of the repeat sequence is very striking. Inspection of the DNA sequence reveals that the sequence is under strong selection. For example, the lysine is completely conserved, or to put it another way, the replacement of any *one* of these lysines results in a protein so much less well adapted to its function that the mutation disappears from the population!? In the case of an enzyme with a unique lysine at the active site, this is easily understood since such a mutation could completely inactivate the enzyme, resulting in a phenotypically significant loss of function. However, in a repetitive protein with a *presumed* structural role, one tends to imagine that any given repeat plays only a small part in the overall structure and function of the protein and, therefore, the alteration of a few residues might be expected to be tolerated. In general, this is what is observed and repeats in structural proteins are usually much less precisely conserved than in this protein, for example in collagen (Burgeson 1988). Silent mutations, on the other hand, have been freely fixed in the TRP encoding gene and the codon usage is close to the overall codon usage of *S. mansoni* (Meadows and Simpson 1989).

In addition to being unusually conserved, the TRP gene is only present in a single copy in the haploid genome of our laboratory strain which is homozygous for the TRP gene. In our original publication, we were using the line of *S. mansoni* maintained in Cambridge (Johnson et al. 1987) and there were two *EcoRI* fragments on a genomic Southern blot using DNA from this strain. Only a single *EcoRI* fragment corresponding to the larger of these two bands is present in our current strain and we have cloned this 12-kb fragment in a bacteriophage lambda vector. There is only one copy of the TRP in this fragment (data not shown). This is in contrast to the GRP genes, which are members of a family of closely related genes (Bobek et al. 1988). We conclude that this protein is something a little out of the ordinary and our subsequent studies have tended to confirm this.

2.3 The TRP Repeat Adopts a Left-Handed α-Helix in Aqueous Solution

We have taken advantage of the highly repetitive nature of the amino acid sequence of the TRP and synthesized short synthetic peptides as models of the larger protein. Using these synthetic peptides and a wide variety of biophysical techniques and computer modeling, we have come to the conclusion that these repeats adopt a left-handed α-helix in aqueous solution (Middaugh et al. 1991). We are not going to present data in support of this observation here since it is not central to the theme of the chapter, and to adequately support this contention would require the whole chapter. After all of our biophysics and our amazing findings, we are in many ways no nearer to understanding the function of the TRP.

That the GYDKY repeat forms left-handed α-helix is amazing but, in itself, it does not offer any immediate explanation for the apparent selective advantage conferred by this sequence.

2.4 Eggshell Precursors Are DOPA Proteins

Herbert Waite and his colleagues have demonstrated, first in mollusks and later in *Fasciola hepatica* (see Rice-Ficht this Vol.), that the precursor proteins of mussel glue and *F. hepatica* eggshells have the majority of their tyrosine residues modified to 3,4-dihydroxyphenylalanine (DOPA) prior to secretion (Waite 1983; Waite and Rice-Ficht 1987). The female schistosome contains such modified proteins, as we discovered serendipitously when we discovered some unusual silver staining properties of these female-specific proteins (Wells and Cordingley 1991a). The modified eggshell precursor proteins actually reduce silver even in the absence of the usual reducing agent (formaldehyde). In the presence of formaldehyde a "redox cycle" is set up resulting in very rapid and intense silver staining of these modified proteins (Paz et al. 1988; Wells and Cordingley 1991a). This staining technique has been invaluable to us in analyzing the eggshell precursor proteins, as the technique allows detection of the eggshell precursors at concentrations well below that needed to detect the proteins by conventional protein staining techniques (Wells and Cordingley 1991a, b).

2.5 Eggshell Precursors Are an Acid-Stabilized Emulsion

In thin sections of vitelline cells prepared for electron microscopy (Wells and Cordingley 1991; Irwin and Threadgold 1970; Erasmus 1987), the vitelline cells contain many secretory vesicles with very electron-dense contents, not as a single inclusion, as is the case in conventional dense core vesicles, but divided into many droplets. These vesicles appear to contain an *emulsion* of eggshell material and our data support this contention. These dense droplets, apparently in the form of an emulsion, are protein, not oil or some similar hydrophobic liquid. Proteins are not usually found in emulsions with this appearance, and we know of no instances except the eggshell precursor vesicles of trematodes as exemplified by *S. mansoni* and *F. hepatica* (Irwin and Threadgold 1970; Erasmus 1987). Typical dense core granules contain a single electron-dense mass with little if any subdivision.

Work in other systems has shown that regulated secretory vesicles are acidic, with internal pHs of approximately 5.5 (Anderson and Pathak 1985; Mellman et al. 1986). The vesicles budding from the *trans* Golgi are acidified at the time of budding and remain acidic in the case of regulated vesicles that are not secreted immediately. In contrast, vesicles budding from the *trans* Golgi, whose secretion is constitutive (i.e., they will be secreted immediately with no regulation), are transiently acidified whilst budding, then the pH returns to above 7 and the vesicles fuse with the plasma membrane and exocytose their contents (Orci et al. 1987). Extrapolating to schistosomes, we guessed that the eggshell precursor vesicles

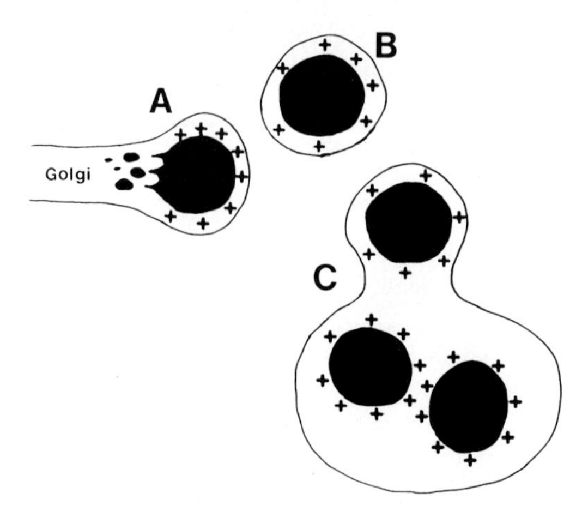

Fig. 3A. A trans-Golgi condensing vesicle is budding from the Golgi containing dense eggshell precursor material. At this stage it begins to acidify. **B** A newly separated vesicle containing eggshell precursor proteins. The vesicle is acidified and a layer of positive charges has formed on the surface of the droplet due to the histidine side chains. **C** Vesicles like the one in **B** fuse together to form larger secretory vesicles. The individual droplets do not fuse together, due to the already formed charge layer

would be acidic until secretion. From a basic knowledge of emulsion technology, we further guessed that the emulsion observed in the vesicles was stabilized by this low pH (Perram et al. 1977). In classical emulsions, such as a Bearnaise sauce, the droplets of oil are stabilized by a layer of charged groups over their surfaces in the same way that detergent molecules stabilize oil droplets. This leads directly to the suggestion diagrammed in Fig. 3. As the eggshell precursor proteins are concentrated in the *trans*-Golgi condensing vesicle, the pH drops until an acidic vesicle buds from the Golgi. The size of these condensing vesicles is approximately the same as the droplets in the mature secretory vesicles, suggesting to us that when these primary condensing vesicles fuse together (as they must) to form the much larger, mature secretory vesicles, then their contents remain separated as a protein emulsion due to the presence of a layer of charges on the surface of the individual droplets.

2.6 Histidine Residues in TRP May Be Responsible for Stabilizing the Emulsion

This raises the obvious question as to what exactly is responsible for these putative pH-dependent charges? Fortunately, one possible answer lies to hand in the predicted histidine-rich tail of the TRP (Johnson et al. 1987). Histidine is perfect for this role, since its side chain is increasingly protonated below pH 6.5, being over 90% protonated at pH 5.5. In contrast, at physiological pH (around 7.4), this charge is essentially gone. This unusually histidine-rich region might plausibly provide the necessary pH-dependent charge layer for the acid stabilized emulsion.

2.7 Raising the Vesicle pH Destroys the Emulsion

If our hypothesis is correct, then treatments which raise the pH in these acidic vesicles will lead to the emulsion being destroyed and the droplets fusing together. This is exactly what we have observed (Wells and Cordingley 1991b) when the live female parasites are cultured in medium containing ammonium chloride or other weak bases such as trimethylamine or chloroquine or with the proton ionophore monensin (Anderson and Pathak 1985). This results in fusion of the eggshell precursor droplets into a single large droplet and, in addition, the cross-linking reactions are triggered, leading to the formation of droplets of cross-linked eggshell inside every secretory vesicle. We shall return to these droplets and their properties later. Thus, we conclude that the apparent emulsion is exactly that and that it is stabilized by a pH-dependent charge layer over the surface of the droplets.

3 The Mechanism of Eggshell Formation and the Role of Mehlis' Gland

3.1 Eggshell Formation in *F. hepatica*

Irwin and Threadgold (1972) published some fascinating and provocative electron micrographs showing eggshell formation in *F. hepatica*. In one micrograph, eggshell can be seen forming in the extracellular space between the vitelline cells and the wall of the ootype. The eggshell is forming, according to the authors, on the interface between the liquid surrounding the vitelline cells and the Mehlis' gland secretions (MGS). There is no "membrane" visible at this interface, it is simply a "discontinuity" between two liquids of slightly different staining properties (Irwin and Threadgold 1972). The droplets of eggshell material exocytosed from the vitelline cells are not fusing together inside this interface, but appear as droplets of the same sizes as those inside the secretory vesicles inside the vitelline cells. However, at this interface they are fusing together and spreading on this "boundary".

3.2 Extending the Picture

Extrapolating from this frozen "snapshot" of eggshell formation in *F. hepatica*, we suppose that more and more droplets will fuse together and spread on the interface until a continuous eggshell results. This agrees with the observation that the eggshell precursor droplets fuse together inside the secretory vesicles when they are alkalinized (Wells and Cordingley 1991b) and therefore the pH inside the interface must be low (5.5) since the exocytosed droplets are observed *not* to fuse together except at the interface between the liquid around the vitelline cells and the

MGS (Irwin and Threadgold 1972). From this we deduce that the interface must be a junction between liquids of different pH, the MGS being around physiological pH (7.4 or higher) and the liquid inside the interface being closer to the pH inside regulated secretory vesicles (5.5).

3.3 Triggering the Cross-Linking

The cross-linking reactions are also triggered when we alkalinize the secretory vesicles (Wells and Cordingley 1991b). This is a perfect final touch to this whole process, since the cross-linking reactions would only be triggered when the droplets fuse with the forming eggshell on the interface. Thus, by spreading the eggshell precursor droplets onto this interface, one achieves the seemingly magical trick of

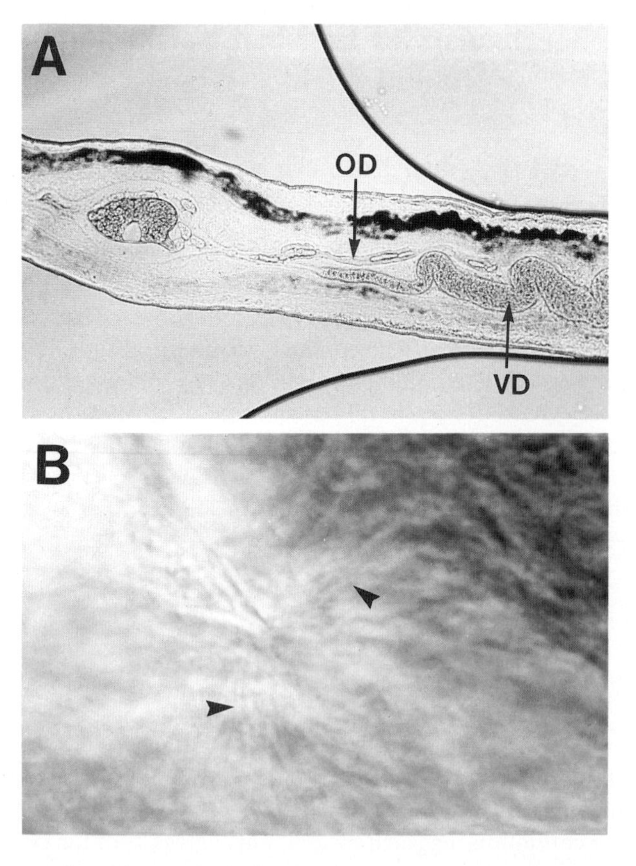

Fig. 4A. Low power photomicrograph of a live female schistosome showing the oviduct (*OD*) and the vitelline duct (*VD*) joining above the ootype to form the common duct which enters the ootype. **B** High-power photomicrograph of the spine of an egg in the opening of the vitelline duct. The *arrowheads* indicate the ducts of Mehlis' gland which empty into the end of the common duct around the spine. At the time of the photograph they do not appear to be actively secreting MGS

making an eggshell over the whole cell mass which ends up being contained within the eggshell.

Figure 4 shows light micrographs of living worms taken with phase contrast optics. In Fig. 4A, taken at low power, one can follow the oviduct and the vitelline duct which join a short distance above the ootype which contains a recently formed egg. Figure 4B is a higher-power photograph of the spine of a newly formed egg in the mouth of the common duct formed by the joining of the oviduct and the vitelline duct. The ducts of Mehlis' gland can be seen emptying into the opening of the common duct around the spine. These two photographs are presented as starting points for the following discussion of the stages leading to eggshell formation.

3.4 The Stages in Eggshell Production

We have observed the formation of eggshell on a very small number of occasions in cultured female worms. This stems from a number of causes. Firstly, the rate of egg-laying slows dramatically in culture so that only one or two eggs may be formed per hour. Secondly, the female worms are in the gynecophoric canal of the male worm and the ootype is usually obscured. On a few occasions, when the orientation of the worm pair was ideal and the female was transiently extended from the gynecophoric canal, we have observed eggshell formation. The cells are introduced by what can only be called injection, being pushed rapidly into the ootype. Then, the ootype is agitated by what we assume to be muscular activity of the female. When the agitation stops, the eggshell appears to be complete. This whole process takes only a few minutes and, even allowing for the possible time distortion in cultured worms, is remarkably quick.

The remaining question is what are the events between egg extrusion from the ootype and the introduction of new cells. The following scheme is suggested as a starting point for discussion and experiment. It is hoped that focusing on these events will assist us in choosing the future directions for our research and also clarify exactly what physicochemical events we need to explain. For example in the past the formation of eggshell *around the cell mass in the ootype* has been unexplained, indeed we cannot remember anyone previously pointing this out as a problem that needed explanation. Without some explicit mechanism to account for the eggshell forming where it does, a more reasonable outcome would be a series of lumps of eggshell of irregular size and shape mixed randomly among the cells in the ootype where they fused together after exocytosis. Any mechanism should also account for the appearance and shape of the spines, both the large subterminal spine and the microspines which cover the outer surface of the eggshell.

Figure 5 is our attempt to make explicit a series of steps that we think are necessary for egg formation. The dimensions and many features of the diagrams are drawn from many hours of observation of living worms and photographs such as those in Fig. 4.

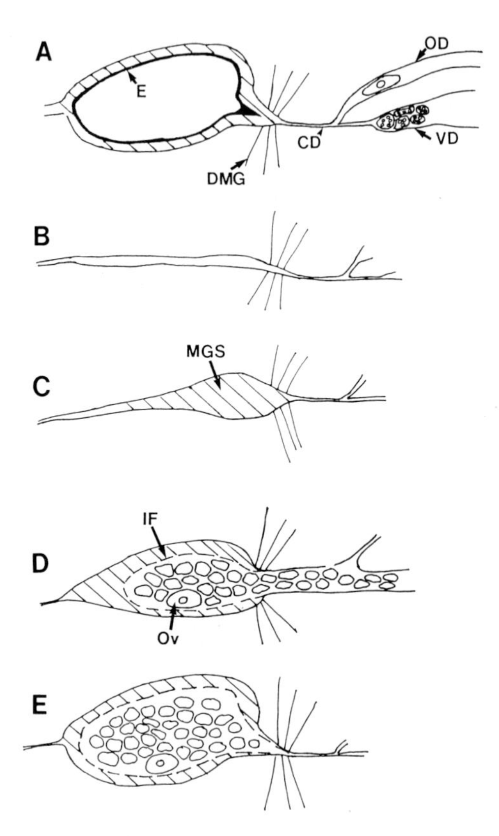

Fig. 5A–E. Sequence of events in eggshell formation. **A** Complete egg in the ootype immediately prior to extrusion from the ootype. Details of the cells inside the egg have been omitted for clarity. **B** The ootype is empty immediately after egg expulsion. **C** MGS begin to accumulate in the ootype. **D** The common duct opens and cells begin to enter the ootype, pushing the MGS into the ootype and spreading the MGS around the ootype walls. The single ovum and the 30 to 40 vitelline cells end up in a droplet of liquid surrounded by MGS. **E** The common duct closes and the exocytosis of the eggshell precursors begins. The ootype agitates to spread the eggshell on the interface. Symbols; *CD* common duct; *DMG* ducts of Mehlis' gland; *E* eggshell; *IF* interface between MGS and the liquid around the vitelline cells; *MGS* Mehlis' gland secretions; *OD* oviduct; *Ov* ovum; *VD* vitelline duct

Figure 5A is a completed egg immediately before extrusion from the ootype and Fig. 5B is the state of affairs immediately following extrusion of the egg. The ootype is "empty", partially collapsed, and presumably contains (if anything) a small amount of residual MGS. At this point, we suggest that Mehlis' gland becomes active, secreting a substantial amount of MGS. This accumulates in the neck of the ootype in the position of the eventual spine (Figs. 4B and 5B). We deduce this from the observed ducts in Fig. 4B, which appear to empty into this region, and the timing is deduced from the observed absence of MGS around the spine, as in Fig. 4B, i.e., at this time Mehlis' gland is not secreting appreciable amounts of material. Then, the cells (one ovum and 30 to 40 vitelline cells) are introduced into, or "through" the MGS in the ootype (however you prefer to imagine it). The MGS is, we suggest (see Sect. 5), quite viscous, and is spread around the walls of the ootype during these events. At these dimensions, viscosity can be a very significant factor and it should be remembered that the interface needs to last for only a few minutes until the eggshell has formed. The vitelline cells are suspended in a nonviscous liquid since they can be observed to move around freely in the vitelline duct as the female worm moves under the microscope. This observation is itself worth comment, since it suggests that the vitelline cells are suspended in a significant volume of liquid as opposed to being tightly packed with very little extracellular fluid. The cell mass comprising one ovum and the vitelline cells is pushed into the ootype as a drop of cells and liquid surrounded by MGS. This is illustrated in Fig. 5D and E. In Fig. 5D, the common duct is open and the cells are being "injected", in Fig. 5E the common duct has shut again and the vitelline cells are beginning to exocytose their eggshell precursor vesicles. We deduce that exocytosis begins after the cells enter the ootype since we never observe eggshell material in the common duct above the sphincter but the spine completely fills the opening on the ootype side of the sphincter, suggesting that eggshell material enters only from the ootype. The spine fills the entrance to the common duct since the injection of the cells must push the majority of the MGS into the ootype and when the sphincter closes there is consequently a conical hole remaining which becomes filled with the eggshell material. An attempt is made to illustrate this in Fig. 5D and E.

MGS become very important in this scenario, and one of our priorities is to characterize this material in more detail.

Another unexpected observation that we made when we alkalinized the secretory vesicles was that the single large eggshell droplets formed within the secretory vesicles were covered in microspines of similar appearance and dimensions to those on the *surface* of authentic eggshell (Wells and Cordingley 1991b). This observation adds weight to our suggestion that MGS is more alkaline than the liquid inside the eggshell, since microspines form only on the outside surface of the eggshell. Also, the microspines form inside the secretory vesicles but quite separate from the vesicle membrane, strongly suggesting that the forming egg is not surrounded by a closely opposed membrane in the ootype, as has been suggested by some previous authors (Clegg 1965), but rather that the spines are forming in a liquid medium, presumably the MGS which surround the egg.

4 The Elusive Phenol Oxidase; Preconceptions, Data, and Suggestions

The previous scenario has deliberately left untouched the mechanism of the cross-linking and, more particularly, the mechanism by which it is triggered. All the reagents are present in the secretory vesicles and the cross-linking is only triggered at the time of eggshell formation. In previous publications, evidence has been presented that suggests, but by no means proves, that the cross-linking is catalyzed by a copper-dependent phenol oxidase (Seed and Bennet 1980; Wells and Cordingley 1991b). This elusive phenol oxidase must be a remarkable enzyme. Consider the following observations. Firstly, *all* of the tyrosine residues in the eggshell precursors are "destroyed" during cross-linking and are not recovered as either tyrosine or DOPA in hydrolysates of eggshells. This is as one predicts if the tyrosine side chains are being oxidized to quinones and reacting to form the cross-links in the eggshell (Smyth and Clegg 1959). However, if a phenol oxidase enzyme is carrying out these reactions, it must be remarkably flexible and adaptable in shape to be able to gain access to *every* tyrosine or DOPA. In the cross-linking eggshell, we would expect steric hindrance to prevent free access for an enzyme to a large proportion, if not most, of the residues. Therefore we are led to suggest that some kind of low molecular weight molecule, that can more readily diffuse through the eggshell, may be involved. Alternatively, the assumption that a *conventional* enzyme is responsible may be mistaken. The observation that copper chelators inhibit the disappearance of the rapid silver staining proteins (due to DOPA side chains) is hardly conclusive evidence for a conventional enzyme. The observation that oxygen consumption is stimulated by tyrosine or DOPA in female worm extracts is also inconclusive, since the acitivity may not be the one responsible for eggshell cross-linking but possibly the activity responsible for the formation of DOPA prior to secretion.

4.1 The Amino Acid Composition of Eggshell

It is possible to account almost entirely for the observed amino acid composition of the eggshell from the sequences of the proteins already identified (GRP and TRP) except for the tyrosine. However, eggshells used in these determinations are isolated by extensive protease digestions from the livers and intestines of mice (Byram and Senft 1979), and we are concerned that one is proteolytically removing part of the eggshell, leaving behind only the most resistant framework of the eggshell. If a phenol oxidase enzyme were present, but only became lightly cross-linked into the matrix of the eggshell, then the proteolysis might easily remove it during eggshell preparation.

Our ability to generate eggshell droplets inside the secretory vesicles of isolated female worms allows us to approach these questions from a different angle. We can isolate eggshell droplets by the simple process of boiling the female worms in sample buffer containing sodium dodecyl sulfate (SDS) and mercaptoethanol after incubation of the female worms in culture medium supplemented with ammonium chloride to alkalinize the vesicles and trigger the fusion and cross-linking (Wells and Cordingley 1991b). In our original publication, we purified the droplets by the same procedure as reported for eggshells to obtain strictly comparable results to those for eggshell (Byram and Senft 1979). In Fig. 6, we present the amino acid analyses for droplets isolated by simply boiling female worms in SDS sample buffer and repeatedly pelleting and washing the droplets by centrifugation. We then digested half of these droplets with collagenase and trypsin (Byram and Senft 1979)

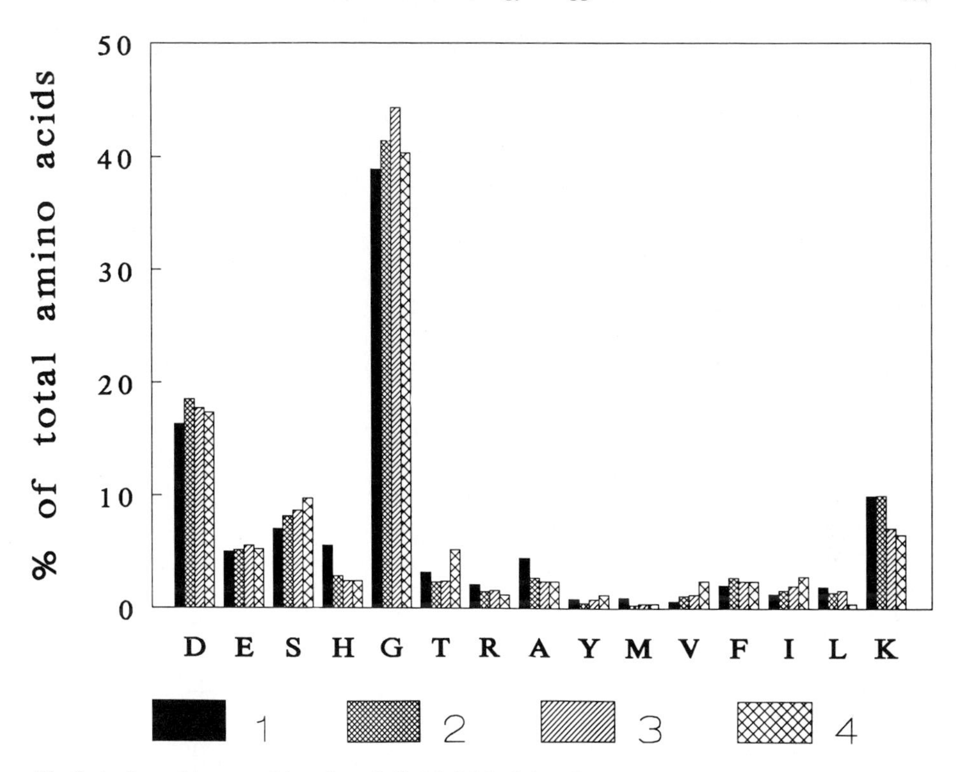

Fig. 6. Amino acid composition of eggshell. *1* Published data (Byram and Senft 1979); *2* composition of eggshell droplets isolated by the same procedure as *1*; *3* composition of eggshell droplets formed in the secretory vesicles after alkalinization. The droplets were isolated by boiling in SDS sample buffer and repeated washing by centrifugation; *4* the same droplets as 3 digested with collagenase and trypsin as in samples *1* and *2*

and performed amino acid analyses on these two sets of droplets. Figure 6 shows there is very little difference between published eggshell compositions and our amino acid analyses of any of the samples prepared from the eggshell droplets formed inside the vesicles. We have repeated these experiments several times with similar results each time. If there had been a phenol oxidase enzyme present that was cross-linked into the eggshell and subsequently removed by proteolysis, then we might have expected it to be represented in the amino acid composition of the "SDS only" samples (3 in Fig. 6) but absent in sample 4 in Fig. 6, which was treated with proteases. We observe very few, if any, significant differences between samples prepared in any of these ways.

Our conclusion from these results is that either (1) the phenol oxidase is present in the vesicles but not cross-linked into the eggshell, or (2) the enzyme is present in amounts too low to be detected by these techniques, or (3) one of the eggshell components already isolated and characterized (the GRP and TRP) actually carries out the function of "catalyzing" cross-linking.

In support of possibility (1), based on histological evidence, Smyth and Clegg (1959) suggested many years ago that the phenol oxidase was around the droplets in the matrix of the vesicles. The lack of any significant electron density in the matrix of the vesicles has led us to doubt this. Possibility (2) is one which can only be confirmed by isolating the phenol oxidase and demonstrating directly its presence in the vesicles, something that has yet to be achieved.

If possibility (3) is accepted as a working hypothesis, then one of the two proteins already identified must be responsible for the cross-linking reactions as both substrate and "catalyst". We should use the expressions "catalyst" or "enzyme" with caution in this regard. The eggshell components are used up irreversibly in eggshell formation so neither definition is really appropriate. They are "reagents" rather than "catalysts".

All these somewhat contradictory ideas illustrate only too clearly that there are several aspects of the process which we currently do not understand correctly. The nature of the "enzyme" or "catalyst" or "trigger" (choose whichever word you prefer) is clearly inadequately accounted for in our current explanations.

In our published work, we have presented evidence that the "phenol oxidase", once it has been triggered, is resistant to boiling in SDS (Wells and Cordingley 1991b). However, if the female worms are boiled in SDS prior to triggering of the phenol oxidase activity, then the amount of activity detectable is dramatically reduced, which leads us to the conclusion that the trigger (a protease?) was SDS-sensitive whereas the phenol oxidase activity, whatever it turns out to be, was not sensitive to boiling in SDS. It is, however, sensitive to copper chelators even after boiling in SDS.

All these observations lead us to the conclusion that we do not understand cross-linking. Our current ideas are just too simplistic to adequately account for our observations or the complex properties of the system.

4.3 Lessons from Chemistry

The formation of chemical polymers between quinones and amines is an instructive comparison (Dagani 1991). The reaction to produce cross-linked polymer requires an excess of quinone. The majority of the quinone is used as oxidizing agent rather than participating in the formation of cross-links directly. The chemists developing this system added calcium hypochlorite as the oxidizing agent in addition to the quinone and amine, thus reducing the amount of quinone required.

In the eggshell, the putative phenol oxidase is invoked to provide the necessary extra oxidizing power by way of molecular oxygen and the active site of the enzyme. However, as outlined above, we see no evidence for a component of the eggshell in addition to those already identified and steric hindrance would prevent an exogenous enzyme from gaining direct access to *all* of the tyrosine side chains. We are forced to the conclusion that there must be a low molecular weight intermediate in the whole process, not necessarily one that participates in the cross-links directly, but one that acts as a carrier of oxidizing equivalents.

4.3 Does Free Tyrosine Play a Role?

Seed et al. (1980) have suggested that free tyrosine has a role in the process of eggshell cross-linking. This is certainly possible, but whatever mechanism and intermediates are hypothesized must account for the total disappearance of the tyrosine side chains of the eggshell proteins. If free tyrosine were being oxidized to DOPA-quinone, and this reacted with amine side chains of proteins, then it still remains to account for the second round of oxidation which is necessary to regenerate the quinone (which is reduced in the first reaction). This is necessary to allow a second reaction with an additional amine or similar nucleophile to generate a cross-link. The same difficulty arises here as with the oxidation of the DOPA residues known to be present on the eggshell proteins. Some oxidizing activity is needed, either a small diffusible molecule or a phenol oxidase enzyme. It is possible that *excess* DOPA-quinone could serve this role, but it remains to be demonstrated, and this leaves the putative role of tyrosine too ill defined to adequately account for the observations in its current form.

5 Observations on Eggshell Formation in Other Platyhelminths

Related platyhelminths make eggs in ways that are instructive when compared with schistosomes. Of the free-living flatworms, egg production has been adequately described for two marine polyclads (Ishida and Teshirogi 1986). In these organisms, the eggs are deposited onto an underwater surface by the worm in a gelatinous mass. The ova are enclosed in this material, which swells and leaves the egg surrounded by a space within the gelatinous material. The eggshell precursor granules are secreted from the ovum itself as opposed to the vitelline cells in Schistosomes. The secreted eggshell granules (or droplets) fuse together on the inside of the gelatinous material and cross-link there. It is very tempting to suggest that the gelatinous material is homologous (in the precise evolutionary sense) to the MGS. The main difference in schistosomes is that the process is occurring inside the female in the ootype rather than in sea water. The eggshell precursors are not divided into an emulsion in the secretory vesicles of the polyclads. This probably correlates directly with the rate of production of *fully cross-linked* eggshell, which is enormously accelerated in parasitic schistosomes. Many of the features of the process in schistosomes have probably evolved to allow the process to be accelerated. High rates of egg laying have long been considered a prime evolutionary advantage for parasites and schistosomes are probably no exception. The subdivision of the eggshell precursors into an emulsion is particularly advantageous as it allows the eggshell to be formed more rapidly, facilitating spreading of the egshell and allowing rapid equilibration of the droplets with their immediate environment when the pH changes.

It seems more than likely to us that the basic mechanism involving the spreading of the eggshell on a viscous or semi-rigid gelatinous capsule has been conserved with appropriate modifications throughout the diverse platyhelminths. The free-living marine turbellarians and schistosomes are unlikely, in our opinion, to have evolved completely different and unique ways of forming their eggshells.

6 Questions for the Future

Instead of simply summarizing what we have been saying, we will close by listing the most important outstanding questions as we see them.

1. What is the chemistry of the cross-links in *S. mansoni* eggshells? The intense autofluorescence of the eggshell suggests that dityrosine is present, but what else, and how much?
2. The nature of the "trigger" that allows all of the eggshell reagents to coexist for days without reacting and then to react rapidly at the time of eggshell formation.
3. What exactly is the role of the TRP? Is it the central reagent in the cross-linking reactions? Or is it just another substrate for the elusive phenol oxidase? The TRP has a very unusual structure and, judging by its conservation, some very important function for the schistosome. Using synthetic peptides as models of the protein, we hope to discover exactly what the significance of this protein is in eggshell formation.
4. What is being made and secreted by Mehlis' gland? This, perhaps more than anything else, lies at the heart of this problem and to find the answer is one of our main goals in the near future. Mehlis' gland is rich in rough endoplasmic reticulum, indicating to us that significant amounts of protein are being secreted. We guess that the main secretory product is one or more glycoproteins which are probably the main component of the secretion.
5. What signals the vitelline cells to exocytose the eggshell precursor vesicles? We guess a small molecule secreted in the ootype or present in MGS.
6. The eggshell has physical properties such as porosity and strength which are very important. We are devising techniques to measure these physical properties, which is a real challenge with the small amounts of eggshell material available.
7. To what extent does the eggshell change size after formation. Some authors say not at all, some say the eggs swell significantly. The stretching of polymer materials is known as "creep" in materials science. Do schistosome eggshells "creep" or not?
8. Last, but by no means least, what is the mechanism of hatching? Once again, authors are divided in their opinions. Some believe osmotic rupture is responsible, others are not so sure. There seems to be a temperature-dependent component which suggests that passive osmotic lysis is not the whole story. Hatching is an extremely important step. We cannot believe it has remained an

entirely passive event throughout evolution and, like everything else about the eggshell, it is going to be much more complex than we currently think when we finally unravel the details. Eggshell formation is just beginning to yield its secrets, perhaps hatching will be next?

Acknowledgments. This work was supported by the National Science Foundation through the EPSCoR program and by the Office of Naval Research grant number N00014-90-J-1997.

References

Anderson RGW, Pathak RK (1985) Vesicles and cisternae in the trans golgi apparatus of human fibroblasts are acidic compartments. Cell 40:635–643

Bobek LA, Rekosh DM, Loverde PT (1988) Small gene family encoding an eggshell (chorion) protein of the human parasite *Schistosoma mansoni.* Mol Cell Biol 8:3008–3016

Bobek LA, Rekosh DM, Loverde PT (1991) *Schistosoma japonicum*: analysis of eggshell protein genes, their expression, and comparison with similar genes from other schistosomes. Exp Parasitol 72:381–390

Burgeson RE (1988) New collagens, new concepts. Ann Rev Cell Biol 4:551–577

Byram JE, Senft AW (1979) Structure of the schistosome eggshell: amino acid analysis and incorporation of labelled amino acids. Am J Trop Med Hyg 28:539–547

Chou PY, Fasman GD (1974) Prediction of protein conformation. Biochemistry 13:222–245

Clegg JA (1965) Secretion of lipoprotein by Mehlis' gland in *Fasciola hepatica.* Ann NY Acad Sci 118:969–986

Dagani R (1991) Novel water-displacing polymers show promise in coating uses. Chem Eng News 69:20–22

Erasmus DA (1987) The adult schistosome: structure and reproductive biology. In: Rollinson D, Simpson AJG (eds) The biology of schistosomes from genes to latrines. Academic Press, London, pp 51–82

Field KG, Olson GJ, Lane DJ, Giovannoni SJ, Ghiselin MT, Raff EC, Pace NR, Raff RA (1988) Molecular phylogeny of the animal kingdom. Science 239:748–753

Irwin SWB, Threadgold LT (1970) Electron microscope studies on *Fasciola hepatica* VIII. The development of the vitelline cells. Exp Parasitol 28:399–411

Irwin SWB, Threadgold LT (1972) Electron microscope studies of *Fasciola hepatica.* X. Egg formation. Exp Parasitol 31:321–331

Ishida S, Teshirogi W (1986) Eggshell formation in polyclads (Turbellaria). Hydrobiologia 132:127–132

Johnson KS, Taylor DW, Cordingley JS (1987) Possible eggshell protein gene from *Schistosoma mansoni.* Mol Biochem Parasitol 22:89–100

Kunz W, Opatz K, Finken M, Symmons P (1987) Sequences of two genomic fragments containing an identical coding region for a putative egg-shell precursor protein of *Schistosoma mansoni.* Nucl Acids Res 15:5894

Meadows HM, Simpson AJG (1989) Codon usage in *Schistosoma.* Mol Biochem Parasitol 36:291–299

Mellman I, Fuchs R, Helenius A (1986) Acidification of the endocytic and exocytic pathways. Annu Rev Biochem 55:663–700

Middaugh CR, Thomson JA, Burke CJ, Mach H, Naylor AM, Bogusky MJ, Ryan JA, Pitzenberger SM, Ji H, Cordingley JS (1991) Evidence for a left-handed alpha helix in an eggshell protein of *Schistosoma mansoni.* (submitted)

Orci L, Ravazzola M, Anderson RGW (1987) The condensing vacuole of exocrine cells is more acidic than the mature secretory vesicles. Nature 326:77–79

Paz MA, Gallop PM, Torrello BM, Fluckiger R (1988) The amplified detection of free and bound methoxatin (PQQ) with nitroblue tetrazolium redox reactions: insights into the PQQ locus. Biochem Biophys Res Commun 154:1330–1337

Perram CM, Nicolau C, Perram JW (1977) Interparticle forces in multiphase colloid systems: the resurrection of coagulated sauce bearnaise. Nature 270:572–573

Rodrigues V, Chaudhri M, Knight M, Meadows H, Chambers AE, Taylor WR, Kelly C, Simpson AJG (1989) Predicted protein structure of a major *Schistosoma mansoni* eggshell protein. Mol Biochem Parasitol 32:7–14

Seed JL, Bennet JL (1980) *Schistosoma mansoni*: phenol oxidases role in eggshell formation. Exp Parasitol 49:430–441

Seed JL, Kilts CD, Bennet JL (1980) *Schistosoma mansoni*: tyrosine, a putative in vivo substrate of phenol oxidase. Exp Parasitol 50:33–44

Simpson AJG, Chaudhri M, Knight M, Kelly C, Rumjanek F, Martin S, Smithers SR (1987) Characterization of the structure and expression of the gene encoding a major female specific polypeptide of *Schistosoma mansoni*. Mol Biochem Parasitol 22:169–176

Smyth JD, Clegg JA (1959) Eggshell formation in trematodes and cestodes. Exp Parasitol 8:286–323

Waite JH (1983) Evidence for a repeating 3,4-dihydroxyphenylalanine and hydroxyproline containing decapeptide in the adhesive protein of the mussel, *Mytilus edulis* L. J Biol Chem 258:2911–2915

Waite JH, Rice-Ficht AC (1987) Presclerotized eggshell protein from the liver fluke *Fasciola hepatica*. Biochemistry 26:7819–7825

Wells KE, Cordingley JS (1991a) Detecting proteins containing 3,4-dihydroxyphenylalanine by silver staining of polyacrylamide gels. Anal Biochem 194:237–242

Wells KE, Cordingley JS (1991b) *Schistosoma mansoni*: eggshell formation is regulated by pH and calcium. Exp Parasitol 73:295–310

Chapter 6

Molecular Architecture of Helicoidal Proteinaceous Eggshells

Stavros J. Hamodrakas[1]

1 Introduction

1.1 Helicoidal Architecture: a Brief Description

Several extracellural fibrous structures are known to have helicoidal architecture. Such structures include arthropod cuticles, vertebrate tendons, plant cell walls etc. The widespread occurrence of the helicoidal structure in spherical shells, such as eggshells, spore walls, cyst walls, and others, and its correlation with the mechanical strength it provides is intriguing.

Excellent reviews on the helicoidal architecture and its appearance in biological systems have been made by Bouligand (1972, 1978a, b) and Neville (1975, 1981, 1986). These works describe in a beautiful and most comprehensive way how helicoids are identified, how widespread they are, the basic molecular principles of their formation and their geometrical, physical, and biological properties.

A brief description of the helicoidal architecture is as follows (Fig. 1).

The helicoidal architecture consists of helicoidally arranged (like the steps of a spiral staircase) parallel planes or sheets of fibrils (mono- or polymolecular). Within individual planes the fibrils are oriented more or less parallel to each other. Between successive planes the fibril direction rotates progressively (through a constant angle — the so called helicoidal twist), thus, giving rise to a helix, with its axis perpendicular to the planes, a sort of multi-directional "plywood" (Fig. 1a).

The helix can be either left- (anticlockwise) or right- (clockwise) handed; most biological systems so far examined have been shown to be left-handed (reviewed in Neville 1975).

The pitch of a helicoid is defined as the minimum distance corresponding to a 360° rotation of the direction of the fibrils. Therefore, the fibrils, are in parallel alignment every half pitch, which is equivalent to one lamella (Fig. 1a).

A helicoidal organization is identified ultrastructurally, usually under a transmission electron microscope, as a lamellar structure, exhibiting in, suitably cut, oblique thin sections, parabolic arrays of fibrils ("arced" patterns), which constitute each lamella (Fig. 1b).

---

[1] Department of Biochemistry, Cell and Molecular Biology, and Genetics, University of Athens, Panepistimiopolis, Athens 157.01, Greece

Results and Problems in Cell Differentiation 19
Biopolymers
Case, S.T. (Ed.)

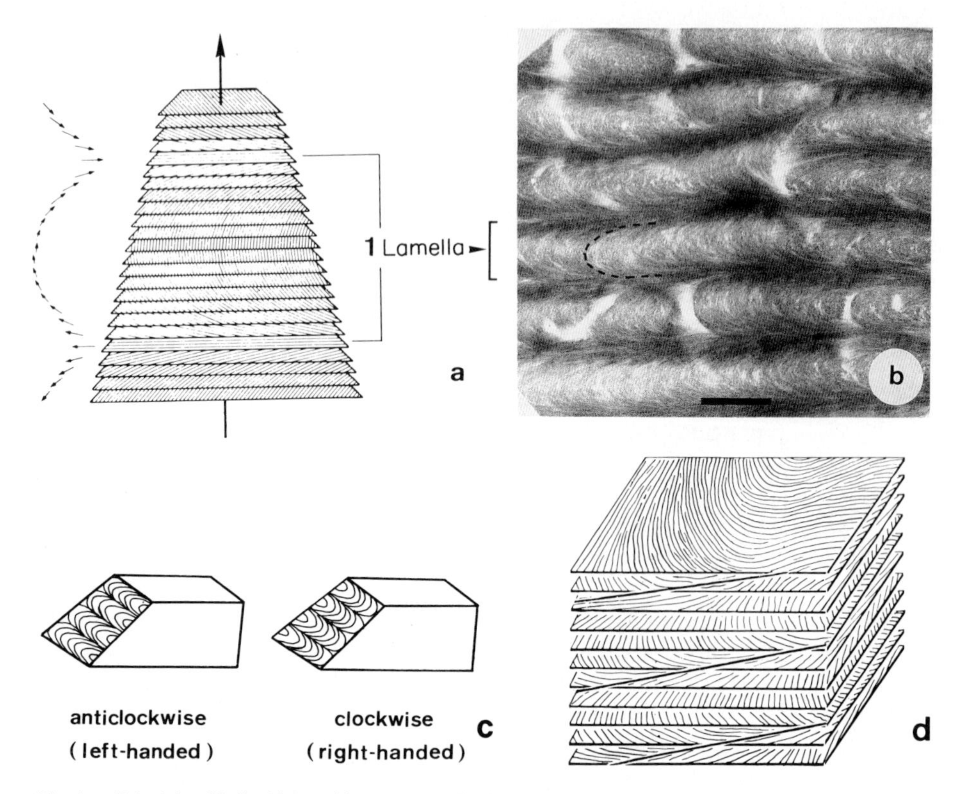

Fig. 1 a. Principle of helicoidal architecture. A helicoidal structure consists of parallel planes or sheets of fibrils. Within individual planes the fibrils are oriented, more or less parallel to each other. Between successive planes the fibril direction rotates progressively through a constant angle, the helicoidal twist, thus giving rise to a helix with its axis perpendicular to the planes. Fibrils are in parallel alignment every half-pitch of the helix (180° rotation of the fibril direction) which is equivalent to one lamella. In oblique thin sections typical parabolic arrays of fibrils ("arced" patterns) are seen to constitute each lamella. **b** Electron micrograph of an oblique section through a helicoidal proteinaceous eggshell of a silkmoth, *A. polyphemus*. Parabolic arrays of protein fibrils are seen in each lamella. *Bar* 0.4 µm. **c** The sense of rotation of a helicoidal structure can be deduced from the direction of the parabolic patterning on a known oblique section. An anti-clockwise (left-handed) helicoid is one in which successive planes of fibrils, twist anti-clockwise in a direction further away from the observer. **d** A generalized variant of the helicoidal model [see text; after Bouligand (1972)]

The sense of rotation (left or right) of a helicoid can be deduced from the direction of the parabolic pattern on a known oblique face (Fig. 1c).

A helicoidal structure should have as components rod-shaped, chiral (optically active) molecules (Bouligand 1972; Neville 1986).

A generalized variant of the helicoidal model may consist of a set of parallel surfaces of any shape, instead of planes, and not necessarily straight fibrils on each surface (Bouligand 1972; Fig. 1d). Two other ways in which helicoids depart from the ideal model are distortions and defects (Mazur et al. 1982 and references therein).

This description emphasizes the analogies of helicoidal fibrous biological structures with true cholesteric liquid crystals (Friedel 1922); the term "cholesteric" was introduced for the simple historical reason that it was first observed in some esters of cholesterol.

The close analogy between the helicoidal structures of (usually extracellular) biological materials and the structure of cholesteric liquid crystals probably suggests that several tissues and organelles are self-assembled according to a mechanism that is very similar to the process allowing materials to form liquid crystals. Apparently, helicoids should pass through a liquid crystalline phase before solidifying. Therefore, it is important to determine in such cases the molecular mechanisms of self-assembly.

Self-assembling systems are important in biology as they are economical in energy terms, requiring neither enzymatic control nor the expenditure of energy-rich bonds. They are particularly appropriate for building extracellular skeletal structures outside the cells which make them (Bouligand 1978b; Neville 1986).

Natural helicoidal composites occur in several combinations: polysaccharide fibers in a polysaccharide matrix (plant cell walls), polysaccharide fibers in a protein matrix, (arthropod cuticle), and protein fibers in a protein matrix (insect and fish eggshells), to mention just a few examples. In all cases principles of molecular recognition should govern the self-assembly mechanisms (Neville 1986).

1.2 Scope of This Chapter

Over the last decade we have been using the silkmoth chorion, the major component of the eggshell, a proteinaceous protective and functional layer surrounding the oocyte, as a model system to study how proteins fold and self-assemble to form complex, physiologically important structures. Silkmoth chorion is an example of a helicoidal composite of protein fibers in a protein matrix. In a previous communication (Hamodrakas 1984) by briefly surveying data collected from the silkmoth *Antheraea polyphemus* (Saturnidae) chorion, we proposed the twisted β-pleated sheet as the molecular conformation which dictates the formation of the helicoidal architecture in proteinaceous eggshells.

Since then several other helicoidal proteinaceous eggshells (both insect and fish) have also been studied, providing both novel information and interesting comparisons. These include eggshells from the domesticated silkmoth *Bombyx mori* (Bombycidae), *Antheraea pernyi* (Saturnidae), the lepidopteran *Manduca sexta* (tobacco hornworm moth, Sphingoidae), and *Sesamia nonagrioides* (an insect harmful to various crops, especially maize, Noctuidae) and the teleost fish *Salmo gairdneri* (trout).

This review will briefly cover the large amount of new information accumulated in the last 7 years which verifies, in the systems studied so far, the existence of a common molecular denominator, the β-pleated sheet, which apparently governs the self-assembly process in helicoidal proteinaceous eggshells, biological analogs of cholesteric liquid crystals.

1.3 Insect and Fish Eggshell: Terminology and Brief Description of the Systems

Hereafter, the terms insect eggshell and fish eggshell will simply denote the Lepidopteran eggshell and the *S. gairdneri* eggshell, respectively, to avoid possible misinterpretations.

1.3.1 Insect Eggshell

The insect eggshell (chorion and vitelline membrane) and the associated follicular epithelium, which secretes its constituent layers, have been the subject of numerous investigations at the cellular and molecular level, providing a model system in several areas of current biological research: physiology of the eggshell layers and morphogenesis of supramolecular structure, control of gene expression in differentiating cells, evolution of multigene families, and structural protein folding and organization.

Work in these areas has been reviewed recently (Kafatos et al. 1977; Hamodrakas 1984; Margaritis 1985; Regier and Kafatos 1985), and, together with the book by Hinton (1981), should be consulted for background information on the biology, biochemistry, physiology, and morphology of insect follicles and eggshells.

The ovary of an adult female lepidopteran usually consists of ovarioles, i.e., strings of follicles attached to each other in linear arrays from least mature to ovulated (Fig. 2a). Hereafter, we shall refer to a certain follicle by declaring its position within the ovariole with respect to the first follicle. With this terminology, the notation 5/22 means the 5th out of 22 choriogenic follicles in an ovariole.

The follicular cells, which surround the oocyte, synthesize and secrete, according to a precise spatial and temporal program, a set of structural proteins onto the surface of the oocyte, which self-assemble to form the multilayered eggshell (Fig. 2b). Protein synthesis occurs over a 2-day period in the silkmoths (Kafatos et al., 1977); in contrast, in *Drosophila melanogaster*, it lasts only 5 h (Margaritis 1985).

The eggshell performs certain functions, permitting sperm entry-fertilization, exchange of the respiratory gases, mechanical and thermal insulation, waterproofing, resistance to external high pressures, exclusion of microorganisms, and hatching (Kafatos et al. 1977; Margaritis 1985).

Chorion, which is largely proteinaceous (over 96% of its dry weight is protein in the silkmoth; Kafatos et al., 1977), accounts for at least 95% of the total eggshell mass in the species studied and it is the part of the eggshell which exhibits a helicoidal architecture.

Hereafter, for the insect eggshell, we shall be using both terms eggshell and chorion to refer to chorion.

1.3.2 Fish Eggshell

The tough eggshells of fish eggs play an essential role in controlling the relations between the external and internal egg environments allowing gaseous diffusion, providing physical protection and provision for sperm entry.

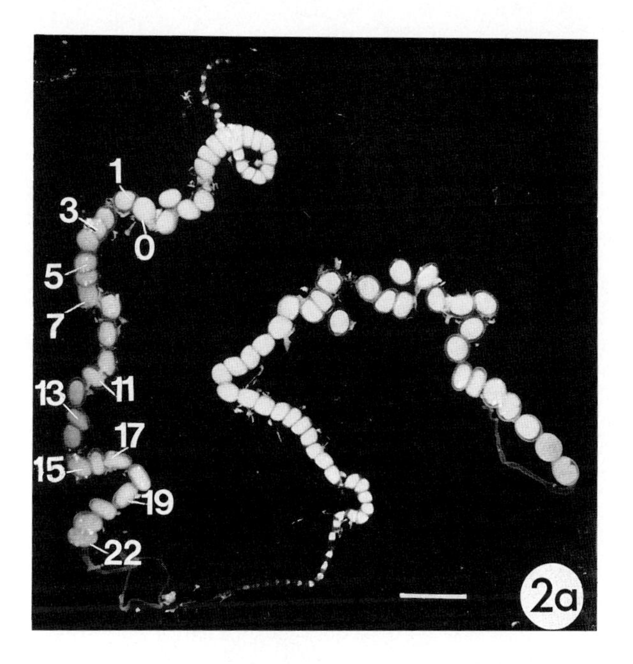

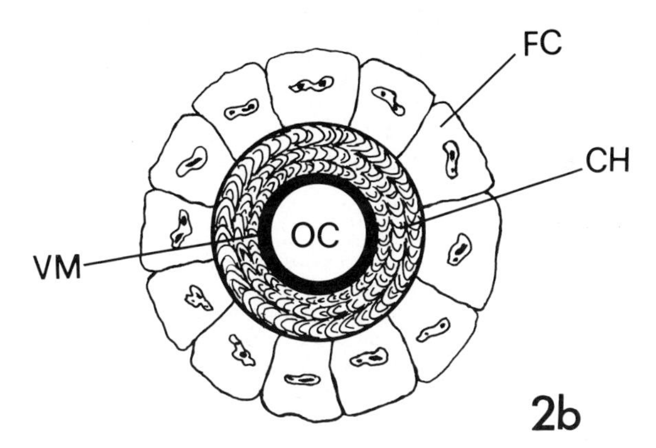

Fig. 2. a Two ovarioles dissected from a developing adult female *B. mori*. The paired ovaries in *B. mori* are each composed of four ovarioles. Follicles in progressively advanced stages of development (*0, 1, 2 . . . , 22*) are interconnected in a linear array within each ovariole; number 0 follicle indicates the beginning of choriogenesis. (Papanicolaou et al. 1986). **b** Schematic diagram of a follicle. *FC* follicle cells; *OC* oocyte. The eggshell consists of chorion (*CH*) and vitelline membrane (*VM*)

There is considerable variation in the nomenclature used to describe the eggshell of fishes (Grierson and Neville 1981; Groot and Alderdice 1985). Commonly used terms for this outer covering of the egg include: zona radiate, zona pellucida, radiate membrane, cortical membrane, vitelline envelope, egg envelope, egg capsule, cortex radiatus, chorion, eggshell.

Hereafter, for the fish eggshell, we have chosen to use the terms eggshell and/or alternatively, "cortex radiatus externus" when referring to its thin protein-polysac-

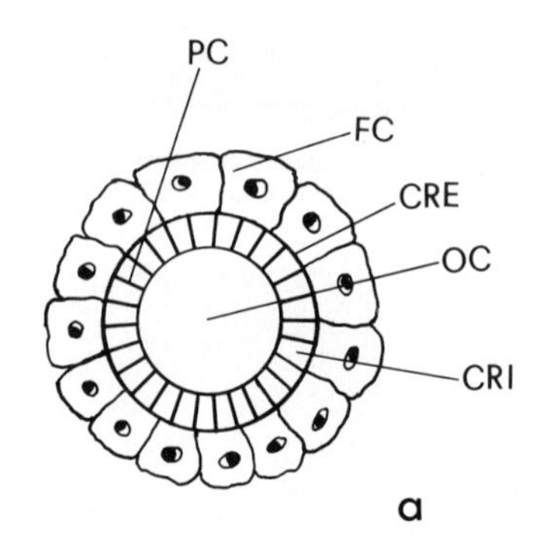

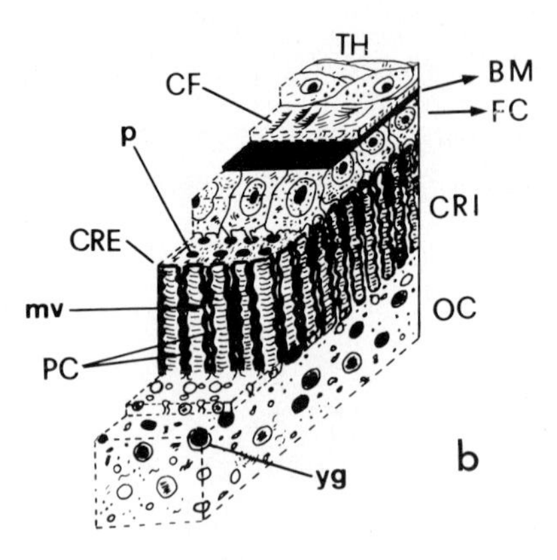

Fig. 3. a Schematic diagram of a fish follicle. *FC* follicle cells; *OC* oocyte. The eggshell consists of *CRE* (outer eggshell; cortex radiatus externus) and *CRI* (inner eggshell; cortex radiatus internus). Pore canals (*PC*) traverse these layers. **b** A diagram of a typical mature *S. gairdneri* follicle. Moving inwards from the periphery the following layers can be seen: *TH* theca layer; *CF* collagen fibers; *BM* basal membrane; *FC* follicle cells; *CRE* outer eggshell (cortex radiatus externus); *CRI* inner eggshell (cortex radiatus internus); *OC* oocyte (*PC* pore canals; *mv* oocyte microvilli; *p* "plugs"; *yg* yolk granules)

charide outermost layer and "cortex radiatus internus" when referring to its largely proteinaceous thick inner layer which has a helicoidal architecture. (Fig. 3a, b).

A wealth of interpretable information has been gathered from numerous studies on the fish eggshell, concerning, its fine structure and development, the structural changes after fertilization and water activation, its chemical composition as well as its degradation by hatching enzyme, its characteristics in relation to the prevention of polyspermy during fertilization and to the ecological significance of variation in eggshell structure. These studies are fully referenced by Groot and Alderdice (1985).

Initially, a few follicle cells are present close to the very young oocytes of fish ovaries (Anderson 1967; Flugel 1967). As the oocyte grows, the follicle cells increase in number, probably by mitosis to constitue a continuous follicular epithelium which retains its single-layered structure throughout oocyte growth in teleosts. The theca layer is composed of collagenous fibers, fibroplasts, blood vessels and large thecal cells that surround the follicular epithelium outside the basal membrane (Hurley and Fischer 1966; Fig. 3a,b).

During oogenesis the oocyte develops microvilli on its surface. Synthesis of the eggshell begins on the oocyte surface at the base of these microvilli and proceeds inwards, towards the oocyte surface. The first layer to be deposited is the thin (0.15 μm) polysaccharide-protein cortex radiatus externus (CRE) followed by the thick (30 μm) proteinaceous, helicoidal cortex radiatus internus (CRI) layer. As the eggshell increases in thickness, the oocyte microvilli increase in length and become enclosed in the presumptive microvillar "pore canals". Follicular cell processes are also found in these pore canals, often in close association with the oocyte microvilli (Fig. 3b) throughout eggshell formation. (Papadopoulou P and Hamodrakas SJ, in prep.; Tesoriero 1977).

Our observations support the view that both oocyte and, perhaps to a lesser extent, the follicular cells contribute to the formation of the eggshell (see also Groot and Alderdice 1985 and references therein). Before ovulation the oocyte microvilli and folicular cell processes withdraw from the pore canals and the outermost (closest to the follicle cells) pore canal openings are blocked with "plugs" of material. However, remnants of the oocyte microvilli and follicular cells processes are, occasionally, found in "plugged" pore canals even after ovulation (our unpubl. observations; for a discussion see also Groot and Alderdice 1985). The important events following water activation and their effects on eggshell structure and chemistry are fully described, discussed, and referenced by Groot and Alderdice (1985).

2 Ultrastructure of Helicoidal Proteinaceous Eggshells

2.1 Macroscopic Surface and Interior Architecture of the Eggshell

Lepidopteran and fish eggs are usually large and extremely hard. Their shape is that of a laterally flattened ellipsoid, with a major diameter ranging from 0.9 mm for *S. nonagrioides*, 1.5 mm for *B. mori*, 2.1 mm for *M. sexta*, 3 mm for *A. polyphemus* and *A. pernyi* to 5.5 mm in *S. gairdneri* and exhibit anterior-posterior polarity (Fig. 4). The dry weight of the eggshell varies accordingly, being approximately 40 μg in *B. mori* and 460 μg in *A. polyphemus* (Kafatos et al. 1977).

In the lepidopteran species, a prominent feature of its surface is a polygonal network of ridges (Figs. 5a, b, 6a). The ridges correspond to the edges of the follicular cells which secrete chorion. They are formed by overproduction of chorionic proteins in the intercellular spaces. Each polygon corresponds to the overlying secretory cell — it is a follicular cell "imprint" — and each ridge to a two-cell junction.

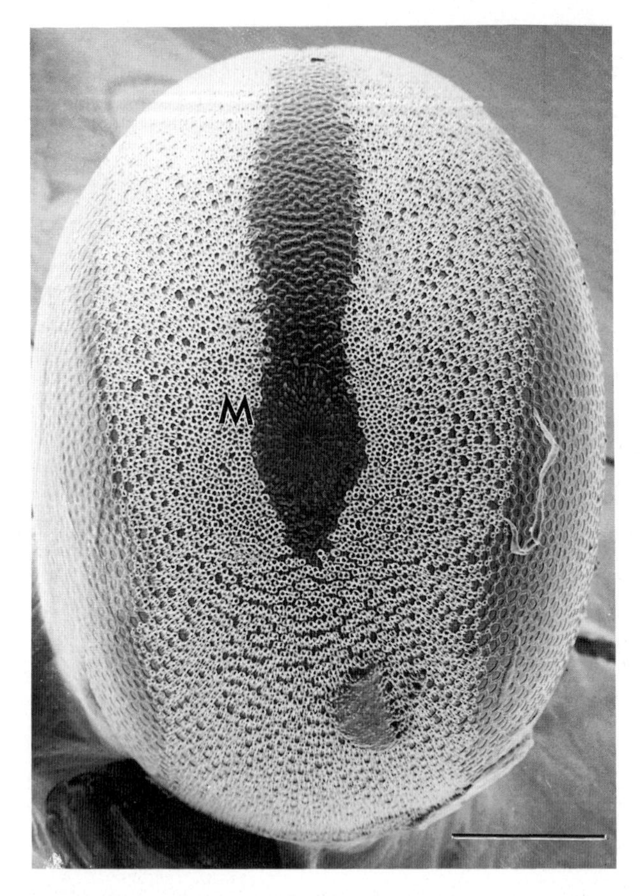

Fig. 4. Scanning electron micrograph of a mature (ovulated) follicle of *A. polyphemus*. At the anterior pole, the micropyle region (*M*) is seen. *Bar* 0.4 mm. (Courtesy of Dr. LH Margaritis)

Each polygonal imprint includes a small number of knobs which varies not only among species, but also among different strains of *Bombyx* (Kafatos et al. 1977 and references therein; Papanicolaou et al. 1986; our unpubl. data).

At ridge corners, aeropyles are found (Figs. 5b, c, 6a). These are round openings leading to internal radial air-channels; they are found normally at the borders of

Fig. 5a–f. Scanning electron micrographs of a purified chorion from a mature *B. mori* follicle, 22/22. **a** There are two planes of symmetry perpendicular to each other, both passing through the longest meridian of the eggshell (one is parallel to the plane of the micrograph). The surface of the chorion shows polygonal imprints (*arrows*), each one created by one follicular (epithelial) cell. The anterior pole (*A*) is slightly flattened. *P* posterior pole. *C* (*arrow*) indicates a ripped region of the chorion. *Bar* 0.4 mm. **b** Closer view of the squared region of the chorion surface shown in **a**. The follicle cell imprints are marked by wide polygonal ridges (*dotted lines*) which correspond to intercellular regions of the follicle cells. Several knobs rest within each imprint. Small pores (aeropyles) are seen at the corners of polygons (*arrows*). *Bar* 20 μm. **c** Side view of the ripped region C of the chorion shown in **a**. The ridges of the imprints become flat and polygons disappear. Aeropyles (*arrows*) are seen on the knobs. *Bar* 20 μm. **d** Detail of the anterior pole (*A*) of the chorion. Polygonal imprints completely disappear (compare with

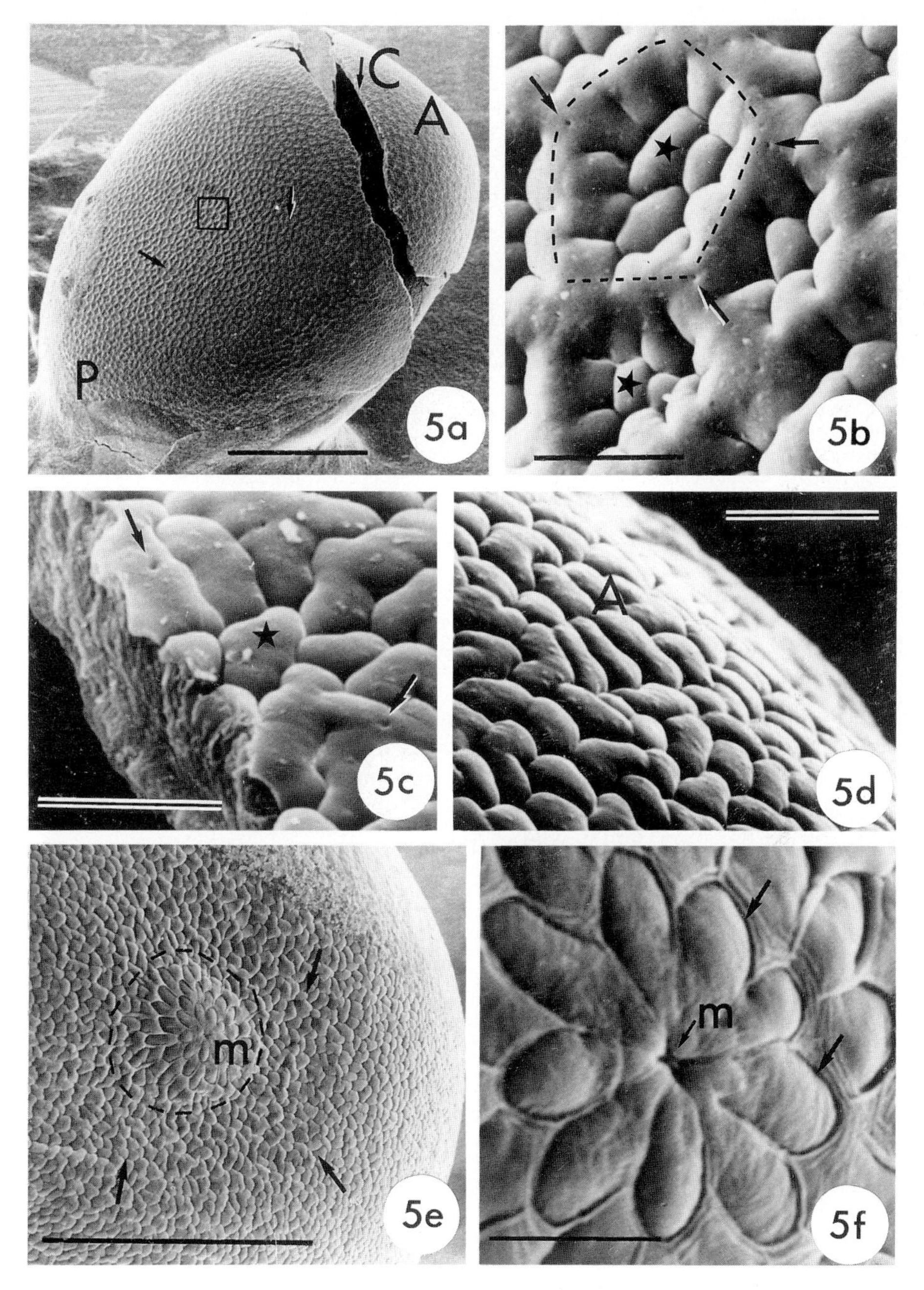

b). The unfocused region is the chorion dome where the micropyle lies. *Bar* 20 μm. **e** Front view of the micropyle dome (*dotted lines*). The micropyle (*m*) is surrounded by four concentric "circles" of cell imprints. The remainder of the surface is covered by knobs (*arrows*). *Bar* 20 μm. **f** Detail of the micropyle (*m, arrow*) shown in **e** where the surrounding cell imprints (*arrows*) take the shape of a rosette. The ridges of the micropyle region are very thin (compare with **b**) and are devoid of aeropyles. The micropyle appears to contain four channels. *Bar* 20 μm. (Papanicolaou et al. 1986)

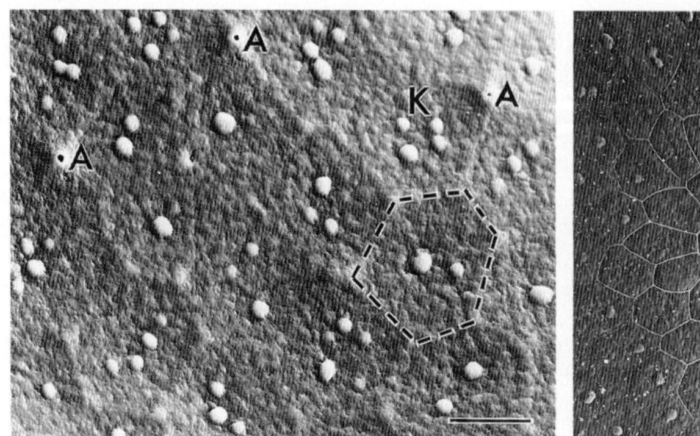

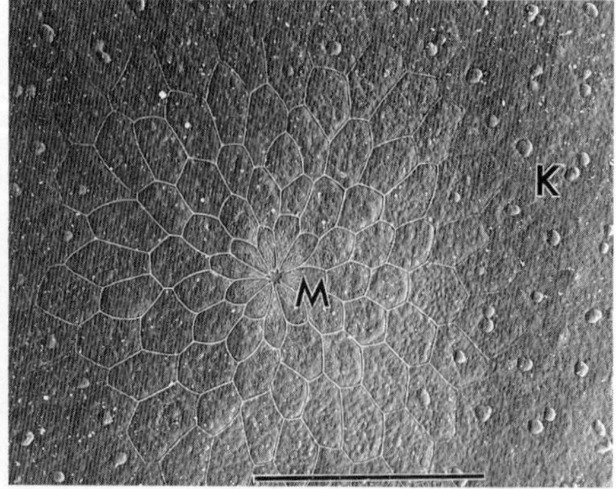

Fig. 6a, b. Scanning electron micrographs showing surface regions of a mature *M. sexta* chorion. **a** Follicle cell imprints are marked by wide polygonal ridges (*dotted lines*). Several "knobs" (*K*) are seen within each imprint. Aeropyles (*A*) are found at the corners of polygons. *Bar* 20 μm. **b** The micropyle (*M*) is surrounded by four concentric "circles" of cell imprints. It is discerned by the fine ridges and the elongated shape of the follicle cell imprints. *Bar* 100 μm

three-cell junctions. The aeropyles may be simple small holes or elaborate chimney-like structures, reflecting the respiratory physiology of the different species (Kafatos et al. 1977).

The anterior pole of chorion contains the micropyle, with the micropylar channels (their number is also species specific), through which sperm entry occurs (Figs. 5d, e, f, 6b). Its external opening is surrounded by a rosette of petal-shaped cell imprints. The ridges are lower than in the bulk of the chorion surface, substantially narrower, and devoid of aeropyles.

In the fish eggshell, the outer surface of the major layer CRI is decorated by numerous "plugs" (Fig. 7a) which block the outer openings of the radial "pore canals" traversing the eggshell (Fig. 7c). An almost hexagonal arrangement of the pore canals openings is revealed, after complete removal of the outer thin CRE layer with a treatment involving a combination of a denaturant (6 M urea) and a reducing agent (1% β-mercaptoethanol) (Fig. 7b).

The pore canal openings retain their regular arrangement in the inner (closest to the oocyte) surface of the eggshell, which shows a fibrous texture (Fig. 7d).

The outer opening of the micropyle, through which sperm entry occurs, is surrounded by a zone devoid of plugs and an outer zone with large plugs (Fig. 7f).

In vertically (to the surface) ripped sections of a lepidopteran eggshell, the first signs of a helicoidal architecture are lamellae lying above a usually thin trabecular layer, the first chorionic layer formed during choriogenesis (Fig. 8a–d).

The number of lamellae varies among species, ranging from 10–20 for *S. nonagrioides* to more than 100 in the moth *Hyalophora cecropia* (Smith et al. 1971). Their orientation relative to the oocyte surface is also variable and, similarly, their thickness and density in different regions of the eggshell. These variations presumably reflect inter-species-specific physiological needs, and/or perhaps different morphogenetic modes and local chemistry.

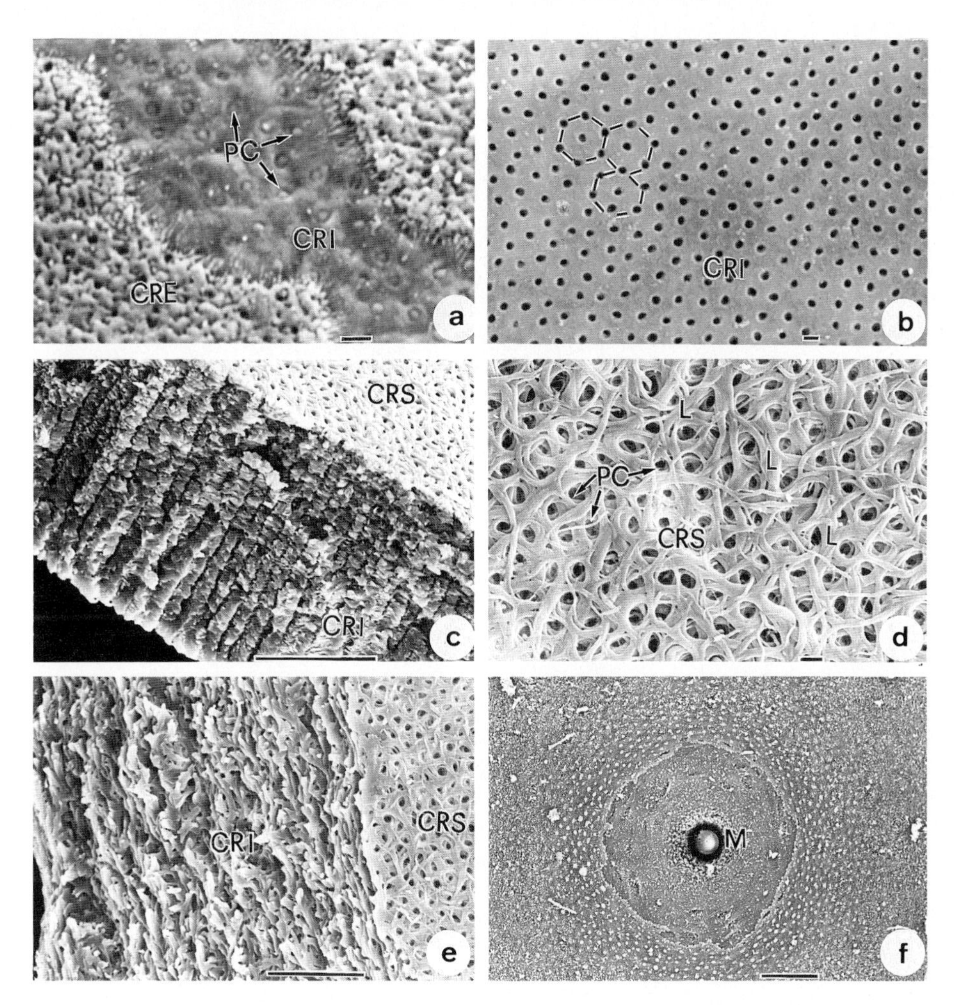

Fig. 7a–f. Scanning electron micrographs showing surface regions and interior architecture of the *S. gairdneri* eggshell. **a** CRE and the outer CRI with "plugs" (*P*) blocking the pore canals are seen (2 days after fertilization). *Bar* 1 μm. **b** Treatment with 6 M urea removes CRE and the plugs, revealing an almost hexagonal arrangement of the pore canals (2 days after fertilization). *Bar* 1 μm. **c** A vertical cross-sectional rip through the eggshell reveals the lamellar organization of CRI. The pore canals are easily discerned traversing the eggshell (18 days after fertilization). *Bar* 10 μm. **d** The inner surface (*CRS*) of the eggshell is seen. Note the network of lamellae, their interconnections and twist around the pore canals (*PC*) (18 days after fertilization). *Bar* 1 μM. **e** An oblique cross-sectional rip through an eggshell showing a different view of the lamellar organization (18 days after fertilization). *Bar* 10 μm. **f** The micropyle, 5.4 μm in diameter, is seen (2 days after fertilization). *Bar* 10 μm

The trabecular layer consists of "columns" oriented perpendicular to the oocyte surface connecting a "roof" and a "floor" (Fig. 8). It contains cavities acting as air-tanks, which are important for respiration (Kafatos et al. 1977; Hinton 1981; Margaritis 1985). Its thickness is also species-specific. It is usually very thin compared to the lamellar part of the eggshell; however, in extreme cases, as in *S. nonagrioides*, it represents almost one third of mature chorion overall thickness

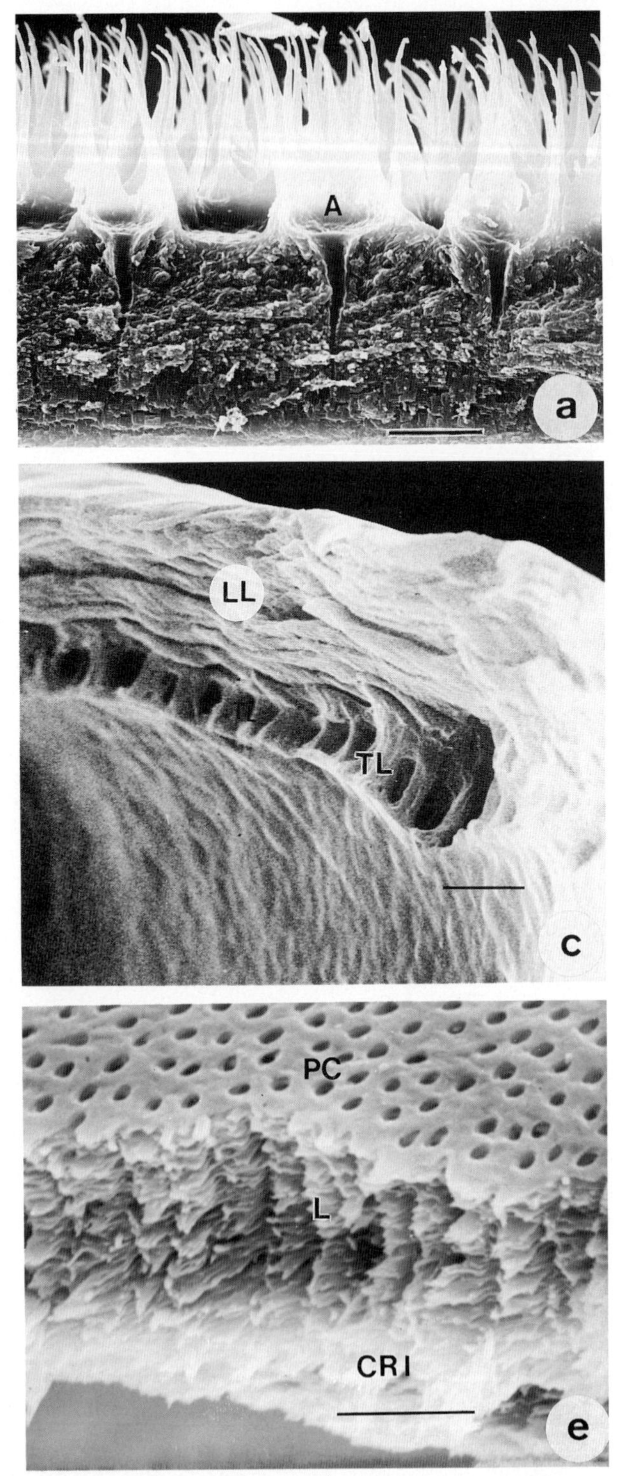

Fig. 8a–f. Scanning electron micrographs of cross-sectional rips through the mature eggshell showing the interior lamellar architecture of: **a** *A. polyphemus*; chimney-like tall aeropyles (*A*) are seen. *Bar* 20 μm. (Courtesy of Dr. LH Margaritis). **b** *M. sexta*; nearest to the oocyte is the trabecular layer (*TL*) which consists of pillars surrounding air-filled spaces. In the inner lamellar layer (*IL*) lamellae are lying parallel

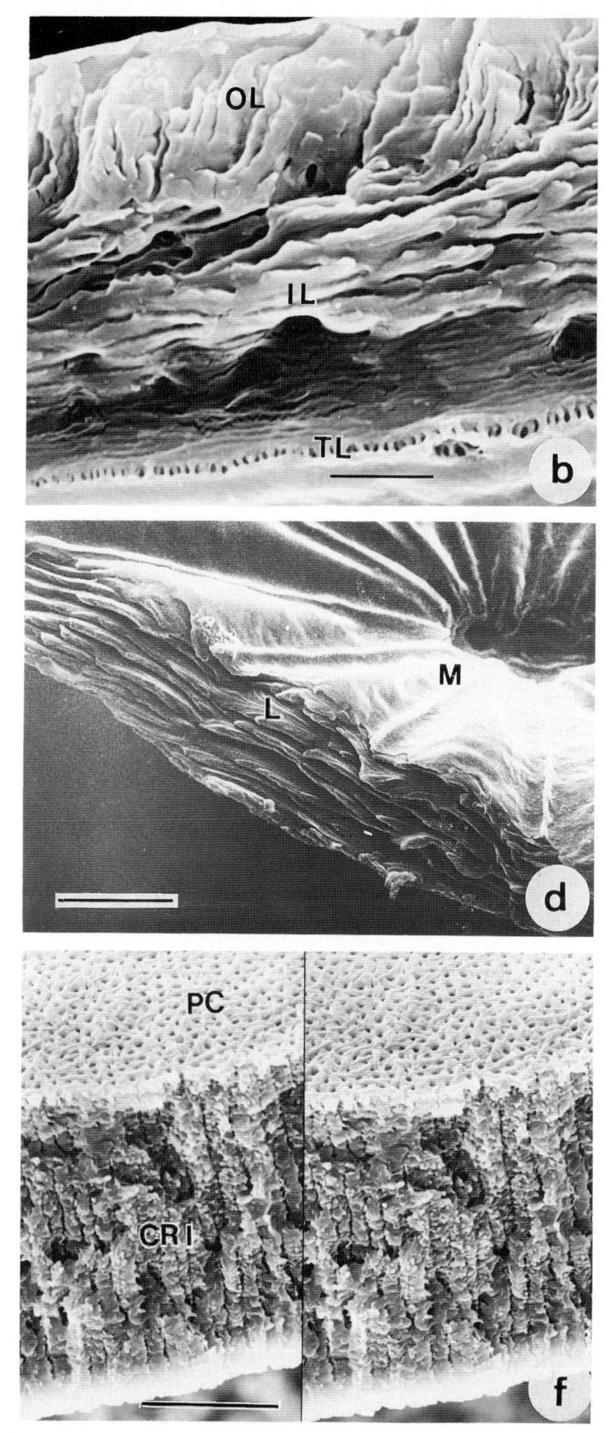

to the oocyte surface, whereas in the outer lamellar layer (*OL*) at an oblique angle to the rest. *Bar* 2.5 μm. **c**, **d** *S. nonagrioides*; the trabecular layer (*TL*) is almost 1/3 of total chorion thickness. The micropyle region is devoid of trabecular layer. In **c** *bar* 2.5 μm and in **d** 5 μm, **e**, **f** *S. gairdneri*; **e** shows an oblique cross-sectional rip, whereas **f** is a stereo-scanning electron micrograph showing the pore canals (*PC*) traversing regularly arranged lamellae. In **e** *bar* 5 μm and in **f** 10 μm

(Fig. 8c). The lamellar structure of fish eggshell can also be seen in vertically to its surface ripped sections (Fig. 8e, f).

2.2 Fine Structure and Morphogenesis of the Eggshell

2.2.1 Eggshell Fine Structure

The eggshells of Lepidoptera are clearly distinguished from those in other orders of insects by their predominantly lamellar ultrastructure (e.g., Fig. 9a, b, c; Smith et al. 1971; Furneaux and Mackay 1972; Kafatos et al. 1977; Hinton 1981; Mazur et al. 1982; Papanicolaou et al. 1986; Fehrenbach et al. 1987; Regier and Vlahos 1988).

A lamellar supramolecular structure also predominates in several fish eggshells (e.g., Fig. 9d; Grierson and Neville 1981 and references therein). Distortions and defects disrupt the regular lamellar arrangement (Mazur et al. 1982) both in lepidopteran and fish eggshells.

In Lepidoptera, under higher magnification, each lamella is seen to consist of fibers, 70–200 Å in diameter, embedded in an amorphous matrix (Fig. 10). The paraboloidal appearance of these fibers in oblique sections clearly suggests a helicoidal architecture (Bouligand 1972). Initially, these fibers are distributed sparsely, but as choriogenesis progresses they thicken and/or increase in number, until they coalesce (Smith et al. 1971; Kafatos et al. 1977; Regier and Vlahos 1988).

In perpendicular sections the paraboloidal "arcs" disappear, leaving alternating bands (zones) of lines and dots. The zones where the fibers are oriented nearly parallel to the plane of section appear darker, giving rise to the lamellar appearance. However, a closer examination of electron micrographs taken from eggshells in the late choriogenetic stages reveals parabolic arrays of thin, 30–40 Å, fibrils confirming the helicoidal model of architecture even at these developmental stages (Hamodrakas 1984; Papanicolaou et al. 1986). Our recent unpublished transmission electron microscopy and optical diffraction data (Hamodrakas SJ and Ottensmeyer FP, in prep.) suggest the existence of 30–40 Å fibrils as constituents of the 70–200 Å thicker fibers throughout choriogenesis.

Applying carefully the technique described by Neville (1975, p. 228), we were able to determine the sense of rotation of the helicoidal structure in lepidopteran eggshells (data not shown): clearly, proteinaceous eggshells in Lepidoptera are left-handed (anti-clockwise) helicoids, which means that successive planes of fibrils twist anticlockwise in a direction further away from the observer. As mentioned earlier, most helicoidal biostructures are left-handed (Neville 1975; Grierson and Neville 1981). Knowledge of the absolute sense of rotation of these helicoids might be crucial in determining the molecular mechanisms of self-assembly (packing of protein fibrils, see below).

The two-phase system (fibers embedded in a matrix) of lepidopteran proteinaceous eggshells appears to be an ideal solution to the problem of constructing eggshells providing strong and elastic mechanical protection and support (Kafatos et al. 1977 and references therein). Our data, however, pose the question whether

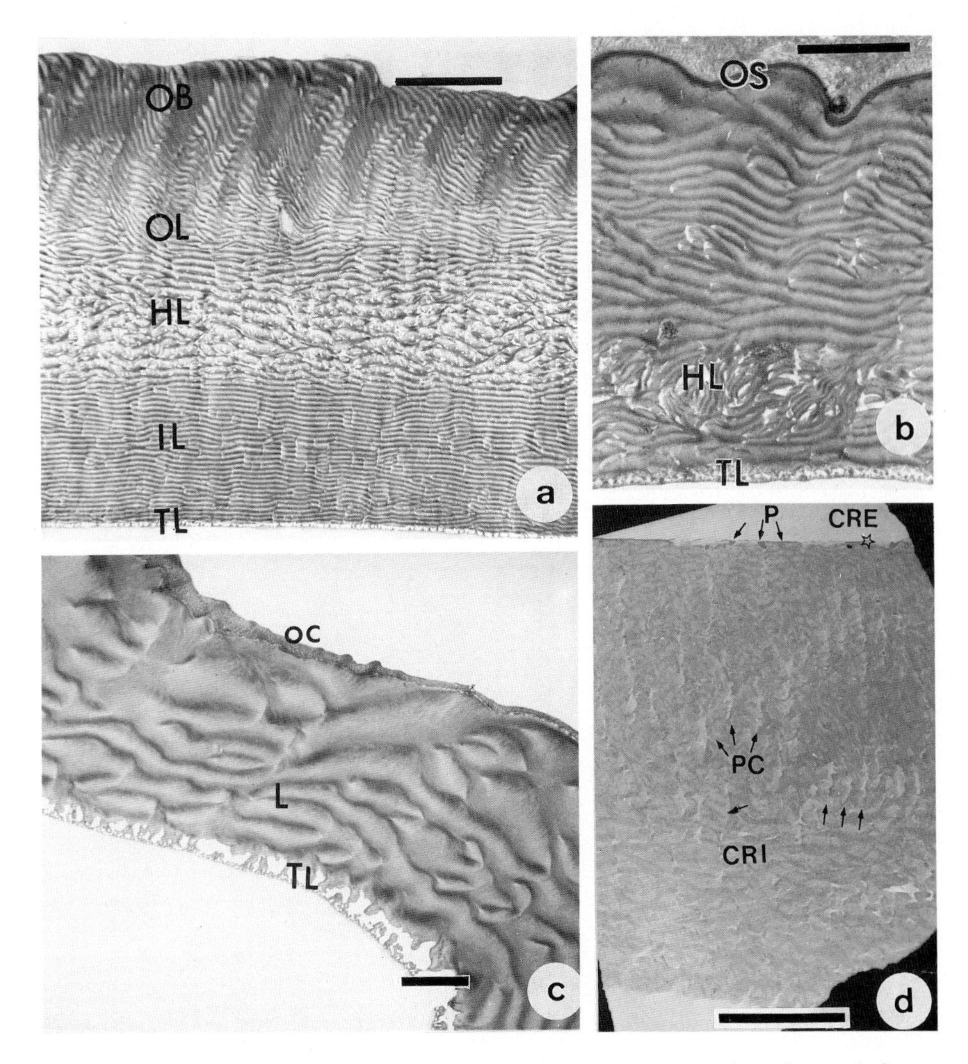

Fig. 9a–d. Transmission electron micrographs of thin sections cut through chorions of: **a** *A. polyphemus*; within the bulk of the lamellar chorion, four types of lamellae can be distinguished: thin lamellae of the inner lamellar layer (*IL*), thick, distorted, spongy lamellae of the holey layer (*HL*), lamellae of the outer layer (*OL*) and, lying at an angle to the rest, the thick lamellae of the oblique layer (*OB*). Nearest to the oocyte is the trabecular (*TL*) layer, which consists of pillars surrounding air-filled spaces. *Bar* 10 μm. (Hamodrakas et al. 1986). **b** *B. mori*; formation of the outer osmiophilic layer (*OS*) containing proteins with a high-cysteine content has just begun; the trabecular layer (*TL*) and the remainder of the chorion have already been laid down. *Bar* 5 μm. (Papanicolaou et al. 1986). **c** *S. nonagrioides*; a "sticky" outer covering (*oc*) has been deposited on the outer surface of a laid egg. It serves as a "glue" sticking together laid eggs. The main bulk of the chorion consists of lamellae (*L*) deposited on a thick trabecular layer (*TL*). *Bar* 2 μm. **d** *S. gairdneri*; the radial pore canals (*PC*) can be discerned (*arrows*) in the lamellar CRI, with a characteristic twisted ribbon structure (Grierson and Neville 1981), implying helicoidal architecture for the eggshell. On the outer surface of the eggshell, the thin osmiophilic *CRE* is the eggshell layer deposited first. "Plugs" can be seen (*P*, *arrows*) blocking pore canals. *Bar* 7 μm. (Hamodrakas et al. 1987)

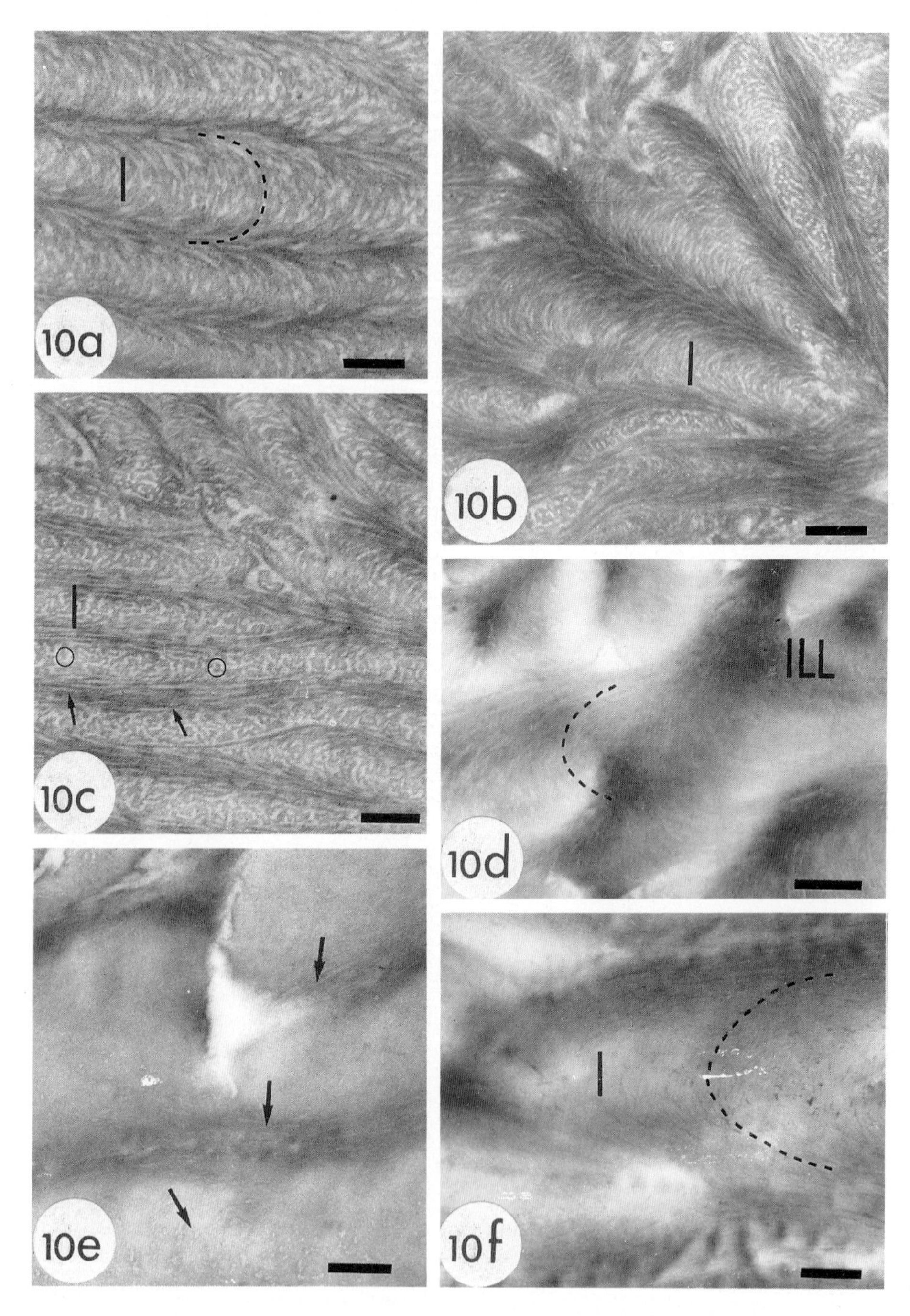

Fig. 10a–f. Transmission electron micrographs of thin sections cut through the chorion of *B. mori*, showing the fibrous ultrastructure of its lamellae. **a** The parabolic pattern of fibers (ca. 110 Å; *dotted lines*) within each lamella (*l*) in an oblique section of follicle 11/22 is seen. *Bar* 0.2 μm. **b**, **c** Fibrous lamellae (*l*) in oblique (**b**) and vertical (**c**) (with respect to the surface of chorion) sections of the 13/22

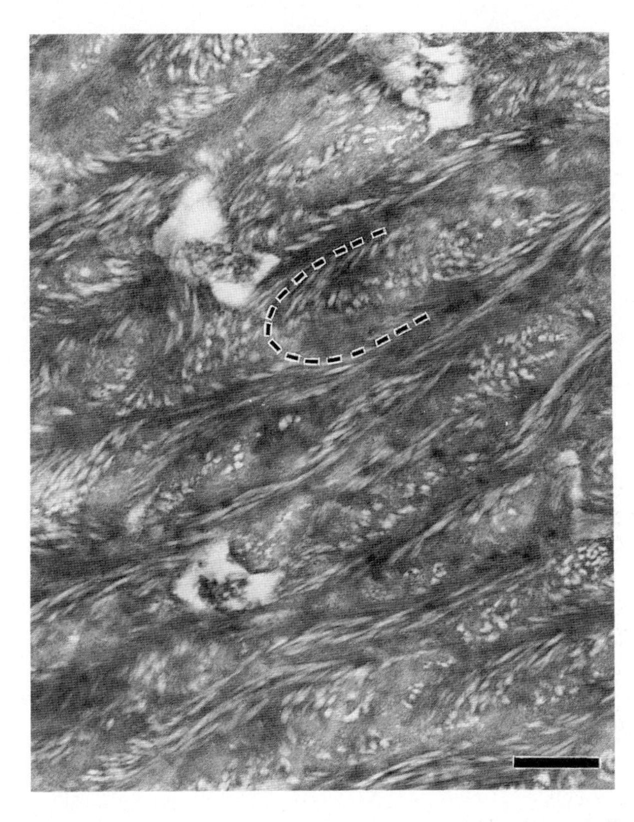

Fig. 11. Thioglycollic acid treatment (after Grierson and Neville 1981) reveals parabolic patterns of fibrils (*dotted lines*) in electron micrographs of oblique, thin sections cut through the eggshell of *S. gairdneri*. *Bar* 0.4 μm

true two-phase systems actually exist in the lepidopteran eggshell: different fixation and/or staining techniques give entirely different images of the fiber-matrix systems (Hamodrakas SJ, unpubl.). Furthermore, the technique of freeze-fracturing (cf. below) does not give any indication of a separate matrix phase. An attractive alternative, might be that the light (not heavily stained) areas of electron micrographs correspond to hydrophobic (not stain-absorbing) parts of the protein molecules constituting the "core" of eggshell fibrils, whereas the dark areas represent hydrophilic, stain-absorbing, portions of the molecules. It might, therefore, be that different portions of the same protein molecule(s) correspond to fiber

chorion. In **c** parabolic arcs disappear; in the lamellar boundaries, fibers cut longitudinally are seen as *parallel lines* (*arrows*) in contrast to those in the center, cut transversely, which are seen as *dots* (*small circles*). *Bar* 0.2 μm. **d** A micrograph showing the inner lamellar layer (*ILL*) of the chorion of follicle 15/22. Fibers ca. 90 Å in diameter form parabolic arcs within individual lamellae (*dotted lines*). *Bar* 0.45 μm. **e** Micrograph of the inner part of chorion (follicle 19/22) where the fibrous ultrastructure is obscured. *Arrows* show fibrillar remnants of the chorion lamellae. *Bar* 0.2 μm. **f** In the mature chorion of the follicle 33/33, fibers of approximately 30–40 Å in diameter can be seen. The *dotted lines* show the arrangement of the parabolic arcs of fibers within one lamella (*l*). *Bar* 0.2 μm. (Papanicolaou et al. 1986)

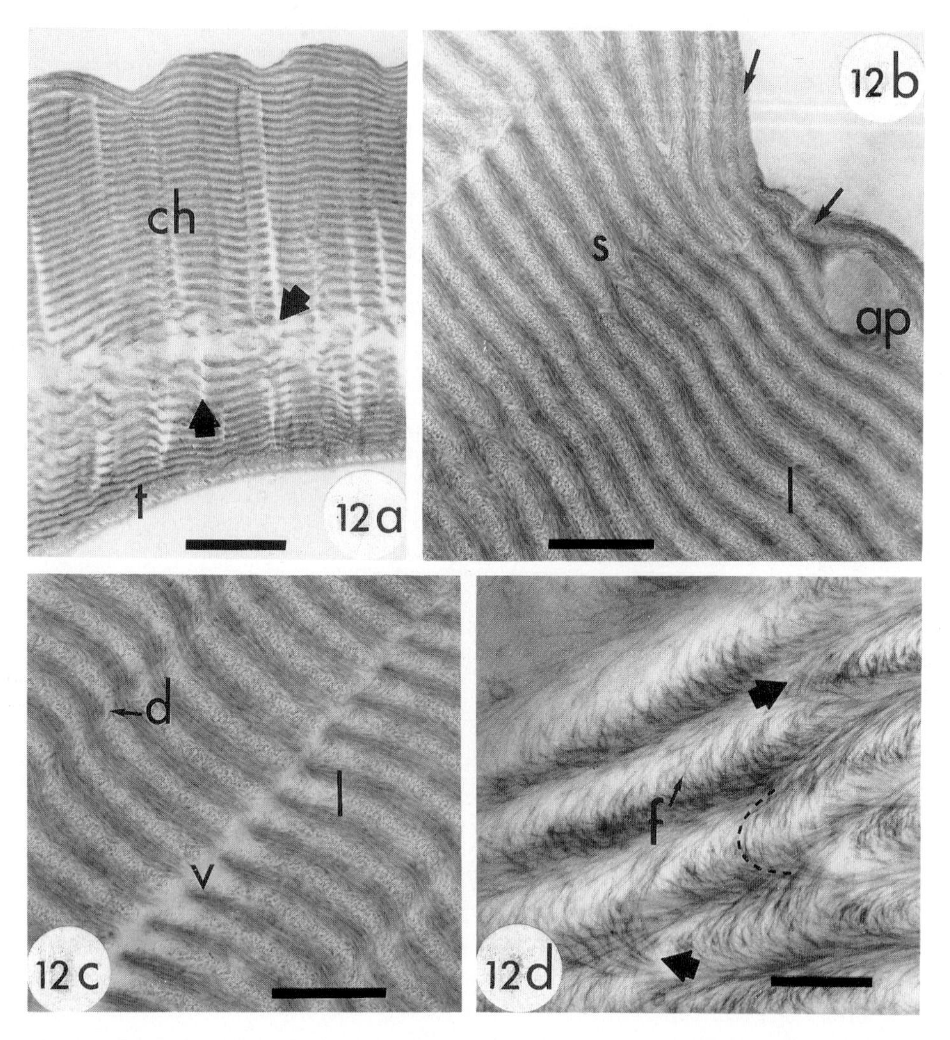

Fig. 12a–d. Transmission electron micrographs of thin sections cut through the chorion of *B. mori* from a number of follicles treated with a denaturing agent before fixation (6 M urea; see Papanicolaou et al. 1986). **a** Low power survey field of the chorion (*ch*) of follicle 19/38. Note the parallel arrangement of the lamellae and the vacancies (*arrows*) between them. *Bar* 5 μm. **b** The outer part of the same chorion at higher magnification reveals an empty aeropyle (*ap*) and a screw dislocation (*s*) (see also Mazur et al. 1982) within the lamellar organization (*l*) Note also the 180° change in the orientation of the parabolic pattern of fibers within individual lamellae, in certain chorion regions. *Bar* 1 μm. **c** Image of the inner part of chorion shown in **a**, illustrating in detail the lamellar organization. The parallel arrangement of the lamellae (*l*) and their fibrous ultrastructure is evident; *d* arrow, and *v* indicate a distortion and a vacancy in the lamellar organization, respectively. *Bar* 1 μm. **d** Fibrous ultrastructure of chorion from follicle 26/38. Thin individual fibers, ca. 30 Å in diameter (*f, arrow*) are clearly resolved, forming parabolic arcs (*dotted lines*). In several cases fibers of adjacent lamellae seem to overlap (*heavy arrows*). *Bar* 0.5 μm. (Papanicolaou et al. 1986)

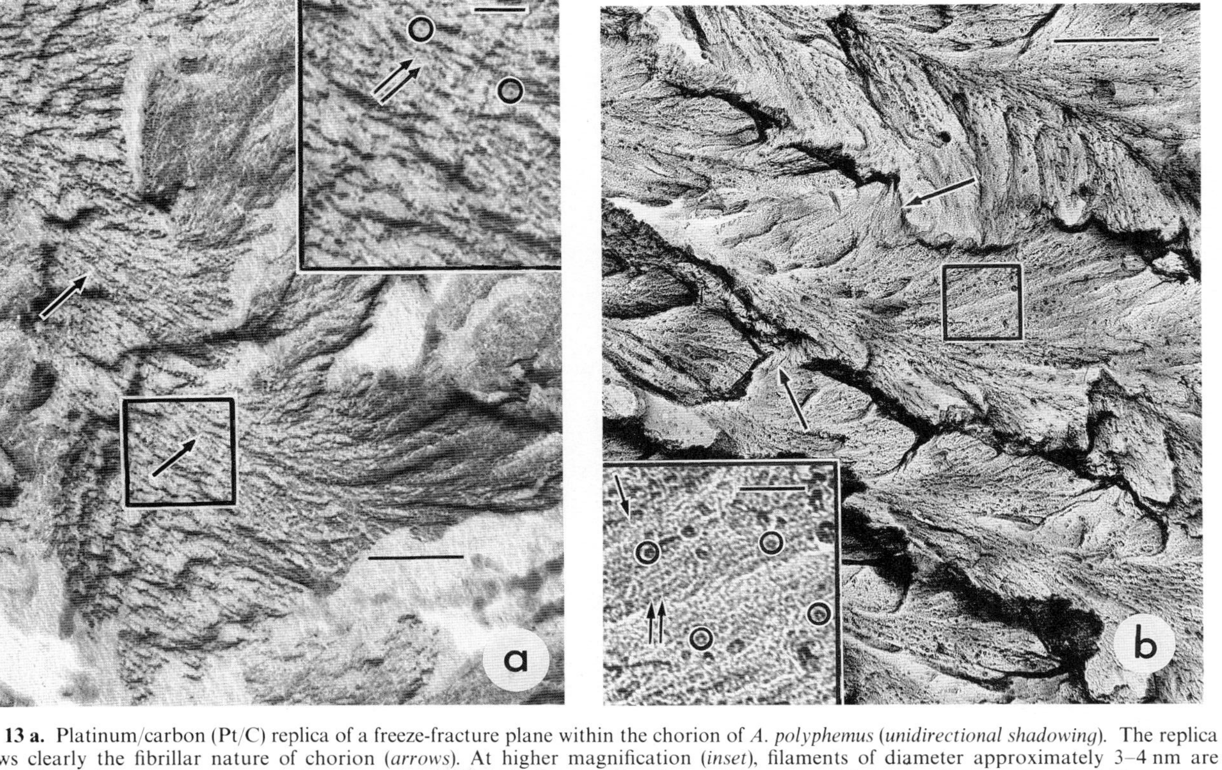

Fig. 13 a. Platinum/carbon (Pt/C) replica of a freeze-fracture plane within the chorion of *A. polyphemus* (*unidirectional shadowing*). The replica shows clearly the fibrillar nature of chorion (*arrows*). At higher magnification (*inset*), filaments of diameter approximately 3–4 nm are longitudinally (*arrow*) or transversely (*circles*) seen. There are indications (*beading*; *multiple arrows*) that these filaments have helical structure. *Bar* 0.2 μm. *Inset*: *Bar* 60 nm. **b** Pt/C replica of a freeze-fracture plane within the chorion of *A. polyphemus* (*rotary shadowing*). The fracture has advanced across successive lamellae producing a series of steps. Therefore the lamellar structure of silkmoth chorion is clearly seen. The helicoidal arrangement of chorion filaments is also obvious (*arrows*). In the inset (*boxed area*), at higher magnification, the helical (?) (*multiple arrows*) chorion filaments are seen either longitudinally (*arrows*) or transversely (*circles*). *Bar* 0.3 μm. Inset: *Bar* 50 nm. (Hamodrakas et al. 1986)

and matrix respectively. In the early stages of choriogenesis (Fig. 10) large stained areas may represent spaces, empty of protein, occupied by stain.

To discern microfibrillar patterns showing characteristic rows of parabolic "arcs", indicative of a helicoidal architecture, in sections of the fish eggshells cut obliquely (Fig. 11), we were forced to pretreat fish eggshells with thioglycollic acid (an agent disrupting disulfide bonds) following a method suggested by Grierson and Neville (1981). The method was first introduced by Filshie and Rogers (1962) to resolve fine structure in feather keratin.

Pretreatment of lepidopteran eggshells with a urea buffer (6 M urea, 0.4 M Tris-HCl, pH 8.5, a denaturing agent) substantially modifies the lamellar organization (Fig. 12), causing extensive rearrangements and texture disruption. Possible reasons for the significant changes observed and their meaning are discussed by Papanicolaou et al. (1986).

Transmission electron microscopy (lamellar organization — parabolic patterns of fibrils) provides only indirect evidence for the existence of helicoidal architecture in lepidopteran and fish eggshells.

Direct visualization of the helicoidal arrangement of fibrils for the formation of eggshell architecture was made possible by freeze-fracturing, utilizing both unidirectional and rotary shadowing (Hamodrakas et al. 1986; Orfanidou C, Hamodrakas SJ, Margaritis LH, and Gulik-Krzywicki T, in prep. Figs. 13, 14).

Apparently, in Lepidoptera, fibrils, hereafter called also "filaments", with a diameter of approximately 30–40 Å are the basic structural units of the eggshell. The beaded appearance of the fibrils along their long axis may be attributed to a helical substructure (e.g., Aebi et al. 1983).

It is perhaps instructive to remember at this point that Rudall, some 35 years ago (1955), had put forward the proposal that helical molecular structures are ideal candidates to serve as basic structural elements of a helicoidal architecture.

The existence of 30–40 Å fibrils as basic structural units of helicoidal proteinaceous eggshells, unraveled by freeze-fracturing, is in good agreement with X-ray diffraction (cf. below; Hamodrakas et al. 1984, 1986) and recent transmission electron microscopy and optical diffraction data (Hamodrakas SJ and Ottensmeyer FP, in prep.).

Freeze-fracturing has been used to reveal the helicoidal organization of microfibrils in the cuticle of a crayfish (Filshie and Smith 1980) and also in studies of the cholesteric liquid crystalline phases of polymer solutions (Livolant and Bouligand 1989) and cholesteric liquid crystalline DNA (Rill et al. 1989).

Fig. 14a. Pt/C replica of a freeze-fracture plane within the chorion of *M. sexta* (*rotary shadowing*). The fracture has advanced across successive lamellae producing a series of steps, revealing the lamellar structure of chorion and the helicoidal arrangement of its constituent fibrils (compare with **b**). In the inset (*boxed area*; ∗), at higher magnification, chorion fibrils are seen either tansversely (*circles*; 3–4 nm in diameter) or longitudinally (*double arrows*). The apparent beading of the fibrils suggests, most probably, a helical fibrillar structure. *Bar* 0.2 μm. Inset: *bar* 50 nm. **b** Stereoscopic view of a Pt/C replica of a freeze-fracture plane within the chorion of *M. sexta* (*unidirectional shadowing*) showing clearly the helicoidal packing modes of its constituent beaded (helical?) fibrils (*arrows*), 3–4 nm in diameter. *Bar* 0.2 μm. **c** Stereoscopic view of a Pt/C replica of a freeze-fracture plane within the chorion of *S. gairdneri* (*unidirectional shadowing*) showing the beaded structure of its constituent fibrils (*arrows*). *Bar* 0.3 μm.

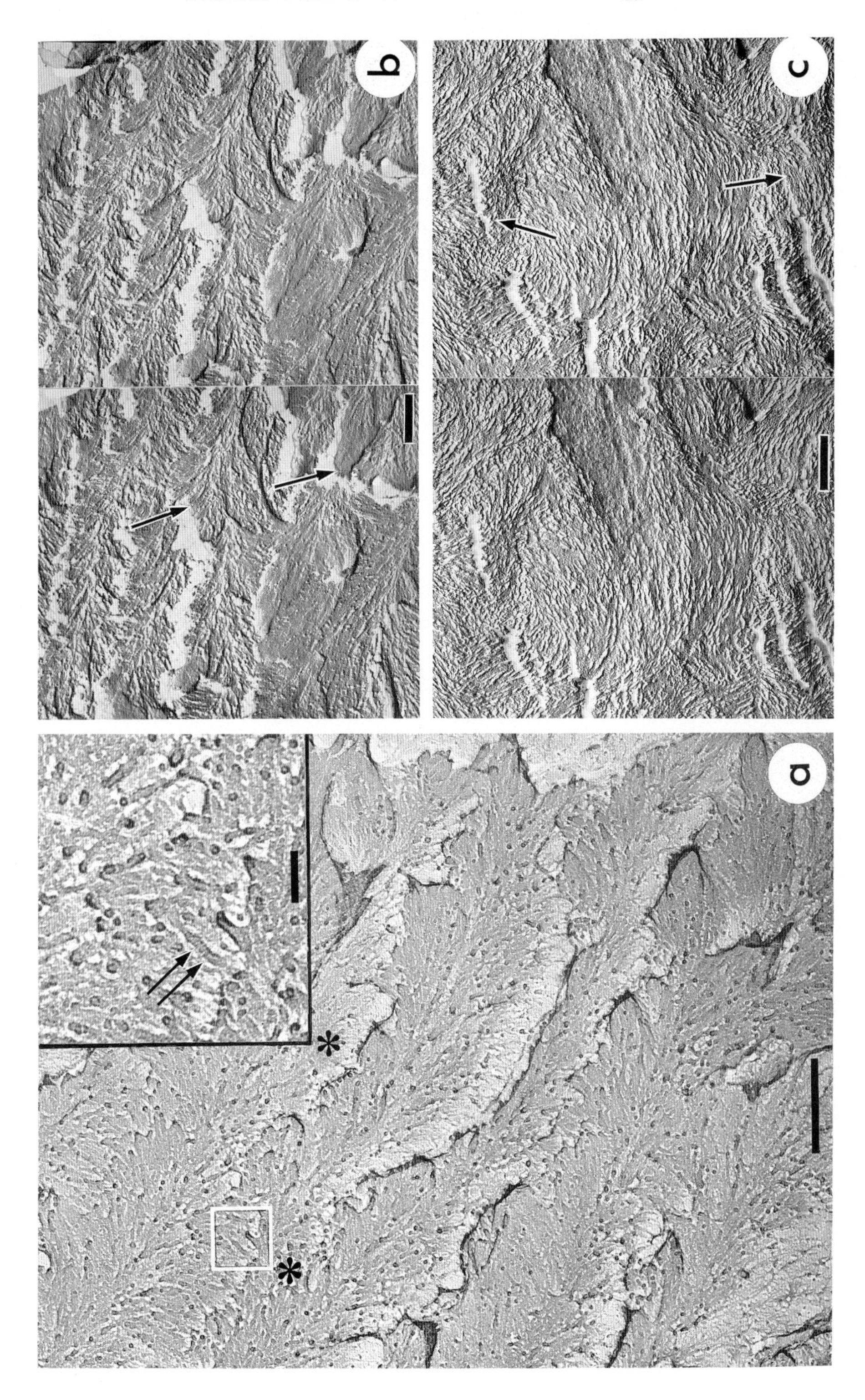

Scanning electron microscopy also provides direct visualization of the helicoid in coelacanth scales (Giraud et al. 1978), the cuticle in *Carcinus maenas* and the test of *Halocynthia papillosa* (Gubb 1975).

Biological structures, in general, are formed by a hierarchy of assembled subunits, each forming the construction unit for the next size level of organization (Crick 1953); this has the advantage that errors are rectified en route by the rejection of faulty components. Helicoidal proteinaceous eggshells are not an exception to this rule; they are typical examples of protein engineering in vivo (Hamodrakas et al. 1988).

The basic structural unit in Lepidopteran helicoidal eggshells appears to be a fibril 30–40 Å in diameter. The next levels of hierarchy are: packing of fibrils into sheets, stacking of sheets one on top of another, helicoid formation.

Questions of interest which remain unanswered to date are: what is the structure of a fibril? Is it formed by one or more protein molecules? How are these protein molecules folded and how do they interact? We shall try to provide partial answers to these questions below. However, it is clear that any simple or heuristic model of self-assembly for helicoidal proteinaceous eggshells should be based on the determination of the folding patterns of individual protein molecules and of protein intermolecular interactions.

2.2.2 Morphogenesis of the Eggshell

Comparison of chorion morphogenesis and protein biochemistry among species offer considerable promise for answering fundamental questions about the relationship between development and evolution (Regier and Vlahos 1988). Unfortunately, to date, a detailed description of Lepidopteran chorion morphogenesis is restricted to only four species: *H. cecropia*, *A. polyphemus*, *B. mori* and *M. sexta* (King and Aggarwal 1965; Telfer and Smith 1970; Smith et al. 1971; Mazur et al. 1980; Regier et al. 1982; Papanicolaou et al. 1986; Regier and Vlahos 1988).

A superb review on silkmoth chorion morphogenesis has been presented by Mazur et al. (1982).

Four different modes of growth appear to occur during formation of the lamellar chorion following the construction of the trabecular layer (Mazur et al. 1982): (1) lamellogenesis or framework formation; most lamellae are deposited early in choriogenesis, (2) permeation or intercalation of newly secreted protein components that leads to expansion of this framework, (3) densification by addition of protein components, and (4) regionalization that leads to apposition of additional lamellae and surface sculpturing.

Species-specific variations are evident (refer to original publications for detailed descriptions) and an evolutionary model has been presented to account for the differences observed (Regier and Vlahos 1988).

Views of sequential events occurring during *B. mori* chorion formation and morphogenesis of chorion components are given as examples in Figs. 15, 16, 17.

Little is actually known about the mechanisms operating during fish eggshell morphogenesis, although the literature concerning fish eggshell formation is relatively rich (cf. Groot and Alderdice 1985). A detailed review of this subject is

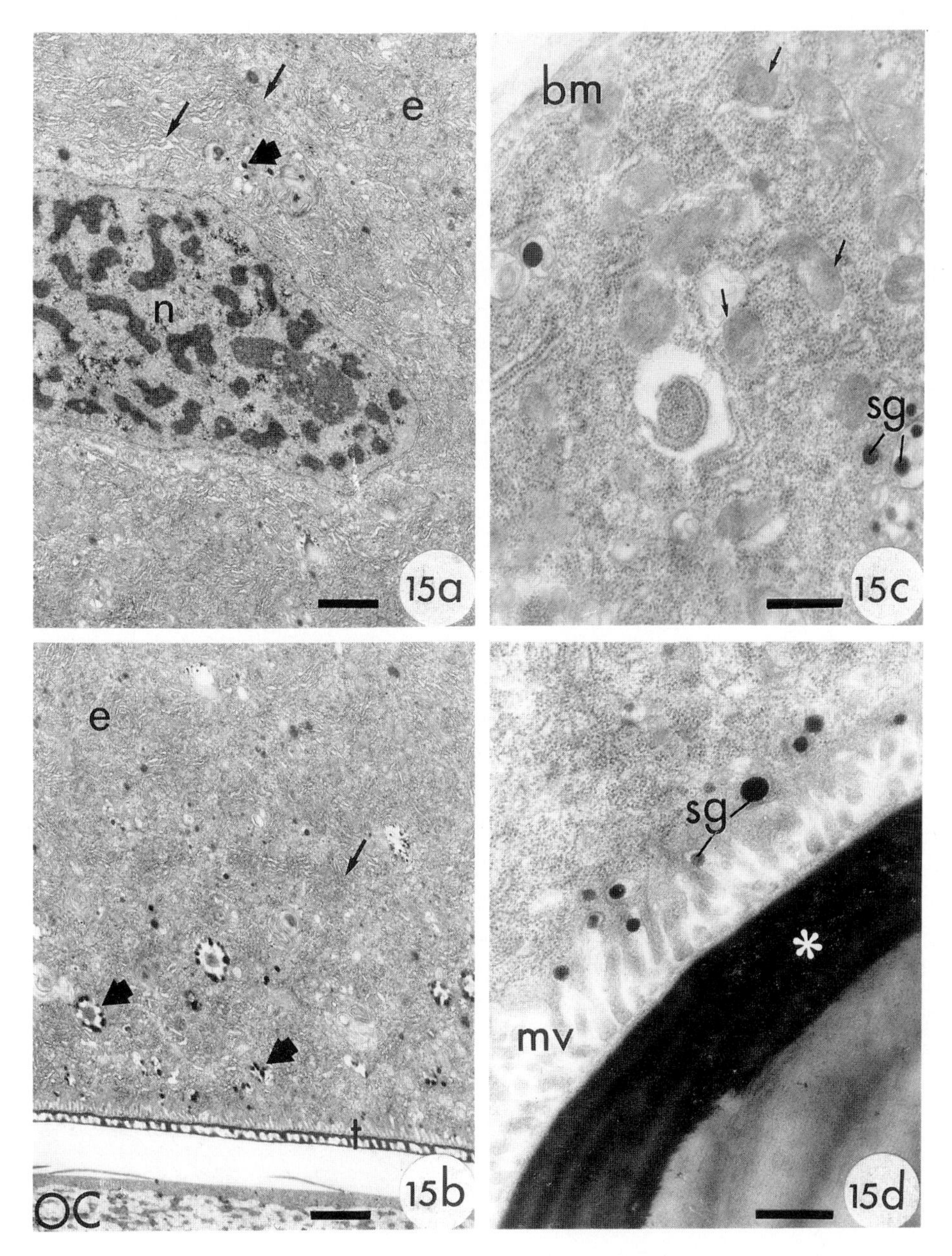

Fig. 15. a, b Transmission electron micrographs of thin sections cut through a follicle cell of *B. mori* (*e* follicle 1/22) showing: **a** its nucleus (*n*), and **b** the cytoplasm close to the secretory region. The high synthetic activity of the cell is apparent by the abundance of RER and Golgi complexes (*arrows*) and also by the osmiophilic secretory granules (*heavy arrows*) in its cytoplasm. *t* Trabecular layer; *OC* oocyte. *Bar* 1 μm. **c**, **d** Transmission electron micrographs of thin sections cut through a follicle cell (follicle 19/22) showing; **c** part of its cytoplasm close to the basal membrane (*bm*) and **d** part of its cytoplasm near the outer surface of the chorion. Secretory granules (*sg*) full of osmiophilic material are seen between, near, or within the microvilli (*mv*) and also close to the basal membrane. A large number of mitochondria (*thin arrows*), RER membranes, and ribosomes are also seen in the cytoplasm; * osmiophilic layer; *Bar* 0.5 μm. (Papanicolaou et al. 1986)

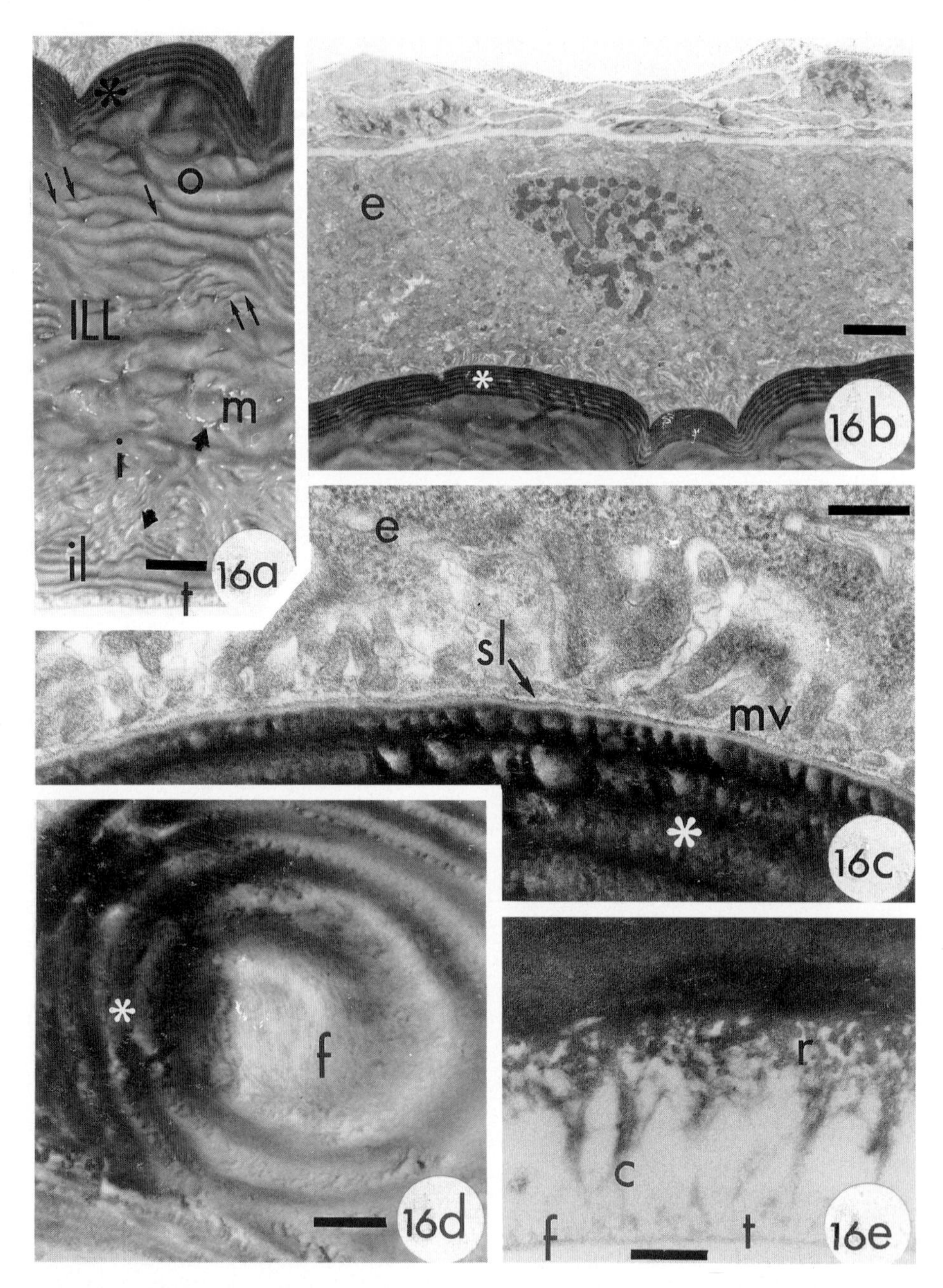

Fig. 16a–e. Transmission electron micrographs of a nearly mature *B. mori* follicle, 22/22. **a** Cross-section of the chorion. Lying closest to the oocyte is the trabecular layer (*t*) and then a thick inner lamellar layer (*ILL*) followed by an osmiophilic outer layer (*). The ILL can be divided into four sublayers, on the basis of their differences in lamellar thickness and orientation; the innermost sublayer (*il*), consisting of two thin, parallel lamellae close to the trabecular layer; the inner sublayer (*i*), consisting of very thin randomly oriented lamellae; the middle sublayer (*m*) with four to five thick lamellae; and the outermost sublayer (*o*) made up of lamellae of moderate thickness. The bulk of the inner lamellar layer

beyond the scope of this chapter, and the reader should consult for details original publications by Anderson (1967), Flugel (1967), Wourms (1976) and our forthcoming article (Papadopoulou P and Hamodrakas SJ, in prep.) as examples.

3 The Biochemistry of Eggshell Proteins

Detailed reviews of silkmoth chorion protein biochemistry have been made by Kafatos et al. (1977) and Regier and Kafatos (1985).

Biochemically, silkmoth chorion is surprisingly complex: as many as 186 protein components have been resolved by two-dimensional gel electrophoresis, from the chorions produced by an individual *A. polyphemus* moth (Regier et al. 1980) and more than 150 different polypeptides in *B. mori* (Regier and Kafatos 1985).

Silkmoth chorion proteins have been classified into four classes called A to D, based on observed clustering in SDS gels. The most abundant are the A's, B's, and C's which together account for approximately 97% of total chorion mass. In *A. polyphemus* the A's, B's, and C's represent about 38, 50, and 9% respectively of the chorion's dry mass, whereas the D's are quantitatively minor. (Kafatos et al. 1977). Molecular weight ranges for A, B, and C proteins are approximately 9000–12 000, 12 000–14 000, and 16 000–20 000 Da, respectively.

A fifth class of minor proteins, the E's (divided into two subclasses, E1 and E2, with molecular weights approximately 15 000 and 85 000 Da, respectively), uniquely assemble to form the "filler" substructure, the bulk of which is localized as little bundles resting on *A. polyphemus* chorion surface, in the so-called aeropyle crown region (Mazur et al. 1980; Regier 1986).

In *B. mori*, another class of species-specific proteins, the Hc's (high-cysteine), are responsible for the formation of the outer osmiophilic eggshell layer (Figs. 15, 16, 17), presumably a very tough weather-resistant covering; *B. mori* embryos are the only silkmoth embryos that undergo prolonged periods of diapause, thus requiring protection. This class is further subdivided into HcA and HcB which are homologous to A's and B's respectively (Rodakis and Kafatos 1982).

shows many small scattered discontinuities (*small arrows*) along with small holes (*heavy arrows*) especially in its inner parts. The outer osmiophilic layer consists of five to six parallel lamellae which are uniform in thickness and show very few or no discontinuities. The osmiophilic layer forms the knobs visible in scanning electron micrographs, in the outer surface of the chorion. *Bar* 2 μm. **b** View of the 11-μm-thick follicular epithelium (*e*) which is attached to the osmiophilic layer. *Bar* 3 μm. **c** Details of the osmiophilic layer (∗) and the microvilli (*mv*) of the follicle cells (*e*). In between lies the thin sieve layer (300 Å), consisting of three minor layers (*sl, arrow*). Its central layer appears as an electron-dense line ca. 150 Å thick. *Bar* 0.2 μm. **d** Spiral arrangement of the osmiophilic layer as seen in a tangential section of a knob (∗). Such spirals were also documented by Bouligand (1972) in the crub cuticle. In the center of the spiral the fibrous (*f*) ultrastructure of the osmiophilic layer is revealed. *Bar* 0.4 μm. **e** Closer view of the trabecular layer (*t*) formed by fibrous columns (*c*) separating two, also fibrous, spongy layers: the roof (*r*) and the floor (*f*). *Bar* 0.3 μm. (Papanicolaou et al. 1986)

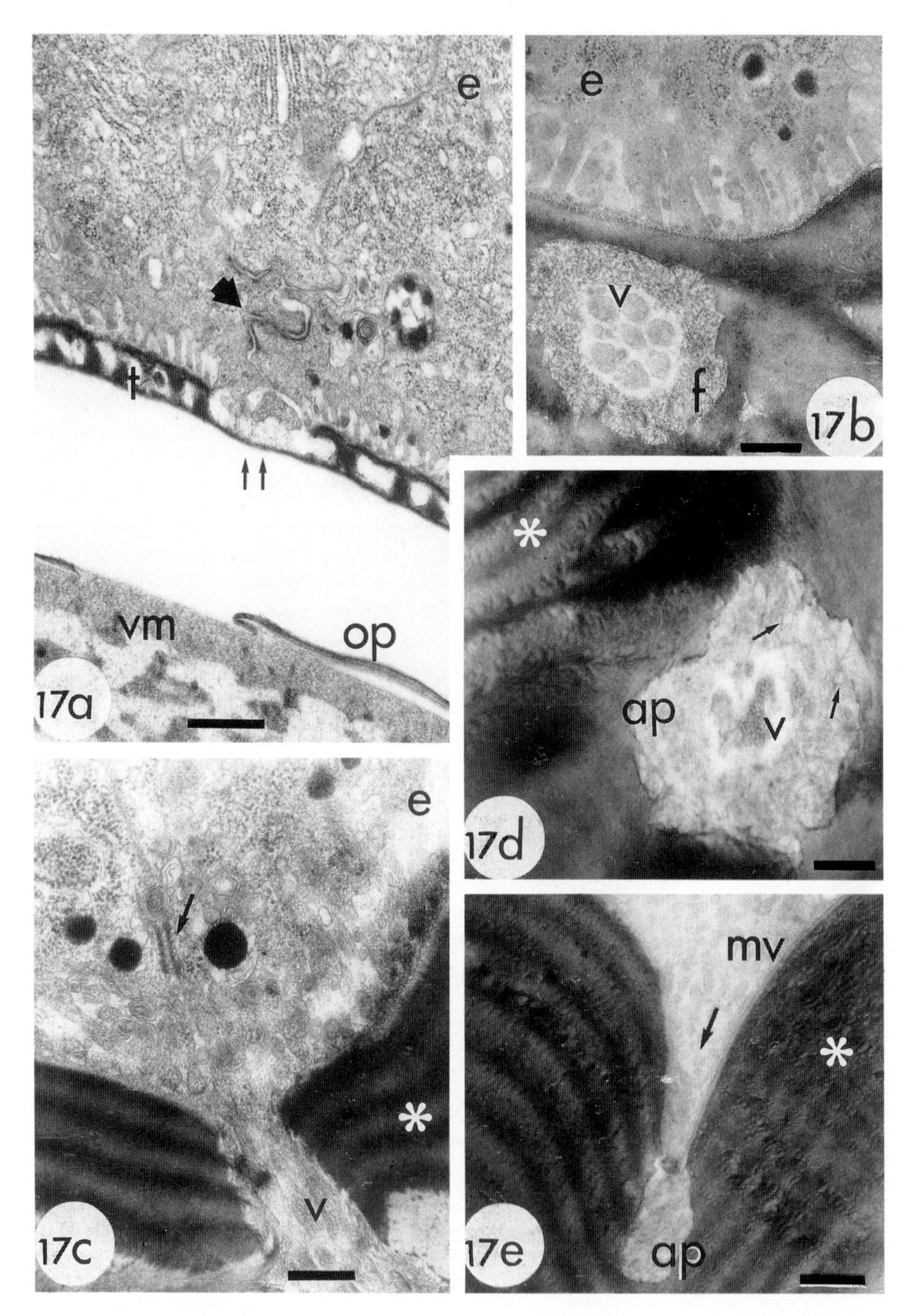

Fig. 17a–e. Transmission electron micrograph of an early stage of choriogenesis (follicle 1/22) showing the formation of an aeropyle in *B. mori*. **a** The aeropyle channel (*arrows*) forms around a bundle of extended microvilli, under a cell junction (*heavy arrow*), during the formation of the trabecular layer (*t*): *e* follicle cell; *vm* vitelline membrane; *op* overlapping plates of the vitelline membrane. *Bar* 0.4 μm.

The C class of proteins, which constitutes the bulk of the "early" proteins, responsible for the formation of the initial chorion framework and perhaps required for the organization of the quantitatively dominant components which are secreted later, can be subdivided further into two subfamilies CA and CB, sharing homologies with A's and B's respectively (Regier et al. 1983; Lecanidou et al. 1986)

It appears that the number of silkmoth genes is at least as high (Eickbush and Kafatos 1982). However, most chorion genes are related: they belong to a small number of gene families (A, B, CA, CB, HcA, HcB ...), each encompassing multiple genes that arose during evolution by reduplication followed by sequence divergence (see, for example, Regier et al. 1978; Jones and Kafatos 1980, 1982). These multigene families are themselves related, and constitute a superfamily, with one branch encompassing the A, CA, and HcA families, and the other the B, CB, and HcB families (Rodakis and Kafatos 1982; Regier et al. 1983; Goldsmith and Kafatos 1984; Iatrou 1984; Rodakis et al. 1984; Eickbush et al. 1985; Burke and Eickbush 1986; Spoerel et al. 1989; Lecanidou et al. 1986). Clear diagrams of sequence relationships and nomenclature in the moth chorion gene superfamily were given by Lecanidou et al. (1986), who called the two branches α and β.

Interestingly, in both *M. sexta* (Regier and Vlahos 1988). and *S. nonagrioides* (Orfanidou C and Hamodrakas SJ unpubl. data), three major protein classes are also resolved on SDS-polyacrylamide gels, the A's, B's, and C's, with similar molecular weights to the corresponding classes of silkmoth chorion. Unfortunately, sequence data are not available to date for members of these classes; however, they are predicted to share sequence homologies with silkmoth chorion proteins (Regier and Vlahos 1988).

Primary sequences have been determined for all six silkmoth gene families and their products (references listed above), comprising a substantial amount of fascinating and, we hope, interpretable information.

Protein sequence comparisons (Fig. 18) and predictions of secondary structure (cf. below) have revealed that chorion proteins have a tripartite structure (Hamodrakas et al. 1982a, b, 1988; Fig. 19). A central domain is highly conserved within each family and can be recognized as highly homologous between families of the same branch (α or β).

A and B central domains show, however, distant similarities, suggesting that the chorion genes constitute a superfamily derived from a single ancestral gene (Lecanidou et al. 1986 and references therein).

---

b Electron micrograph of a thin cross-section cut through an aeropyle channel in the outer chorion of follicle 15/22. The extended villi (*v*), which are used as a mold for the formation of the channel, are closely packed and surrounded by a loose fibrous matrix (*f* filler); *e* follicle cell. *Bar* 0.3 µm. **c** Thin section of the top of an aeropyle of 19/22 follicle. A bundle of microvilli (*v*) traverses both the distupted sieve layer and the osmiophilic layer (*) forming the basis for a future air channel. The aeropyle is formed under a cell junction (*arrow*) of the follicular epithelium *e*. *Bar* 0.3 µm. **d** Thin section of an aeropyle (*ap*) right under the osmiophilic layer (*) of follicle 22/22. Inside the villi (*v*) of the channel, numerous cytoplasmic ribosomes can be resolved. Note also the loose fibers (*arrows*) of the channel filler. *Bar* 0.3 µm. **e** A micrograph indicating that the furrow (*arrow*) between the osmiophilic knobs (*) of the outer surface of the chorion is a site where an aeropyle (*ap*) is formed. *mv* microvilli (follicle 21/22). *Bar* 0.4 µm. (Papanicolaou et al. 1986)

a

pc18 YGCGCGCGLGGYGGLGYGGLGYGGLGYEGTGACLG | EYGGTGIGNVAVAGELPVAGKTAVGGQVPIIGAVGFGGTAGAAGCVSI ¦ AGRC | GGCGCGCGRGIY

pc609 VCRGGLAAPACGCGGLGYEGLGLGALGYEGIGYGAGWAGTACG | SYGGEGIGNVAVAGELPVAGKTAVPGQVPVIGAVDFCGPATAGGCVSI ¦ RGRC | CGCGCGCGSSYAY

pc292 VCRGGLGLKGLAAPACGCGGLGYEGLGYGALGYDGLGYGAGWAGPACG | SYGGEGIGNVAVAGELPVAGTTAVAGQVPIIGAVDFCGRANAGGCVSI ¦ GGRC | TGCGCGCGSSYPY

m2774 ...GGGWNGWNGLGGGWNGLGVGWSGLDGGYGGGCG | SYGGEGIGNVGVADELPVGGVTAVGGRVPIIGGVEYGGPARAAGAVSI ¦ CGHC | APTCGCGRAGLGGYY

m1911 QCCGCGCGGGCGCG | CYGGEGDGNVNVCGELPVCGETCVCGRVPICGSVCYGGPACASGCVSI ¦ CGRC | CGCGCGGCGGCGCGCGGCGGCGCGCGGCGGCGCGRRCCCC

b

B e 2G12 QYLSRVGCGCGGVGNGLGYGGIGYNGYVGNDIGAAGALGA | SNGGCLNVVSSSAAPTS ¦ LGVASENSYEGTVGVCGNLPLLGTAIVTGEFSTGGLGGIN ¦| YGCGNGAVGITAEERAGIGYAGGLGYGSGYGLGYGGYAGNGCGCGGAY

B m 2807 GLGGGCGRGF | S.GGGLPVATASAAPTG ¦ LGIASENRYEGTVGVCGNLPFLGTADVAGEFPTAGIGEID ¦| YGCGNGAVGITREGGLGYGAGYGDGYGLGYGGYGGGYGLGYGGYGGCGCGCG

B m 1768 GGLGYGGLGGGCGRGF | S.GGGLPVATASAAPTG ¦ LGIASENRYEGTVGVSGNLPFLGTADVAGEFPTAGIGEID ¦| YGCGNGAVGITREGGFGYGAGYGDGYGLGFGGYGGGYGLGYGGYGGCGCSWGY

B Hc-B12 TGCGCCCRGCGCGCGGCGC | GCCENFRVCSNSAAPTG ¦ LSICSENRYKGDVCVCGEVPFLGTADVCGNMCSSGCGCID ¦| YGCGNGCVGITRSCGGCGCGCGGCGCGCGGCGCCGCSCGRSCCGVY

B pc401 QCLGRWGPGLGRCGGCGGCDGWGGRLGYGAGIGEIGLGCGLEA | SYGGGLGVASASAVPV (P) ¦ GLGVASENMYEGCVGVAGNLPFLGTAGVEGVFPTAGAGVIN ¦| YGCGDGAVGITSEGGYGGLGYGGLGYEGVGGYGLGYGGYGLGGCGCGCGRYL

C pc404-H12 IGREAIVGAGLQGPFGGPWPYDALSPFDMPYGPALPAMSCGAGSFGPSSGFAPAAA | YGGGLAVTSSSPIST (P) ¦ GLSVTSENTIEGVVAVTGQLPFLGAVVTDGIFPTVGAGDVW ¦| YGCGDGAVGIVAETPFASTSVNPAYGYGGAIGGGVPYNSYGPIGYGGCGYNALY

Fig. 18. a Sequence of typical A and HcA proteins determined from complementary DNA clones: pc18, pc609, and pc202 are members of the A family from *A. polyphemus* (Jones and Kafatos 1982), m2774 is an A component from *B. mori* (Rodakis et al. 1982) and m1911 is an HcA component from *B. mori* (Rodakis and Kafatos 1982). *Unbroken vertical lines* flank the portions of the sequences considered in this chapter, and a *broken line* indicates a slightly different border for the previously defined central domain (Jones and Kafatos 1982). *Dots* indicate that the m2774 sequence is incomplete. (Hamodrakas et al. 1985). **b** Sequences of typical B, HcB, and CB proteins. *Unbroken lines* mark the borders of the central domain and *broken lines* flank the portions of the sequences considered in this work. Proteins Be 2G12, Bm2807, and Bm1768 are representative proteins of the B family of *B. mori*, B Hc-B12 of the HcB family of *B. mori*, B pc401 of the B family of *A. polyphemus*, and C pc404-H12 of the CB family of *A. polyphemus*. (Hamodrakas et al. 1988)

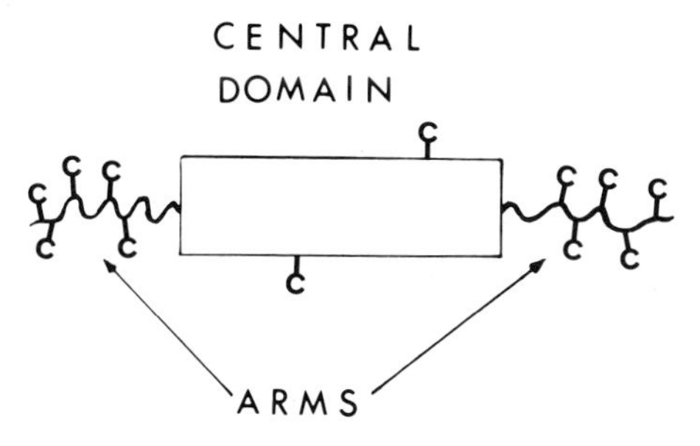

Fig. 19. Schematically, the tripartite structure of the silkmoth eggshell proteins. A central highly conservative and regularly structured domain and two, more variable, flanking "arms", particularly enriched in cysteine (*C*), constitute each protein. (Hamodrakas 1984)

The flanking amino- and carboxyl-terminal domains or "arms" are more variable and are marked by the presence of characteristic tandemly repetitive peptides that do not appear in the central domain. Nevertheless, "arms" also show similarities both within and among chorion families.

Ultimately, we want to relate the primary sequences and tripartite composition of chorion proteins to the fibrous structures they assume, and the regular assemblies of these fibers. Although chorion is very complex, this undertaking is facilitated by the evolutionary relatedness of the components. Limited evolutionary variations amount to information that can help specify further the fundamental molecular properties of these proteins.

Lepidopteran chorion becomes insoluble by disulfide bond formation between its constituent proteins (Kawasaki et al. 1971). This process has been called cementing (Blau and Kafatos 1979), and has been demonstrated to occur in a relatively short period (8 h) around the time of ovulation (Hamodrakas et al. 1982b, Regier and Wong 1988).

In contrast, developing chorions isolated from all stages of choriogenesis prior to ovulation can be almost completely dissolved with protein denaturing agents only (e.g. guanidinium chloride, SDS, 6 M urea) without a sulfhydryl reducing agent (e.g., dithiothreitol or β-mercaptoethanol) (Kafatos et al. 1977; Regier and Wong 1988; Orfanidou C and Hamodrakas SJ, unpubl.).

Several important findings for silkmoth chorion assembly emerged from the work of Regier and Wong (1988), who have tried to find in vivo patterns of disulfide bond formation. These can be summarized as follows:

During choriogenesis, low molecular weight disulfide-bonded chorion protein multimers form that are soluble in the absence of a reducing agent.

Multimers generally contain two to six protein subunits, with dimers and trimers the most abundant. Within multimers, only a small portion of cysteine residues per protein are involved in disulfide bond formation. These multimers, that assemble inside follicle cells, within minutes after protein synthesis, are, apparently, secreted normally (secretion rate is the same as that of monomers), and remain intact (their sizes and concentration do not increase) throughout choriogenesis.

Only a relative minor fraction of the total chorion (5–10% of total) proteins participate in multimer formation. Even so, all classes of chorion proteins are present in multimers. Furthermore, certain linkage patterns are preferred. In particular, class B proteins cross-link abundantly with A's and with themselves, but A's do not abundantly cross-link with themselves.

This evidence suggests that multimers represent normal intermediates in the pathway of chorion protein assembly.

Our opinion is that they may serve as "nuclei" in the "crystallization" process of self-assembly, "guiding", perhaps, the modes of packing of monomers by non-covalent weak interactions, and presumably playing an important role in the "cementing" process as well.

However, multimer formation during choriogenesis and "cementing", which occurs around the time of ovulation, appear to represent quite distinct phenomena, despite the fact that they both depend on formation of disulphide bonds.

Biochemical data on fish eggshell proteins are unexpectedly poor. Apart from papers determining the amino acid composition of fish eggshell proteins, which indicate the relative abundance of Glu, Asp, Pro, and of certain hydrophobic and polar residues (e.g., Leu, Ala, Thr) in these proteins (Ohzu and Kusa 1981; Kobayashi 1982), to our knowledge, no other significant work on fish eggshell protein biochemistry has been published.

Apparently, attempts to study fish eggshells are hindered by the fact that isopeptide bonds between glutamic and/or aspartic acid and lysine/arginine residues (peptide bonds formed between the side chains of these residues, e.g., Glu-Lys) render fish eggshells insoluble to normal denaturing and reducing agents and/or their combinations. It has been shown conclusively that these strong covalent bonds are primarily responsible for the process of "hardening" in the fish eggshell, a result of the so-called cortical reaction thought to occur after water activation (Hagenmaier et al. 1976; Lonning et al. 1984; Davenport et al. 1986).

We, ourselves, have experienced great difficulties in solubilizing eggshells isolated from *S. gairdneri* eggs and characterizing their constituent proteins on SDS-polyacrylamide gels. Recently, we have been able to obtain electrophoretic patterns, which suggest the presence of a complex set of proteins in the fish eggshell, with molecular weights in the range 12–100 kDa, (with prominent bands at approximately 12, 18, 32, 35, 43, 48, 54, 70, 85 and 96 kDa some of which might be arising from multimers), but, these data should only be considered as preliminary (Papadopoulou P and Hamodrakas SJ, unpubl.).

4 Elucidation of Eggshell Protein Structure

Between the detail revealed by low resolution (20–30 Å) electron microscopy and the wealth of information available in the determined amino acid sequences of silkmoth chorion proteins there is an information gap. This is a well known problem in fibrous proteins (e.g., keratins, collagen, etc.) and their supramolecular assemblies; it has been proved to be an extremely difficult task to bridge this gap with the use of existing experimental methods (Fraser and McRae 1973; Squire and Vibert 1987).

Our work on silkmoth chorion and similar-in-structure proteinaceous eggshells has been focused in attempts to partially "fill" this gap.

Ideally, this can be achieved by isolating, crystallizing, and solving the three dimensional crystal structure of individual protein components by X-ray crystallographic methods near atomic resolution, and, in the next step, determining the modes of protein-protein interactions for the formation of fibrils and of higher order helicoidal structure. Unfortunately, this approach seems to be unrealistic at present, since it requires the use of suitable single crystals of fibrous chorion proteins which cannot be produced easily (if at all!) and for various other obvious reasons.

Alternatively, a promising approach would be to isolate individual protein components, from suitable fibers and study these fibers by X-ray diffraction to determine the major structural features of the molecules. However, this is a difficult task because of the substantial complexity of chorion proteins. Nevertheless, it remains one of our future aims although our preliminary attempts towards achieving this goal have failed.

Therefore, we decided to perform an "anatomy" of silkmoth chorion protein sequences, hoping to determine folding patterns and modes of protein-protein interactions, utilizing information "hidden" in the primary sequences. We considered it likely that this undertaking might be facilitated by the possible existence of periodicities (repetitions or tandemly repeating motifs) in the primary sequences of silkmoth chorion proteins, a characteristic and common feature of fibrous protein structure (Fraser and McRae 1973).

In the last two decades, the amino acid sequences of >3000 proteins have been determined (Barker et al. 1986), and also the three-dimensional structure of >300 proteins has been elucidated mainly by single crystal X-ray crystallographic methods and more recently by NMR methods as well, at atomic or near atomic resolution (Bernstein et al. 1977). Since experimental findings indicate that all the necessary information for a protein to fold into its native structure is coded into its amino acid sequence (Anfinsen 1973), several attempts have been made to predict three-dimensional structure from sequence (Taylor 1987 and references therein). In these attempts the correct prediction of elements of secondary structure, frequently, plays a key role, since these elements may represent the initial nuclei in the process of protein folding.

For various other reasons, not mentioned here but described extensively in recent reviews (see Hamodrakas 1988 and references therein) it is also important to predict correctly the secondary structure of proteins from their sequence alone.

In the case of chorion proteins, there is also another reason which makes valuable the correct prediction of their secondary structure elements: the packing of protein molecules, which dictates self-assembly, is known to depend on the interactions of their secondary structure elements (α-helices and β-sheets). Therefore, the first step towards unraveling the modes of interaction of protein molecules for the formation of helicoidal proteinaceous eggshells is the determination of eggshell protein secondary structure.

Our approach was both theoretical (secondary structure prediction) in cases where amino acid sequences were available and experimental (X-ray diffraction and laser-Raman, infrared and circular dichroism spectroscopy) to: (1) verify theoretical results and (2) determine secondary structure when protein primary sequence information was lacking.

In the course of our investigation an attractive idea emerged: the possibility of the existence of a universal folding pattern which dictates the formation of the helicoidal architecture in proteinaceous eggshells, biological analogs of cholesteric liquid crystals. All evidence, both theoretical and experimental, in the systems studied so far, favors this notion.

4.1 Secondary Structure Prediction

Although amino acid sequence dictates native conformation, secondary structure predictions based on primary sequences should be undertaken with full awareness of their limitations. Even in the case of globular proteins, for which they were initially developed and applied, the accuracy of prediction methods is limited (Chou and Fasman 1978). The methods fail altogether when short precise repeats are present, and thus have been applied only rarely to structural proteins in which internal repeats are widespread. However, comparisons of evolutionarily related sequences or imprecise internal repeats are invaluable in overcoming these limitations: limited variation should reduce the "noise" and help identify consistent structural features (e.g., Hamodrakas and Kafatos 1984; Hamodrakas et al. 1989; Aggeli et al. 1991).

Several prediction algorithms have been published and can be classified into two categories — statistical or stereochemical — but their success has been rather limited. It has been claimed that combined prediction schemes provide a higher degree of accuracy than individual prediction methods and on this basis we have constructed and published an algorithm which combines six of the most popular and successful prediction methods. The prediction package runs on microcomputers and requires as input only the amino acid sequence of the protein(s) under study (Hamodrakas 1988).

The algorithm was applied to the determined amino acid sequences of the A, B, and C families of silkmoth chorion proteins. Detailed results on representative, predominant members of these families are shown in Figs. 20, 21, 22 (Hamodrakas et al. 1982a; Regier et al. 1983).

For each protein individual predictions for α-helix (α), β-pleated sheet (β) and β-turns (T) were made according to six prediction methods. Joint prediction histograms were then constructed by tallying individual predictions since they are more dependable than individual prediction schemes. Structures predicted by three or more methods (out of six, or out of five in the case of turns) were considered as most probable and are shaded. To make secondary structure prediction plots most comparable sequences within each family or sequences from different homologous families were aligned.

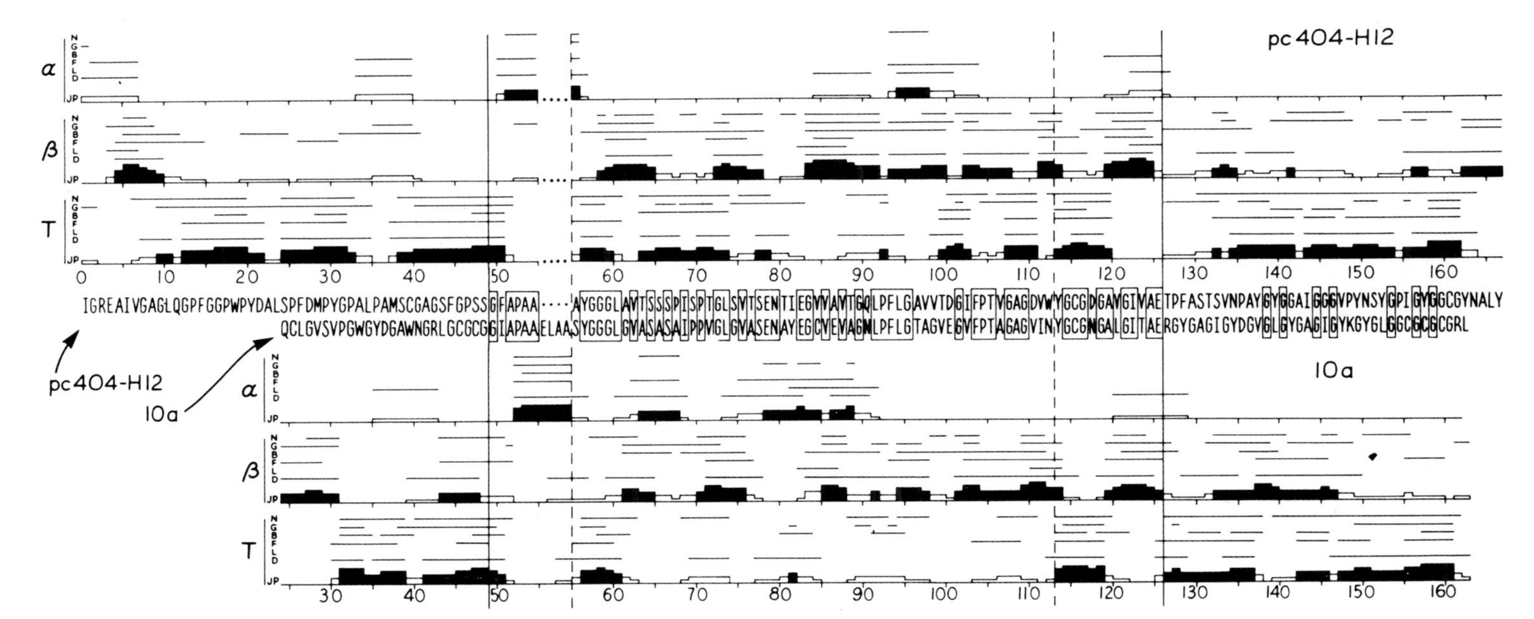

Fig. 20. Secondary structure predictions for protein sequences of the pc404-H12 (CB family) and 10a (B family) chorion components. Individual predictions for α-helix (α), β-sheet (β), or β-turn (*T*) are shown by *horizontal lines*, as derived according to Nagano (*N*), Garnier et al. (*G*), Burgess et al. (*B*), Chou and Fasman (*F*), Lim (*L*), and Dufton and Hider (*D*). See Hamodrakas et al. (1982a) for a complete listing of references. Joint prediction histograms (*JP*), constructed by tallying the individual predictions, are also shown. The most probable structures, predicted by three or more methods, are *shaded*. Sequences are numbered from the amino- to carboxyl-terminus of pc404-H12. Identical residues are enclosed in *boxes*. A gap between positions 55 and 56 in pc404-H12 is indicated by *dots*. The *solid vertical lines* indicate the borders of the clear interfamily homologies; the *dashed vertical lines*, the conservative central region of B protein sequences. (Regier et al. 1983)

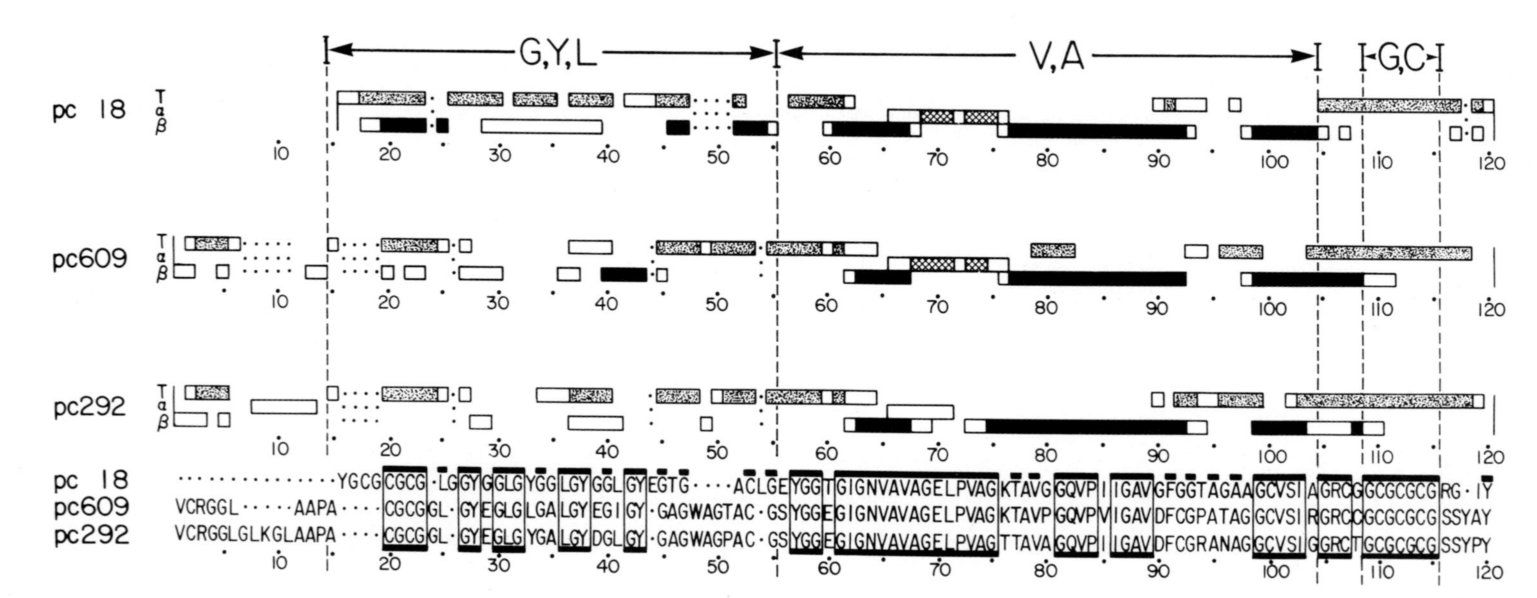

Fig. 21. Summary of joint prediction histograms for secondary structure in proteins of the A family of silkmoth chorion proteins. Structures predicted by two methods are shown as *open rectangles*, while the most probable structures predicted by three or more methods are shown *solid* (β-sheet), *hatched* (α-helix), or *stippled* (β-turn, *T*). The amino- and carboxyl-termini are indicated by *short vertical lines*, and regions enriched in certain amino acids are outlined by *dashed lines* and named accordingly (*G*, *Y*, *L*; *V*, *A*; *G*, *C*). Sequences are aligned and numbered, and necessary gaps are indicated by *dots*. The actual sequences are presented at the bottom, according to the IUPAC-IUB one-letter code, as follows: *A* Ala; *C* Cys; *D* Asp; *E* Glu; *F* Phe; *G* Gly; *H* His; I Ile; *K* Lys; *L* Leu; *M* Met; *N* Asn; *P* Pro; *Q* Gln; *R* Arg; *S* Ser; *T* Thr; *Y* Tyr; *V* Val; *W* Trp. *Blocks* outline two or more residues which are invariant in these and four additional related sequences (18b, 18c, 292a, 292b); individual invariant residues are *overlined*. (Hamodrakas et al. 1982a)

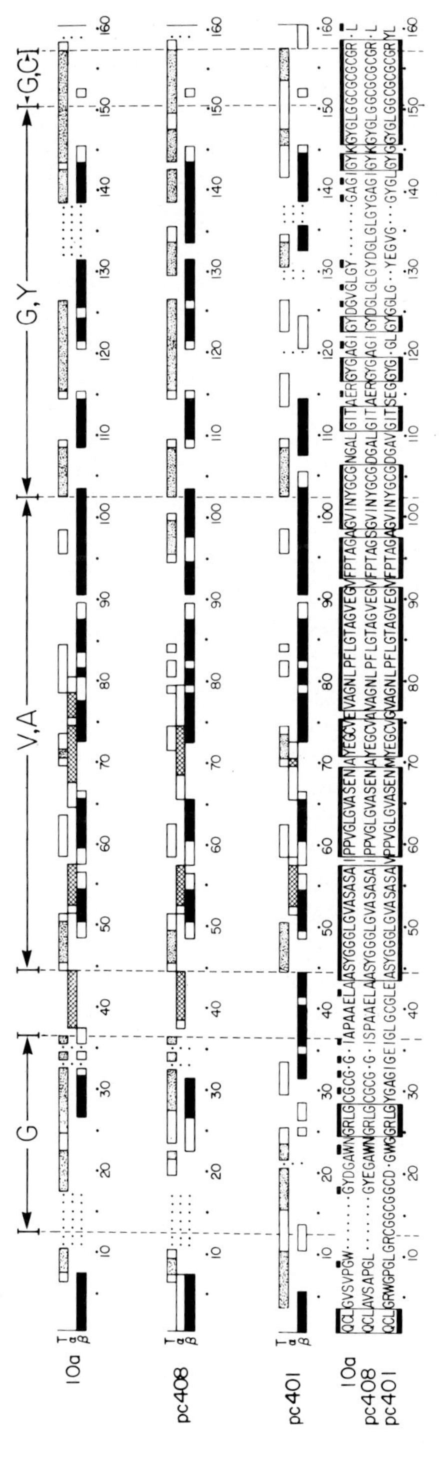

Fig. 22. Summary of joint prediction histograms for secondary structure in proteins of the B family. Details as in Fig. 21. Blocks and overlinings refer to four additional sequences (pc10, 10b, 401a, 401b). (Hamodrakas et al. 1982a)

The diagrams are accompanied by the actual protein sequences presented in the one letter code and displayed in a manner emphasizing regions rich in certain amino acids and also sequence conservation (Jones and Kafatos 1982).

The prediction results clearly indicate that in members of the A, B, and C families β-sheet structure predominates, whereas α-helix is almost totally absent. β-Turns are predicted frequently, and may often connect short, presumably antiparallel β-sheet strands. The evidence that the β-sheets are antiparallel has been provided from laser-Raman and infrared spectroscopy data (see Sect. 4.3) and for purely topological reasons (short connections between β-strands).

The proteins can be divided into a number of distinct regions or "domains" according to the degree of evolutionary constancy in sequence, the amino acid composition and secondary structure features. They appear to have a tripartite structure (Figs. 19, 20, 21, 22): a central domain ("core") highly structured into β-sheet structure, evolutionarily highly conserved, rich in Val, Ala, and two more variable flanking "arms" (or "tails"), varying in length, particularly enriched in Cys in the most abundant families A and B (and in their homologous Hc's in *B. mori*), which appear to be less structured and presumably reflect protein and/or family specific functions.

The most intriguing feature of prediction, which, however, was the clue that led us to the proposal of the detailed protein structural models presented below (cf. Sect. 5) was an apparent periodicity of β-sheet strands in the central domain evidenced by the nearly periodic occurrence of maxima for β-sheet predictions (Fig. 20; Hamodrakas et al. 1982a). In between these maxima, β-turns were also predicted. This feature, which is obscured in the summaries (Figs. 21, 22), indicated a regular model of alternating β-strands-β-turns, in the central domain.

It is interesting to note that insect eggshells which do not exhibit a helicoidal architecture contain proteins with a tripartite structure (e.g., *D. melanogaster*; *Ceratitis capitata*; Hamodrakas et al. 1989; Aggeli et al. 1991). A tripartite structure is also a characteristic feature of several other fibrous proteins, e.g., avian — feather and scale — keratins (Fraser and McRae 1973, 1976; Gregg et al. 1984).

On the basis of partial sequencing data, Regier et al. (1978b) pointed out that, as in avian keratin, in silkmoth chorion proteins, cysteines are preferentially localized near the two ends of the polypeptide chain. The complete chorion sequences confirmed this feature and the tripartite structure which emerged after secondary structure prediction further emphasizes the analogy with avian keratins.

In both feather and scale keratin, a fiber-matrix texture can be detected under the electron microscope when appropriate stains are used. Amino- and carboxyl-terminal ends, which are rich in cysteine, are thought to provide the stain-absorbing matrix, whereas it was suggested that the regular β-sheet central, conservative portion of keratin constitutes the fibrils (Fraser and McRae 1976).

In analogy with avian keratins, we proposed that the central, highly structured, evolutionarily conserved domains of silkmoth chorion proteins constitute the fibrils and the cysteine-rich protein "arms" the matrix seen in electron micrographs of silkmoth chorion (Hamodrakas et al. 1982a). This proposal appeared to be valid for a number of years and yet might finally prove to be correct during several developmental stages of choriogenesis. However, it should be slightly revised to take into account novel information, provided by X-ray diffraction, freeze-fracturing (Hamodrakas et al. 1986) and protein model-building studies (Hamodrakas et al. 1988) (see also Sects. 2.2.1 and 5).

The structure of silkmoth chorion protein "arms" deserves special attention. Protein arms were for a long time thought to be less structured than the central domain, as prediction results indicate (Hamodrakas et al. 1982a). However, since they contain tandem repeats and prediction methods fail in such cases, this interpretation may be entirely erroneous! On the contrary, the periodical structure of the peptides constituting the arms (see, e.g., Lecanidou et al. 1986) strongly favors a regular type of structure. The question, of course, is: what type of structure? To provide a plausible answer, let us examine briefly the primary structure features of the "arms".

Thus, most members of the abundant A and B families contain reduplications of Gly-Cys (G-C) repeats which are thought to serve for cross-linking chorion proteins via disulfide bonds. These are predicted mostly as β-strands. The alternation of "small" (G) and "bulky" (C) residues is reminiscent of the β-sheet structure in silk fibroin (Marsh et al. 1955), with small and bulky residues pointing to opposite sides of the β-sheet. In these and other parts of the molecules (even in the central domain), especially in sequences predicted as β-sheet strands, relatively small residues (G, A, or T) tend to alternate with bulky residues (V, L, I, Y). This might be important in chorion for the packing of successive β-sheets (Hamodrakas et al. 1982a).

In both the left and right arms, "late" *B. mori* HcA and HcB sequences are highly similar, consist almost entirely of glycine and cysteine, and contain almost exclusively G-G-C and G-C subrepeats which form a $(GGC)_2$-$(GC)_2$ major repeat (Burke and Eickbush 1986).

The "early" CA and CB protein "arms" are proline-rich and cysteine-poor (Lecanidou et al. 1986).

However, the most striking feature in the arms of the abundant A and B proteins is the existence of tandem repeats of the pentapeptide GYGGL (or LGYGG) or its variants (Regier and Kafatos 1985). The repeat array is frequently predicted as β-sheet, although an alternative prediction of a series of β-turns GYGG separated by a single residue (L) is possible (Hamodrakas et al. 1982a). In scale keratin, a repetitive sequence of the type $(GGX)_n$ (where X = F, L or Y) exists (Walker and Bridgen 1976) and it is thought to form β-sheet structure, possibly facilitating the characteristic hexagonal packing of scale keratin (Stewart 1977). Therefore, it appears that, the fiber-matrix systems of feather and scale keratins are analogous to silkmoth chorion at all structural levels: primary and secondary structure of protein components and also ultrastructure (Hamodrakas et al. 1986). For historical reasons, we should perhaps mention at this point that the term "ichthulokeratin" was introduced by Young and Inman (1938) to describe the eggshell proteins of Salmon eggs.

The protein sequence data banks NBRF and SWISSPROT were searched for the possible occurrence in other proteins of the tandem motifs (e.g., LGYGG) appearing in silkmoth chorion protein arms, utilizing the commands SCAN and MATCH of the Protein Sequence Query (PSQ) program. This is the main access and retrieval program of the Protein Identification Resource (PIR; Orcutt et al. 1983). The findings of this search were unexpected and require further investigation: the pentapeptide LGYGG and its variants are constituents of proteins (or portions of proteins) exhibiting a fiber-matrix texture of ultrastructure. Thus, it is found repeatedly at the carboxyl-terminal end of scale keratin, which consists of 4×13 amino acid residue repeats, most probably adopting a characteristic antiparallel β-sheet structure comprising the matrix (Gregg et al. 1984), and also in the amino-terminal sequences of several cuticular matrix proteins from the locust *Locusta migratoria* (Hojrup et al. 1986). Insect cuticle is an extracellular layer, exhibiting a helicoidal architecture with a fiber-matrix texture, surrounding the whole animal. The cuticle consists of fibers of chitin embedded in a protein matrix (Neville 1975).

Furthermore, a study of the literature shows that the protein oothecin, secreted by the left colleterial gland of the cockroach *Periplaneta americana* to form the eggcase, consists, almost entirely, of tandem repeats of this pentapeptide (Pau 1984). Similar peptides appear also in the sequence of *Schistosoma mansoni* eggshell proteins and are predicted to form a β-sheet type of structure (Vanderlei et al. 1989) and also in the β-sheet structure of spider silk as well (Xu and Lewis 1990). Homologies of small portions of these proteins are summarized in Table 1.

In all the proteins (or protein segments) of Table 1, the peptide LGYGG and its variants, apparently, adopt a β-sheet type of structure. Its appearance in the structure of proteins which exhibit a fiber-matrix texture most probably indicates that this peptide is well tailored to play its role in the matrix of fiber-matrix systems: it may be a common architectural feature of matrix, or, if the alternative interpretation of fiber and matrix is taken (cf. Sects. 2.2.1 and 4.2), its presence perhaps ensures a better packing of the system components. To gain further insight, we performed a comparison of silkmoth chorion proteins with scale keratins. A sample of such a comparison performed with the FASTP program of Lipman and Pearson (1985) is shown in Fig. 23. To our surprise, it was found that the abundant silkmoth chorion protein Bpc401 (JBA041) shares a 41% similarity with chicken scale keratin (KRCHS) over a segment of 87 residues. This relatively high homology awaits further interpretation, in terms of both structure and evolution.

Summarizing the evidence presented above, we propose that the predominant secondary structure of chorion protein "arms" is that of β-sheet.

Table 1. Homologies between segments of proteins containing the pentapeptide GYGGL. Gaps necessary for alignment are represented with dashes. References are given in the text

Protein	Sequence segment
Silkmoth chorion Bpc401	. . . G L G Y G G – Y G L G G . . . .
Silkmoth chorion Bpc401	. . . G L G Y G G L – G Y G G . . . .
Silkmoth chorion Apc18	. . . G L G Y G G L – G Y G G . . . .
Scale keratin	. . . S L G Y G G L Y G Y G G . . . .
Locust cuticle Pr37	. . . G L G Y G G – Y G Y G . . . . .
Spider silk	. . . Q G G Y G G L G G Q G . . . . .
Oothecin	. . . G L G Y G G L – G Y G G . . . .

```
                80          90          100         110         120         130
JBA041     GLGVASENMYEGCVGVAGNLPFLGTAGVEGYFPTAGAGVINYGCGDSAVGITSEGG
                     ...:  ... :..:.....   .. : : . ::.:....: ..  :
KRCHS      FPGPILSSFPQDSVVGSSGAPIFGGSSLGYGGSSLGYGGL-YGYGGSSLGYGGLYG
             50          60          70          80          90          100

                  140         150         160         170
JBA041     YGG--LGYGGLGYEGVGGYGLGYGG-YGLG-GCGCG-SGRYL
           :::  :::::: : : :: .::::: :: : ..:.: :..:
KRCHS      YGGSSLGYGGL-Y-GYGGSSLGYGGLYGYGRSYGSGYSSPYSYRYNRYRRGSCGPC
                   110         120         130         140         150
```

Fig. 23. Sequence similarities between silkmoth chorion protein Bpc401 (*JBA*041) and chicken scale keratin (*KRCHS*). The alignment optimized by the program FASTP (Lipman and Pearson 1985) is denoted by a *colon* for an identity and a *dot* for a conservative replacement. Insertions are marked with a *dash*. A z value of 3.69 standard deviations above mean was calculated for the initial score and a z value of 4.88 standard deviations above mean was calculated for the aligned score with the program RDF (Lipman and Pearson 1985). Fifty random sequences were generated. z values of initial and optimized scores >3 are considered as possibly significant, whereas z values >6 are probably significant according to Lipman and Pearson (1985)

Having predicted the secondary structure of silkmoth chorion proteins, experimental evidence was necessary to verify predictions. Experimental studies of eggshell protein structure were not confined, however, to silkmoths. Since they do not require previous knowledge of protein primary structure, they were extended to include helicoidal eggshells of other Lepidoptera (*M. sexta* and *S. nonagrioides*) species and of fish (*S. gairdneri*) as well. We shall briefly describe below the results of these studies.

4.2 X-Ray Diffraction Studies of Eggshells

X-ray diffraction studies of fibrous proteins usually provide information about their secondary structure and the modes of orientation and packing of protein molecules; frequently, these studies lead to the proposal of a protein structural model (Fraser and McRae 1973). However, samples should fulfill certain requirements: they should have intrinsic high order and diffract X-rays strongly to give interpretable X-ray diffraction patterns. Our experience showed that, unfortunately, both requirements are not easily met by the proteinaceous eggshells; it proved to be difficult to obtain useful X-ray diffraction patterns (Hamodrakas et al. 1983, 1986). Samples used for X-ray diffraction experiments were either hemispherical half-chorions, or stacked arrays of almost flat chorion fragments. High and low angle diffraction patterns were taken with the geometries shown in Fig. 24. Full details of the experimental procedures are given in the original publications.

In the "in-plane geometry" (Fig. 24a) the beam was parallel to the chorion surface and the planes of the stacked fibrils. In this case, the vertical axis of the diffraction pattern (meridian) corresponded to the radial axis of the chorion, i.e., revealed order along the axis from the inner to the outer surface; the horizontal axis (equator) of the diffraction pattern corresponded to the lateral axis of the chorion, i.e., revealed order within planes parallel to the surface. In the perpendicu-

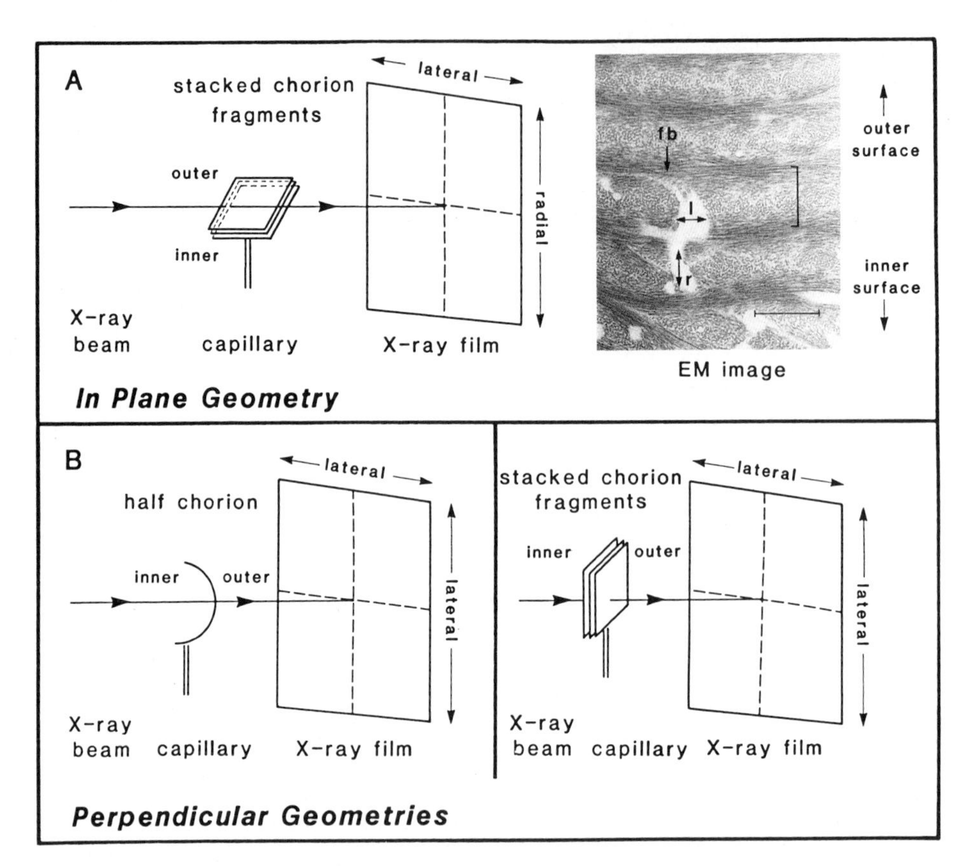

Fig. 24A, B. Geometries employed in the high angle X-ray diffraction experiment. In **A**, a stacked array of almost flat fragments, obtained from the hollow chorion sphere, is irradiated with the beam in the plane of the fragments, i.e., parallel to the outer and inner surfaces of the chorion. Note that the horizontal axis of the film is parallel to the fragments and thus reveals lateral periodicities, whereas the vertical axis of the film reveals radial periodicities (along the axis between inner and outer chorion surfaces). The electron micrograph is from an immature chorion and corresponds to the cut face that would be encountered by the beam. Fibrils (*fb*) are seen in orientations that vary with the plane; the *bracket* outlines one lamella (180°) rotation in fibril orientation), and the lateral (*l*) and radial (*r*) axes are indicated, as are the outer and inner chorion surfaces. *Bar* 500 nm. In **B**, the beam is oriented along the radial axis of the chorion hemisphere, or of stacked chorion fragments; consequently, both the horizontal and the vertical axes of the film reveal lateral periodicities. (Hamodrakas et al. 1983)

lar geometry (Fig. 24b) the beam was perpendicular to both the chorion surfaces and the planes of the stacked fibrils. In this case, both the vertical and the horizontal axes of the diffraction pattern revealed lateral order, within these planes.

In both orientations the high angle diffraction patterns had certain features in common (Fig. 25a,b). However, asymmetrical texture (scattering in an ellipse) was only observed in the in-plane geometry (Figs. 25a, 26, 27), indicating preferential orientation of the molecular chains. According to Kakudo and Kasai (1972, p. 404), scattering in an ellipse denotes scattering by stacked lamellar crystals with cylindrically symmetrical orientation, which is (locally) the case of helicoidal structure. Therefore, X-ray diffraction provides a direct proof of the existence of a helicoidal architecture.

Predictions of secondary structure based on known sequences of major silkmoth chorion proteins led us to the suggestion that β-sheet structure is predominant in the silkmoth chorion. The preponderance of β-sheet structure was also supported by the results of laser-Raman and infrared studies (Hamodrakas et al. 1982b, 1984, 1987; cf. Sect. 4.3). The X-ray diffraction studies further support this notion.

Both perpendicular and in-plane geometries yielded broad, nearly uniform rings, centred at 0.464 nm. This spacing is typical of antiparallel β-sheet structures

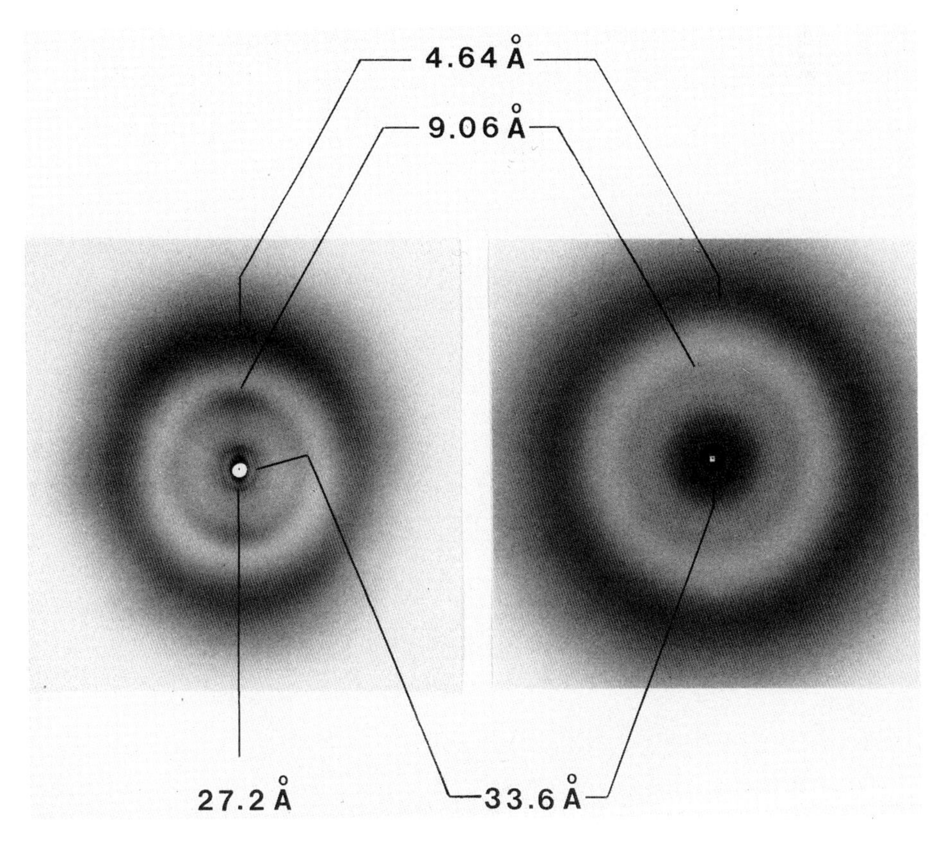

Fig. 25. High-angle X-ray diffraction patterns from mature silkmoth chorions of *A. polyphemus*. Pattern (*left*) is from the in-plane geometry (cf. Fig. 24a) and (*right*) from the perpendicular geometry (cf. Fig. 24b; half chorion). Note the presence of 0.464 nm, 0.906 nm, and 3.36 nm reflections which form rings in the perpendicular geometry. In the in-plane geometry, the same reflections occur but the 0.906 nm repeat is stronger and predominantly oriented along the radial axis. A 2.72 nm radial repeat is also seen, whereas the 3.36 nm repeat is confined to the lateral axis. (Hamodrakas et al. 1983)

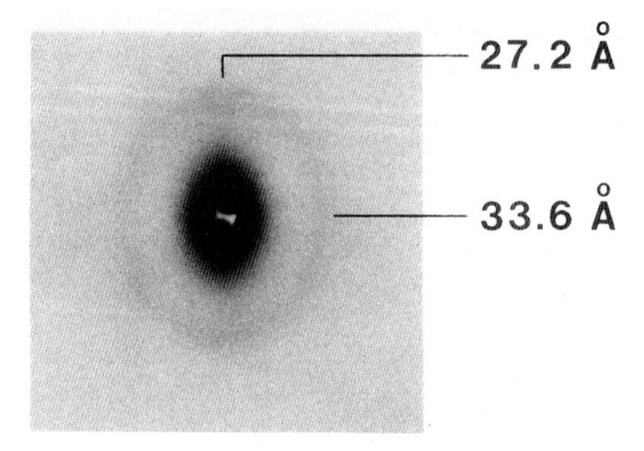

Fig. 26. Low-angle X-ray diffraction pattern from mature silkmoth chorion of *A. polyphemus*. Elliptical scattering, obtained using the in-plane geometry, suggests a helicoidal organisation (see text). This pattern confirms the results seen in the low-angle region of Fig. 25a; there is a 2.72 nm radial repeat and a 3.36 nm lateral repeat. (Hamodrakas et al. 1983)

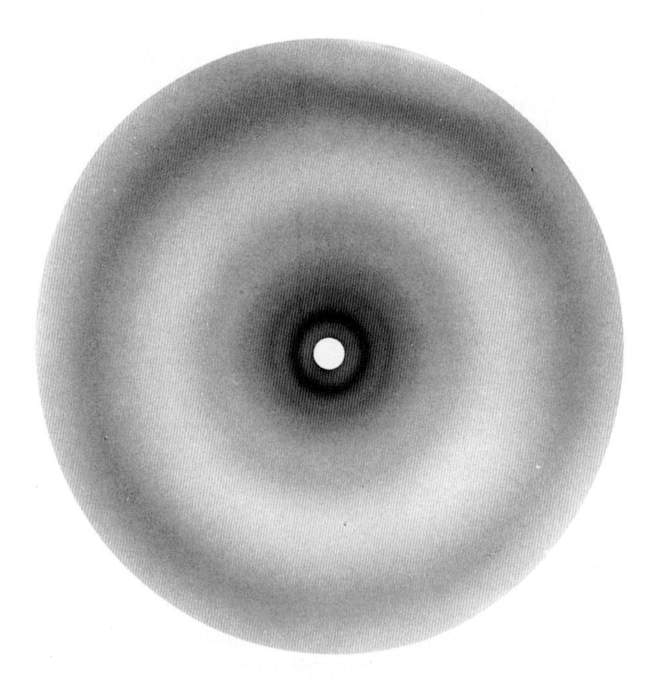

Fig. 27. X-ray diffraction pattern from mature silkmoth chorion of *A. polyphemus*. Incident beam parallel to the chorion surface which is horizontal (in-plane geometry). The plane of the film is vertical. The chorion sample was treated with 2% uranyl acetate for 60 min before irradiation. The concomitant significant increase in the intensity of the ca. 3 nm reflection can be seen. A toroidal camera was employed. Specimen to film distance was approximately 70 mm. (Hamodrakas et al. 1986)

corresponding to the interchain distance, between hydrogen bonded peptide chains of each sheet. A second ring was observed, at a 0.906-nm spacing, which may be attributed to the intersheet packing distance between regularly packed adjacent β-sheets. In the perpendicular geometry this ring was uniform and relatively weak (Fig. 25b). By contrast, in the in-plane geometry it included strong and broad reflections along the radial axis (Fig. 25a).

The X-ray diffraction patterns do not contain a ca. 0.54 nm reflection which could be representative of α-helix, confirming the suggested paucity of α-helical structure in the chorion (cf. 4.1). A third ring corresponded to spacings of ca. 3 nm. In the perpendicular geometry it was circular and corresponded to 3.36 nm. In the in-plane geometry it was oblong (elliptical), and in some patterns it was clearly resolved into a strong 2.72 nm reflection along the radial axis and a weaker 3.36 nm reflection along the lateral axis (Fig. 25a).

Low-angle X-ray diffraction patterns (Fig. 26) confirmed the existence of oriented 2.72 and 3.36 nm spacings in the in-plane geometry. These patterns revealed no other discrete reflections in the low angle region; diffuse central scatter was seen at very low angles, starting at 8 nm along the radial axis and 11 nm along the lateral axis.

We wish to draw particular attention to the remarkable similarities between the diffraction patterns presented here, and those previously reported for chicken scale keratin. These similarities include the prominence of interchain and intersheet reflections (0.47 and 0.94 nm^{-1} respectively, in the case of scale keratin); the presence of oriented reflections in the in-plane but not the perpendicular geometry; and the presence of oriented reflections in the vicinity of 3 nm^{-1}, in both high and low angle X-ray diffraction patterns obtained using the in-plane geometry (2, 2.3 and 3.5 nm^{-1} in the case of scale keratin).

When considered together with the electron microscopic evidence (Fig. 24a), the X-ray diffraction patterns suggest substantial orientation of the β-sheets relative to the fibrils. Since the 0.906 nm^{-1} reflections are most prominent along the radial axis (in-plane geometry), and weak along the lateral axis (both in-plane and perpendicular geometries), it would appear that β-sheets tend to be stacked across the fibril, occupying planes parallel to the chorion surface (rather than say, being stacked along the long axis of the fibril). The weak 0.906 nm^{-1} reflections in the lateral axes could be ascribed either to the disrupted and oblique lamellae, or to stacking of β-sheets in nonradial orientations even in fibrils which are parallel to the surface. It might also arise if the β-sheets are not "flat", but twisted β-sheets (with the majority of their strands more or less parallel to chorion surface; cf. also Sect. 5). Twisted sheets and disrupted or oblique lamellae would explain the paradoxical observation that in the in-plane geometry the 0.464 nm^{-1} (interchain) reflection is not preferentially observed at a 90° angle relative to the 0.906 nm^{-1} (intersheet reflection) and/or nonstacked sheets or sheets stacked in various orientations. Oriented 0.94 nm^{-1} and unoriented 0.47 nm^{-1} reflections are also observed in the in-plane patterns derived from scale keratin [see Fig. 2b in Stewart (1977)].

Prominent reflections at ca. 3 nm^{-1} are also observed in the chorion patterns. In the perpendicular geometry they are unoriented and correspond to 3.36 nm, whereas in the in-plane geometry oriented 3.36 nm and stronger, 2.72 nm reflections are observed. One possible explanation for these reflections is that the 8–20 nm fibers seen by electron microscopy are aggregates of oriented ca. 3 nm filaments. The shape and stacking of these filaments would be asymmetrical in cross-section, resulting in the 2.72 nm radial and 3.36 nm lateral periodicities. Similar filaments, measuring ca. 3 nm in diameter, and spaced at distances of 3 to 3.5 nm, have been seen in both feather rachis and scale keratin (Stewart 1977).

To examine whether the ca. $1/3\ nm^{-1}$ reflections appearing in the X-ray diagrams correspond to periodicities between fibrous elements of chorion, a simple experiment was carried out, similar to that of Fraser and McRae (1959) in studies of feather keratin: the silkmoth chorion was treated with 2% osmium tetroxide or 2% uranyl acetate, which results in considerable deposition of osmium or uranium within the structure, as evidenced by the brown and yellow coloration, produced respectively. The high angle diffraction pattern is unaffected by these treatments and it may be concluded as in feather keratin, that deposition of osmium or uranium occurs mainly between, rather than within, the fibrous elements of the structure. At low angles, the principal effect is a dramatic intensification of the $1/3\ nm^{-1}$ reflection (Fig. 27) which indicates periodic fluctuations of considerable amplitude in the density of osmium and uranium in the structure, of the same order of magnitude.

If the $1/3\ nm^{-1}$ reflection is considered as indicating packing distances between chorion filaments, then the dramatic increase in the intensity of this reflection may be easily explained, since the contrast between the filaments and the matrix is considerably increased, if osmium tetroxide or uranyl acetate bind to the matrix intervening between the filaments. The concept that osmium tetroxide or uranyl acetate bind preferentially to the matrix is supported by the proposal put forward by Hamodrakas et al. (1982a), that the "less structured", high in cysteine content, variable arms of chorion proteins constitute the matrix. The high affinity of osmium tetroxide and uranyl acetate to cysteine is well known. However, an alternative explanation is that the filaments constituting chorion are formed by the hydrophobic cores of the folded proteins and that the matrix corresponds to the hydrophilic exterior of the proteins, with osmium tetroxide and uranyl acetate binding to the polar groups of the protein surface; groups of the protein surface include the side chains of glutamate, aspartate, and cysteinyl residues (Hamodrakas et al. 1985) to which uranyl acetate and osmium tetroxide preferentially attach (Blundell and Johnson 1976).

Supporting and conclusive evidence for the existence of ca. 3–4 nm fibrils in silkmoth chorion was provided by freeze fracturing studies (Hamodrakas et al. 1986; cf. Sect. 2.2.1) and recent transmission electron microscopy and optical diffraction data (Hamodrakas SJ and Ottensmeyer FP, in prep.).

Similar, but very weak X-ray diffraction patterns were obtained also from eggshells of the fish *S. gairdneri* (Papadopoulou P and Hamodrakas SJ, unpubl.) and *M. sexta* (Orfanidou C and Hamodrakas SJ, unpubl.).

In conclusion, X-ray diffraction studies confirm the prevalence of β-sheet structure in the chorion, reveal a degree of regular orientation (preferred packing) of β-sheets relative to its surface, provide a direct proof of its helicoidal architecture and suggest the existence of fibrils ca. 3–4 nm in diameter as its basic structural elements.

4.3 Laser-Raman and Infrared Spectroscopic Studies of Eggshells

Laser-Raman spectroscopy is a relatively new technique for examining changes in the frequency of emitted radiation due to molecular vibrations. It is related to, but distinct from, infrared (IR) spectroscopy, which examines absorption of radiation due to the same molecular vibrations. Both techniques have been empirically demonstrated to be quite sensitive to protein conformation; they are very powerful and reliable techniques for the determination of protein secondary structure. Laser-Raman spectroscopy can also provide a wealth of useful information about the state of certain amino acid residues in proteins and protein systems, thus offering distinct advantages compared to IR spectroscopy. The weak emission as opposed to absorption of radiation by water is another important advantage of Raman spectroscopy for studies of biological materials. (Parker 1971; Frushour and Koenig 1975; Spiro and Gaber 1977; Yu 1977; Carey 1982).

The eggshell is a relatively favorable structure for Raman spectroscopy studies. It consists almost exclusively of protein and functions as an essentially dry shell; thus it meets the criterion of high protein density which is required for adequate sensitivity in Raman studies. The absence of major admixtures, such as chitin or other carbohydrates, minimizes interference from other than protein vibrations. Furthermore, we have encountered only limited interference from fluorescence in our studies. These features have permitted analysis of the intact eggshell structure, as opposed to protein extracts, ensuring that the structural features observed reflect a physiological state.

Serious difficulties were encountered in our initial attempts to obtain eggshell samples [in the form of KBr pellets, containing about 2% (w/w) material, which was thoroughly ground in a vibrating mill before mixing with KBr] suitable for IR spectroscopy. This was due to the toughness, relative opacity and elasticity of the eggshells. Nowadays, however, useful IR spectra from eggshells are obtained in an almost routine manner.

To date, Raman spectra have been obtained from eggshells of *A. polyphemus* (Hamodrakas et al. 1982b), *B. mori* (Hamodrakas et al. 1984), *A. pernyi* (Hamodrakas SJ and Petrou A, unpubl.), and *S. gairdneri* (Hamodrakas et al. 1987; Papadopoulou P, Kamitsos EI, and Hamodrakas SJ, in prep.) and IR spectra from *A. polyphemus*, *B. mori*, *A. peryi*, *S. gairdneri*, *M. sexta*, and *S. nonagrioides* (Orfanidou C and Hamodrakas SJ, in prep.). Representative laser-Raman spectra from *A. polyphemus*, *B. mori*, and *S. gairdneri* eggshells are shown in Fig. 28a, b, and c, whereas Fig. 29 shows a representative IR spectrum from *S. gairdneri* (Hamodrakas et al. 1982b, 1984, 1987).

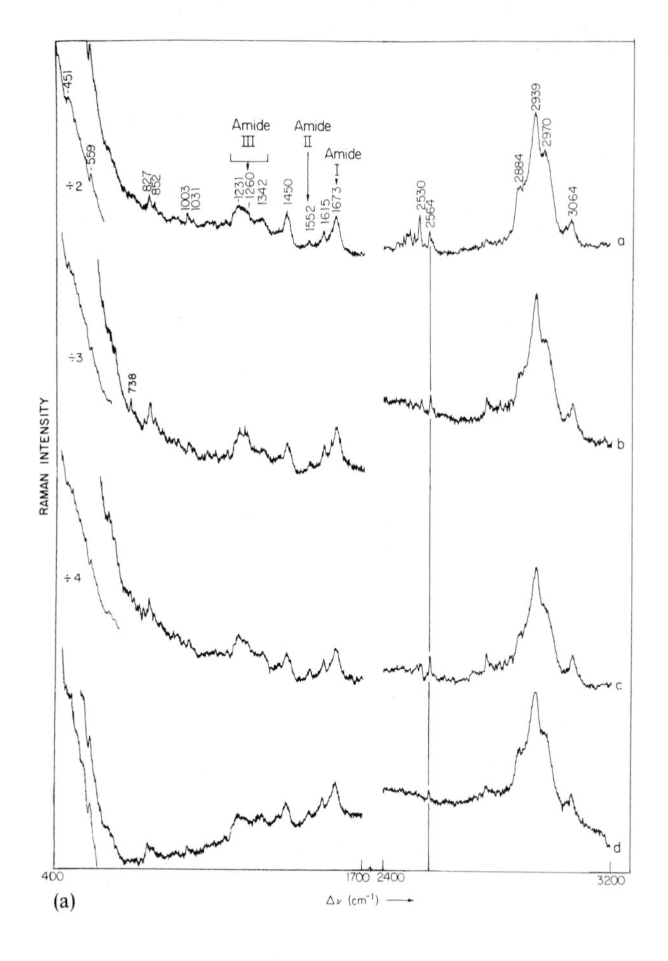
RAMAN INTENSITY
Amide III
Amide II
Amide I
451
559
827
852
1003
1031
1231
1260
1342
1450
1552
1615
1673
2530
2564
2884
2939
2970
3064
738
÷2
÷3
÷4
a
b
c
d
400
1700 2400
3200
Δν (cm⁻¹)

(a)

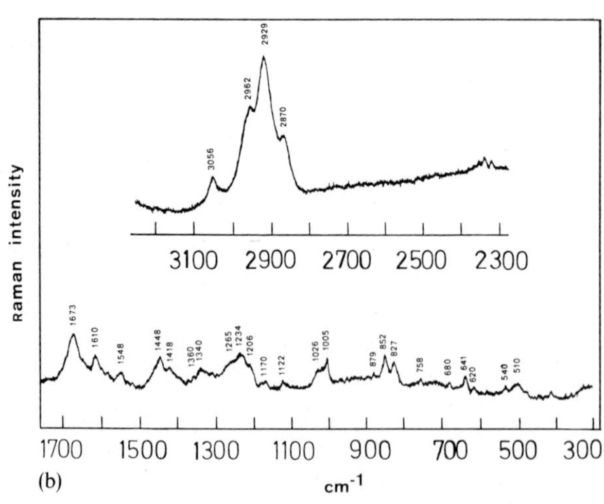
Raman intensity
2929
2962
2870
3056
3100 2900 2700 2500 2300
1673
1610
1548
1448
1418
1360
1340
1265
1234
1206
1170
1122
1026
1005
879
852
827
758
680
641
620
540
510
1700 1500 1300 1100 900 700 500 300
cm⁻¹

(b)

Fig. 28a, b

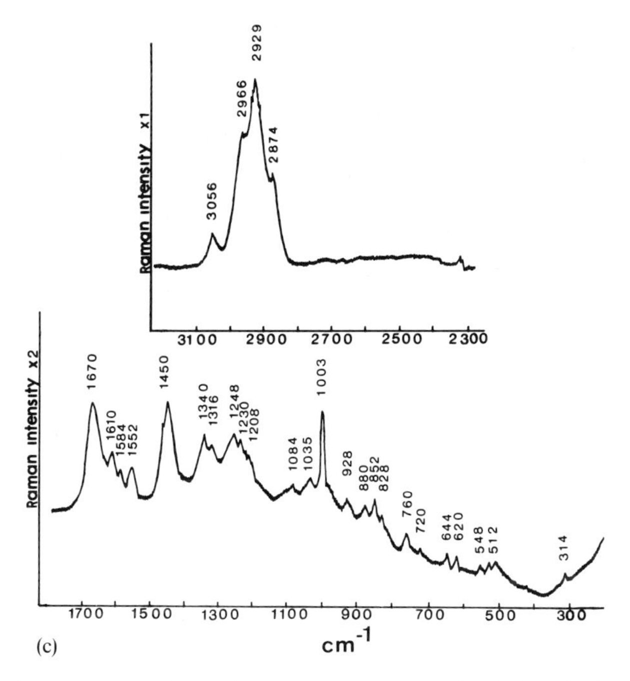

Fig. 28 a. Laser Raman spectra of *A. polyphemus* chorion. Early developing chorion (*a*) or ovulated chorion (*b–d*) samples were used. *a–c* were examined tangentially to the surface, and *d* in the radial orientation along a cut edge. In *c*, the inner structure was examined tangentially, after scraping off the outer layers of the chorion. Several Raman bands are identified, including those discussed in the text. A change in scale was necessary for plotting the low end of the spectrum in *a*, *b*, and *c*. Instrumental conditions: excitation wavelength = 5145 Å; average slit width = 5.2 cm^{-1}; integration time = 1 s; laser power at the sample = 100 mW. (Hamodrakas et al. 1982b). **b** Laser Raman spectrum of the eggshell of the silkmoth *B. mori*. A 90° scattering geometry was employed with the laser beam hitting the eggshell surface tangentially (outer lamellar layer). Instrumental conditions: excitation wavelength 514.5 nm; scanning speed = 10 cm^{-1} min^{-1}; time constant = 2 s; spectral resolution 5 cm^{-1}; laser power at the sample = 100 mW. (Hamodrakas et al. 1984). **c** Laser Raman spectrum of the eggshell of the fish *S. gairdneri*. A 90° scattering geometry was employed, with the laser beam hitting the eggshell surface tangentially. Instrumental conditions: excitation wavelength, 514.5 nm; scanning speed, 10 cm^{-1} min^{-1}; time constant, 2 s; spectral resolution, 5 cm^{-1}; laser power at the sample, 100 mW (Hamodrakas et al. 1987)

4.3.1 Raman and Infrared Spectra: Secondary Structure of Eggshell Proteins

According to extensive theoretical and experimental studies, the frequencies of amide I, II, and III bands in Raman spectra are useful indicators of protein and polypeptide secondary structure (Carey 1982). The amide I bands (which occur in the region 1630–1690 cm^{-1}) appear to have contributions from C–O stretching (approximately 70%) and C–N stretching (approximately 16%). The generally weak amide II Raman bands (1510–1570 cm^{-1}), and the amide III bands (1220–1330 cm^{-1}) have significant contributions from N–H in plane bending and

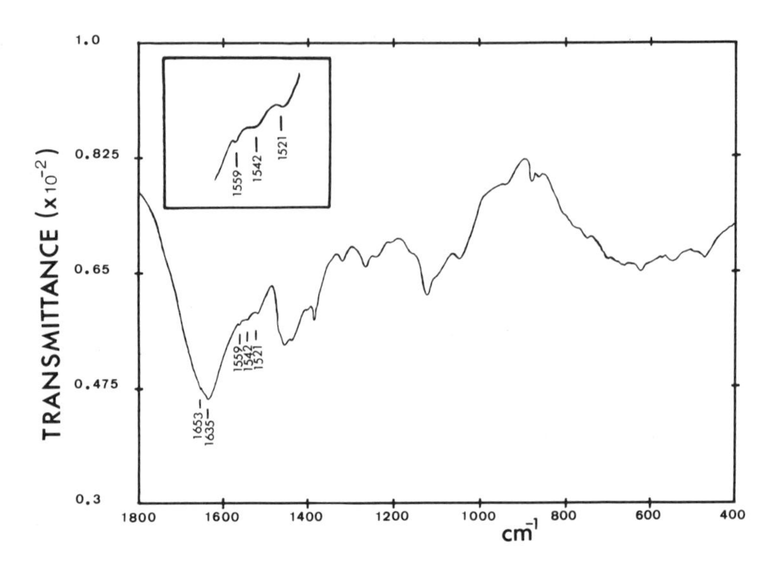

Fig. 29. Fourier transform infrared spectrum of the eggshell of the fish *S. gairdneri*. The spectrum is the result of signal averaging of 100 scans, at 2 cm^{-1} resolution. Samples were in the form of KBr pellets, containing about 2% wt. material, thoroughly ground in a vibrating mill, before mixing with KBr. *Inset* shows, on an expanded scale, the 1500–1600 cm^{-1} spectral region used for identifying the bands at 1521, 1542, and 1559 cm^{-1}. (Hamodrakas et al. 1987)

C–N stretching. Table 2 summarizes the diagnostic locations of these bands for α-helical, β-sheet, and β-turn structures, and lists the corresponding frequencies observed in laser Raman spectra of *A. polyphemus*, *B. mori*, and *S. gairdneri* chorion samples (Fig. 28a, b, c).

Table 3 gives the frequencies and our tentative assignments of the bands appearing in the laser-Raman (Fig. 28b) spectrum of the *B. mori* eggshell. Additional bands are resolved but not tabulated, because insufficient data are available for unambiguous assignments. The IR spectra exhibit a wealth of information. However, in such complicated proteinaceous systems it is difficult to assign all absorption bands to certain vibrations of defined chemical groups. Therefore we limit our attention to identifying amide I, II, and III bands suggesting a certain type of secondary structure.

The Raman spectra clearly indicate that the β-sheet conformation is predominant in proteinaceous eggshells (Fig. 28; Tables 2, 3): The bands at ca. 1673 cm^{-1} (*A. polyphemus*), 1673 cm^{-1} (*B. mori*), 1670 cm^{-1} (*S. gairdneri*) [amide I] and 1231 cm^{-1} (*A. polyphemus*), 1234 cm^{-1} (*B. mori*), 1230 and 1248 cm^{-1} (*S. gairdneri*) [amide III], can best be interpreted as resulting from abundant antiparallel β-pleated sheet structure in eggshell proteins.

The amide I band of the spectra was analyzed following the method of Williams and Dunker (1981) to estimate the percentage (%) of secondary structure of eggshell proteins. The method is described fully by Hamodrakas et al. (1984). The

Table 2. Summary of diagnostic laser-Raman amide bands and their observation in proteinaceous eggshells (Fig. 28)

Band nature	Bands characteristic of			Observed in chorion
	α-helix	β-sheet	β-turn	
Amide I	1650 to 1660	1665 to 1680	1665, 1690	1673 *A. polyphemus* 1673 *B. mori* 1670 *S. gairdneri*
Amide II	1516, 1545	1535, 1560	(I) 1550 to 1555, 1567 (II) 1545, 1555, 1560	
Amide III	1260 to 1290	1230 to 1240	1290 to 1330	1231 *A. polyphemus* 1234 *B. mori* 1230, 1248 *S. gairdneri*

Table 3. Wavenumbers and tentative assignments of bands in the laser-Raman spectrum of the eggshell of *B. mori* (Fig. 28b)

Wavenumber (cm^{-1})	Tentative assignment
510	S–S stretch
540	S–S stretch
620	Phe
641	Tyr
680	C–S stretch? Trp?
758	C–S stretch? Trp?
827[a](+)	Tyr
852[a](+)	Tyr
879	Trp
1005[a](+)	Phe or C–C stretch (β-sheet)
1016	Phe-Trp
1026	Phe
1122	C–N stretch
1170	Tyr
1206	Tyr, Phe
1234[a](+)	Amide III (antiparallel β-sheet)
1265[b](+)	Amide III (β-turns? cross-β? α-helix? coil?)
1340	Amide III (β-turns) or Trp
1360	Trp
1418	Trp
1448[a](+)	CH_2 deformation
1548	Amide II (β-turns) or Trp
1610	Tyr, Phe, Trp
1673[a](+)	Amide I (antiparallel β-sheet)
2800–3100	C–H stretch

[a] (+) A strong peak.
[b] (sh) A shoulder.

analysis suggests that the proteins of silkmoth chorion consist of 60–70% antiparallel β-pleated sheet and the remainder 30–40% of β-turns (a 2:1 ratio). For *S. gairdneri*, it is estimated that 50–60% is antiparallel β-pleated sheet, 30–40% β-turns and 10% α-helix (Papadopoulou P, Kamitsos EI, and Hamodrakas SJ, in prep.).

The distribution of the phi and psi angles in the β-sheets appears to be rather narrow (in other words the β-sheets exhibit a uniform structure), since the amide I band at ca. 1670 cm^{-1} is sharp: its half-width is approximately 40–45 cm^{-1} compared to the 27 cm^{-1} observed in the very uniform silk fibroin and the 76 cm^{-1} in the less uniform β-keratin (Frushour and Koenig 1975).

Supporting evidence for the prevalence of antiparallel β-pleated sheet in the proteins of helicoidal eggshells was supplied by Fourier transform IR spectroscopy: For the *S. gairdneri* eggshell proteins, the observation of a very intense absorption band at 1635 cm^{-1} (amide I) and of a weak band at 1521 cm^{-1} (amide II) in the IR spectrum (Fig. 29) strongly indicate an antiparallel β-pleated sheet conformation. Similar results were obtained from IR spectra of *A. polyphemus*, *B. mori*, *A. pernyi*, *M. sexta* and *S. nonagioides* (Orfanidou C and Hamodrakas SJ, in prep.). In at least two cases, *A. polyphemus* (Fig. 28a) and *S. gairdneri* (Papadopoulou P and Hamodrakas SJ, in prep.) a study of eggshell protein structure has been performed at different developmental stages, utilizing laser-Raman spectroscopy. The relative invariance, during development, of the bands indicative of antiparallel β-sheet suggests the preponderance of this structure in eggshell proteins throughout choriogenesis and is significant in terms of eggshell morphogenesis.

4.3.2 Raman Spectra: Side Chain Environments

Raman spectra yield useful information on amino acid residues of Cys, Tyr, Phe, and Trp (Carey 1982). We shall focus our attention here only on Cys and Tyr.

Bands in the 500–550 cm^{-1} region are typically associated with the S–S stretching mode of the C–C–S–S–C–C structural unit of disulfide bonds. Following Sugeta et al. (1972), bands at 510, 525 and 540 cm^{-1} may be assigned to S–S bridges in *g-g-g*, *g-g-t* and *t-g-t* (*g* and *t* denote *gauche* and *trans*) conformations, respectively. Bands in the 2530–2580 cm^{-1} spectral region are typically associated with the –S–H stretching mode (Yu 1977). Our studies on *A. polyphemus* eggshells have shown that the highly localized cysteines of chorion proteins are apparently found in diverse environments during chorion development, as can be judged by the presence of multiple bands, assigned to free sulfhydryls, in the 2530–2580 cm^{-1} region in developing eggshells and are cross-linked by disulfide bonds at/or near ovulation. This is strongly supported by the evident suppression and even disappearance of these bands in the ovulated samples (Fig. 28a; Hamodrakas et al. 1982b).

In the outer osmiophilic layer of a mature *B. mori* eggshell formed by the Hc proteins, the great majority of the cysteines are cross-linked by disulfide bonds in *g-g-g* and *t-g-t* conformations, as can be judged by the presence of bands at 510 and 540 cm^{-1} and the absence of significant features in the 2530–2580 cm^{-1} spectral region (Fig. 28b; Hamodrakas et al. 1984). These bonds harden and waterproof the eggshell during the unusually long diapause periods, which is essential for the oocyte survival (Kafatos et al. 1977).

Determination of the exact time of formation of disulfide bonds in *S. gairdneri* eggshells does not seem to be necessary; apparently, its cysteines are cross-linked via disulfide bridges throughout development (Papadopoulou P,

Hamodrakas SJ, and Kamitsos EI, in prep.). All possible types of disulfide conformation exist in mature eggshells, as can be judged by the presence of bands at ca. 510, 525, 548 cm^{-1} (Fig 28c; Hamodrakas et al. 1987).

The intensity ratio of the tyrosine doublet at 850 and 830 cm^{-1}, $R = I_{850}/I_{830}$, is sensitive to the nature of hydrogen bonding, or to the state of ionization of the phenolic hydroxyl group: basically, if tyrosine functions as a strong hydrogen-bond donor to a negative acceptor in a hydrophobic environment, the ratio R is low, perhaps 0.3–0.5. Hydrogen bonding in which the phenolic oxygen serves as a weaker donor or as an acceptor yields a higher ratio (Siamwiza et al. 1975). In the *A. polyphemus* spectra (Fig. 28a), $R = 0.3 \pm 0.1$, suggesting that the tyrosines are buried in a hydrophobic environment and strongly hydrogen bonded. This is significant, since the tyrosine residues are highly localized within the "arms" of chorion proteins. In *B. mori* spectra (Fig. 28b), $R = 1.2 \pm 0.1$, which probably suggests that most tyrosines act as much weaker hydrogen bond donors or as acceptors. In *S. gairdneri* (Fig. 28c), $R = 2.2 \pm 0.1$, an unusually high value (Yu 1977).

Summarizing, all evidence to date, both theoretical and experimental, strongly suggests that uniform antiparallel β-pleated sheet is the predominant molecular conformation of helicoidal eggshell protein structure during development. Apparently, this conformation dictates protein self-assembly for the formation of higher order helicoidal architecture. The follicle cells (or the oocyte) do not play a direct role in the process, since self-assembly takes place extracellularly; (1) the proteins interact to form the three dimensional structure of the eggshell at some distance from the points of their secretion (Smith et al. 1971); (2) secreted protein molecules pass through the porous sieve layer (Fig. 16) to reach their destination: this ensures minimum follicle cell involvement in the self-assembly process, apart from secretion of multimers, which may act as "nuclei" in the self-assembly process. The question now arising is: how is self-assembly done in molecular terms? To answer this question, models of protein structure are needed. Obviously, our description will be confined to silkmoth chorion proteins with known primary structure.

5 Structural Motifs in Proteins of Helicoidal Eggshells and Protein Structural Models

Tandemly repeating peptides have been found in the sequences of most fibrous proteins and play an important role in the formation of the fibrous structure (Fraser and McRae 1973; Parry 1979). Individual repeat units tend to be conformationally equivalent. If the equivalence is exact, a helical structure results, if not, the local conformations of the repeat units are likely to be similar (Parry et al. 1979). An important question which always arises in such cases is what type of structure is formed by these repetitive peptides.

Our analysis clearly established that silkmoth chorion proteins have an unusual tripartite structure: a central domain, highly conserved within each family and recognizably homologous between families of the same branch, and two variable flanking domains or "arms", marked by the presence of tandemly repetitive peptides (Figs. 18–22) that are not apparent in the central domain. The evolutionary conservation and length invariance of the central domain, in each branch (α or β) of chorion protein families, suggest that this domain assumes a precise and functionally important three-dimensional structure.

5.1 Structure of the Central Domain

The existence of tandemly repeating peptide motifs as a characteristic feature of fibrous protein structure and the observation of periodicities of β-sheet maxima in the central domain, alternating with β-turns, in secondary structure prediction histograms (cf. Sect. 4.1), led us to search for "hidden" periodical patterns of residues in the central domain.

5.1.1 Definition of the Sequences Considered

Figure 18 presents typical chorion sequences of the α (A, HcA, CA) and β (B, HcB, CB) branches and shows the borders of the central domains analyzed (Hamodrakas et al. 1985, 1988). The analysis has been performed on the sequences shown; our conclusions apply to all other available sequences, published and unpublished.

5.1.2 Hexad Periodicities

The region considered is highly conserved among subfamilies of the α and β branches, and has not undergone deletions or insertions over more 50 million years (Rodakis et al. 1982). For preliminary analysis of periodicities a Fortran program was written (Hamodrakas SJ, unpubl.), which revealed a sixfold periodicity for Gly. Accordingly, sequences were written out in rows of six residues (Figs. 30a, 31a) and the significance of the nonrandom distribution of residues in the six columns thus generated, was analyzed with the method of McLachlan (1977), calculating the pattern strength, P. This measure is the difference between observed and randomly expected unevenness in the distribution of amino acid residues over the columns, divided by the standard deviation. Therefore, values of P greater than 3.0 are highly significant. The analysis established clear sixfold periodicities for various types of residues (Table 4) which may be summarized by declaring that the periodically repeating hexapeptides have the general form:

α branch: Gly-X-large hydrophobic-Y-large hydrophobic-hydrophobic

β branch: Gly-X-large hydrophobic-Y-large hydrophobic-Z,

with X usually a β-turn former residue and Y, Z of a more general type.

Similar hexapeptide periodicities were detected by Fourier transform analysis of the sequences. Fourier transforms were obtained essentially as outlined by

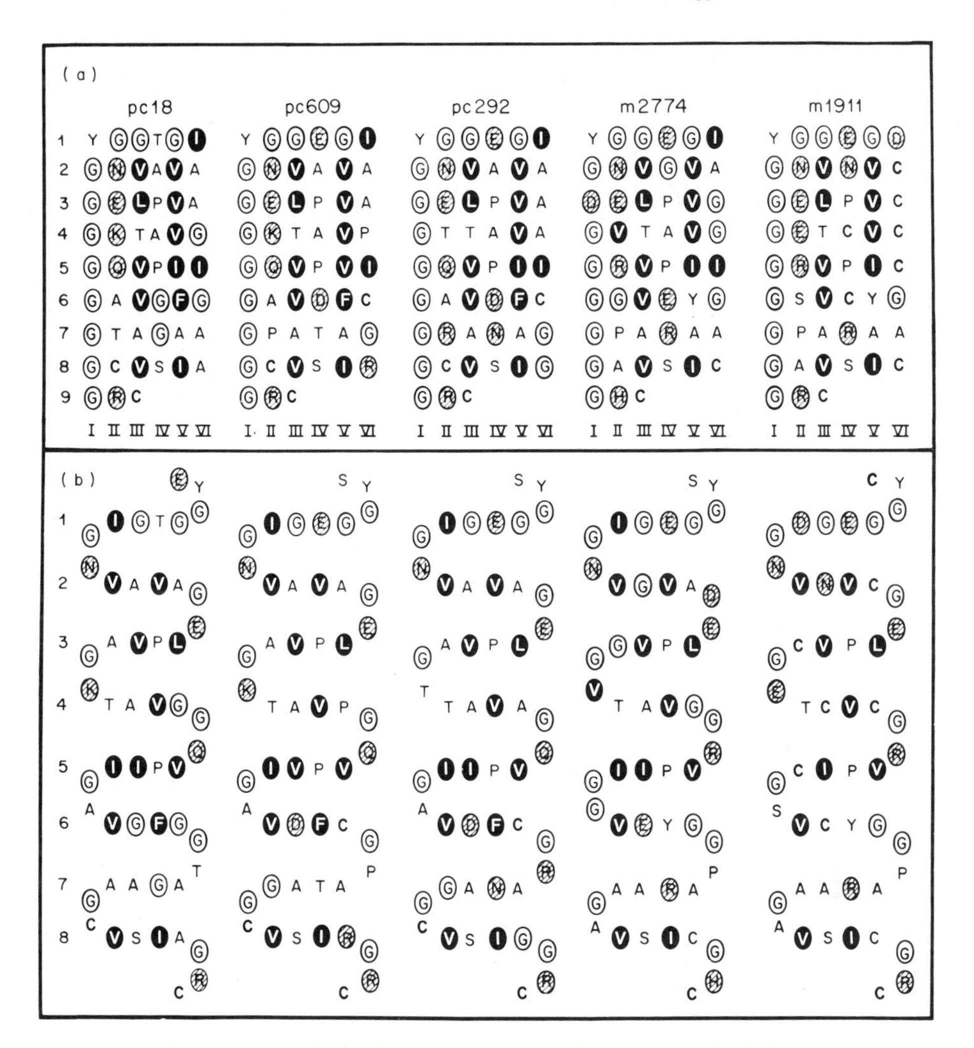

Fig. 30. a Regular amino acid distribution within the central domain of the A/HcA chorion proteins (cf. Fig. 18). To reveal the sixfold periodicities, sequences have been written in rows (numbered *1* to *9*) of six residues each; they should be read *left to right, top to bottom. Vertical columns* (numbered *I* to *VI*) can thus be seen to have nonrandom prevalence of certain types of residues, which are distinctively marked (see text). **b** Anti-parallel β-sheet model for the central domain. Sequences should be read continuously, beginning at the top. Each row of **a** corresponds to a vertical β-turn and a similarly numbered horizontal β-sheet strand. For further details, see text. (Hamodrakas et al. 1985)

McLachlan (1977) and McLachlan and Stewart (1976), using a Fortran 77 computer program: each sequence of N residues was represented as a linear array of N terms, with each term given a value of 1 or 0, according to whether the condition considered (e.g, presence of a Gly residue) was or was not satisfied. To increase resolution, this array was embedded in a larger array of zeros. A summary of the Fourier analysis results for the A/HcA proteins is shown in Table 4.

a

	B e 2G12	B m 2807	B m 1768	B Hc-B12	B pc401	C pc404-H12
1	L G V A	L G I A	L G I A	L S I C	L G V A	L S V T
2	S E N S Y E	S E N R Y E	S E N R Y E	S E N R Y K	S E N M Y E	S E N T I E
3	G T V G V C	G T V G V C	G T V G V S	G D V C V C	G C V G V A	G V V A V T
4	G N L P L L	G N L P F L	G N L P F L	G E V P F L	G N L P F L	G Q L P F L
5	G T A I V T	G T A D V A	G T A D V A	G T A D V C	G T A G V E	G A V V T D
6	G E F S T G	G E F P T A	G E F P T A	G N M C S S	G V F P T A	G I F P T V
7	G L G G I N	G I G E I D	G I G E I D	G C G C I D	G A G V I N	G A G D V W
	I II III IV V VI	I II III IV V VI	I II III IV V VI	I II III IV V VI	I II III IV V VI	I II III IV V VI

b

B e 2G12	B m 2807	B m 1768	B Hc-B12	B pc401	C pc404-H12
L G V A	L G I A	L G I A	L S I C	L G V A	L S V T
S	S	S	S	S	S
E	E	E	E	E	E
E Y S N	E Y R N	E Y R N	K Y R N	E Y M N	E I T N
G	G	G	G	G	G
T	T	T	D	C	V
V G V C	V G V C	V G V S	V C V C	V G V A	V A V T
G	G	G	G	G	G
N	N	N	E	N	Q
L L P L	L F P L	L F P L	L F P V	L F P L	L F P L
G	G	G	G	G	G
T	T	T	T	T	A
A I V T	A D V A	A D V A	A D V C	A G V E	V V T D
G	G	G	G	G	G
E	E	E	N	V	I
G T S F	A T P F	A T P F	S S C M	A T P F	V T P F
G	G	G	G	G	G
L	I	I	C	A	A
G G I N	G E I D	G E I D	G C I D	G V I N	G D V W

II′ I′

Fig. 31. a Regular amino acid distribution within the central domain of the B, HcB, CB silkmoth chorion proteins. To reveal the sixfold periodicities, sequences have been written in *rows* (numbered *1–7*) of six residues each; they should be read *left to right*, *top to bottom*. *Vertical columns* (numbered *I* to *VI*) can thus be seen to have nonrandom prevalence of certain types of residues. A distinct pattern of Gly-X-large hydrophobic-Y-large hydrophobic-Z emerges, where X is usually a β-turn former and Y, Z other types of residues. **b** Antiparallel β-sheet model for the portion of the central domain considered. Sequences should be read continuously, beginning at the top. Tentative I′ and II′ β-turns alternate with four-residue β strands. For further details, see text. (Hamodrakas et al. 1988)

5.1.3 Interpretation of Hexad Periodicities

The following arguments, described in detail by Hamodrakas et al. (1985), led us to interpret the hexad periodicities appearing in the central domains of the α and β branches of proteins, by the alternating β-turn/β-strand model of an antiparallel β-pleated sheet shown in Figs. 30b, 31b:

1. Secondary structure prediction indicates a regular alternation of β-sheet maxima with β-turns in the central domains of silkmoth chorion proteins (cf. Sect. 4.1).
2. Analysis of the amide I band of the laser-Raman spectra suggests a 2:1 ratio of antiparallel β-sheet/β-turns in chorion proteins (cf. Sect. 4.3).

Table 4. Residue periodicities in the central domain of silkmoth chorion A proteins

Pattern strength analysis

Type of residue	pc18	pc609	pc292	m2774	m1911
G	9.02	12.60	11.41	6.34	12.60
β-Turn[a]	6.62	6.70	6.17	4.86	9.46
V	4.07	4.82	4.07	2.97	4.07
β-Sheet[b]	7.19	6.11	6.48	5.97	4.94

Fourier transform analysis

Type of residue	pc18	pc609	pc292	m2774	m1911
G	3.48	6.41	7.61	5.07	6.60
	(98°)	(113°)	(110°)	(114°)	(111°)
β-Turn[a]	3.11	4.55	4.49	2.88	3.31
	(144°)	(131°)	(139°)	(115°)	(109°)
V	3.40	3.43	3.40	2.36	3.20
	(− 86°)	(− 75°)	(− 87°)	(− 96°)	(− 87°)
β-Sheet[b]	4.04	4.49	5.11	3.23	7.76
	(− 47°)	(− 51°)	(− 44°)	(− 50°)	(− 45°)

Top:
Pattern strength (P) values are shown for a periodicity of six residues; values are expressed in standard deviation units (σ).

Bottom:
Fourier transforms for a periodicity of 5.82 residues (a value giving consistent maxima at approximately six residues). Each entry includes a intensity (I) value and a phase angle in parentheses. Intensity values are scaled as recommended by McLachlan and Stewart (1976); the probability of observing by chance an intensity, I, at any particular periodicity is exp(− I).

[a] β-Turn formers, G, P, D, N, S, C, K, W, Y, Q, T, R, H, E; Chou and Fasman (1978).
[b] β-Sheet formers, V, L, I, F, W, Y, T, C; Chou and Fasman (1978).

3. Strong periodicities were observed for groups of β-sheet former and β-turn former residues with a period of approximately six (6) residues and a phase difference of approximately 180° (out-of-phase; Table 4).

In this model, the horizontal rows, each containing four residues, represent antiparallel β-sheet strands and consist of the residues shown in columns III to VI of Figs. 30a and 31a. The end residues of these short strands also participate in β-turns, together with the vertically displayed dipeptides (columns I and II of Figs. 30a and 31a; residues Gly-X in the hexapeptide motifs). The latter represent the central residues i + 1 and i + 2 of the β-turns, respectively.

The model is somewhat reminiscent of Silver Gull feather keratin (Fraser and McRae 1976), where an eight-residue periodicity in β-sheet propensities and a similar but out-of-phase periodicity of random coil propensities reflects the structure of a β-sheet consisting of short antiparallel strands. Similarly, the model is reminiscent of the cross-β-sheet structure in the shaft of the adenovirus fiber protein (Green et al. 1983); in that case, the antiparallel β-sheet strands are usually five or six residues long and together with two β-turns that punctuate them, result in an observed 15-residue periodicity.

5.1.4 β-Turn Type Determination and Modeling

To construct realistic models of the proposed antiparallel β-pleated structure for all chorion proteins, it was necessary to determine the types of the β-turns (Chou and Fasman 1977) which give the silkmoth chorion protein central conservative domains their characteristic fold (Figs. 30b, 31b).

Since the conformation of residues i, i + 3 is constrained to have values corresponding to those of an antiparallel β-sheet (these residues are parts of the four-residue β-strands), this means that we had to determine the conformation of residues i + 1, i + 2, of each β-turn.

A similar analysis has been performed by Geddes et al. (1968) to determine the types of β-turns which generate the cross-β conformation in proteins. These authors arrived at two types of turns to be the most favorable for the formation of the cross-β conformation, which they denoted as fold A and fold B. They essentially correspond to the II′ and II types of Chou and Fasman (1977), respectively. In these folds, a hydrogen bond is formed between the –NH group of residue i + 3 with the –C=O group of residue i.

Our analysis, which in general terms resembled the analysis of Geddes et al. (1968), but was carried out with our own Fortran program, was done as follows:

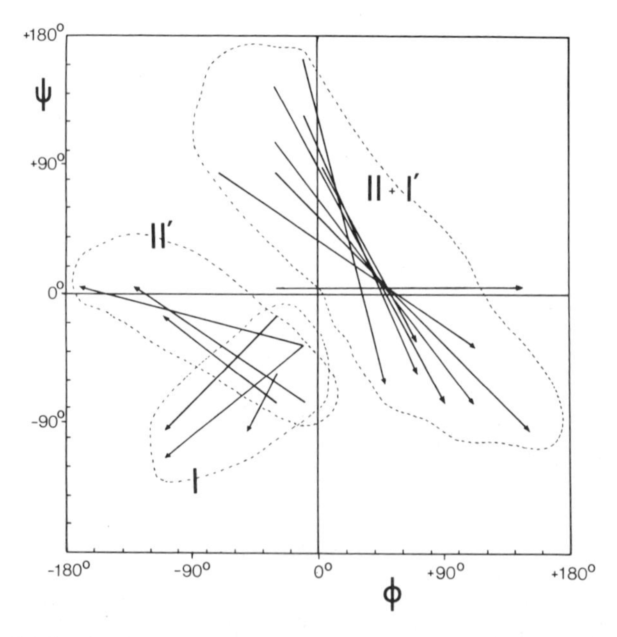

Fig. 32. Phi-Psi plot showing the conformation of the turn residues i + 1, i + 2 by *arrows*, as were found by our conformational analysis of the decapeptide VAVAGELPVA, as described in the text. The *beginning of each arrow* marks residue i + 1 and the end, residue i + 2. Three groups of possible β-turns were found: group 1, which corresponds to distorted types of II and I′ turns (Chou and Fasman 1977), group 2, corresponding to distorted types of II′ turns, and group 3, which corresponds to distorted types of III β-turns. Presumably, these distorted turns are necessary in the model because the phi, psi angles of the sheet residues were kept constant. (Hamodrakas et al. 1988)

1. Representative decapeptides of the type forming the characteristic structure of Figs. 30b, 31b were chosen, which contain two consecutive β-strands linked by a β-turn.
2. Backbone dihedral angles of the β-strand residues were set to phi = − 120° and psi = + 135°. These values correspond to the centre of the allowed region for β-pleated sheets and they generate β-strands suitable for the formation of twisted β-pleated sheets (Schulz and Schirmer 1978). We have chosen these values since most known β-sheets, both in globular and structural proteins are twisted (Fraser and McRae 1976; Richardson 1981; Lotz et al. 1982), and since experimental findings (Hamodrakas et al. 1984, 1986) indicate that silkmoth chorion proteins contain twisted sheets.
3. The phi and psi angles of residues i + 1, i + 2 of the turns were varied in a systematic way, in order to generate allowed conformations without steric hindrance, and such that the −NH and =CO groups of the β-strands can create hydrogen bonds.

Several possible conformations were found from this analysis which are mostly distorted types of well known β-turns (Fig. 32). They can be classified into three major classes: Type I′ (or II), type II′, and type III β-turns. However, from model building, we selected β-turns I′ (or II), and II′, which, provided they alternate along the structure shown in Figs. 30b and 31b, create satisfactory models for the conservative domains of silkmoth chorion proteins. In these models, favorable hydrogen bonds are formed between the NH group of residue i + 3 with the CO group of residue i.

5.1.5 Structural Models of the Central Domain

The models obtained from this stereochemical analysis, assuming β-turns of the I′ (or II) and II′ type, alternating with four-residue β-strands along the sequence, were refined in detail on an interactive graphics system. They are shown for a representative A class protein in Fig. 33.

The β-pleated sheet twist of these models is possibly exaggerated. This might be a consequence of the proline residues in two strands of each sheet, whose phi angle has been constrained to about −60° in the models and of the phi and psi angles of the remaining strand residues being held constant at −120° and +135°, respectively.

A structure closely resembling the models presented in Figs. 30b, 31b, and 33 seems almost inevitable in view of all the evidence for the central domain of silkmoth chorion proteins. Perhaps, there are still ambiguities for a few minor points. For example, what is the exact type of β-turns which generates this structure? Unfortunately, the ambiguities cannot be resolved experimentally (Hamodrakas et al. 1988).

It was very interesting to note that a recent systematic search of β-turns connecting antiparallel β-sheet strands in globular proteins (Sibanda and Thornton 1985) has revealed that the types II′ and I′ turns clearly predominate. Out of the 29 β-turns observed experimentally, 15 were type I′, 10 type II′ and only 4 type

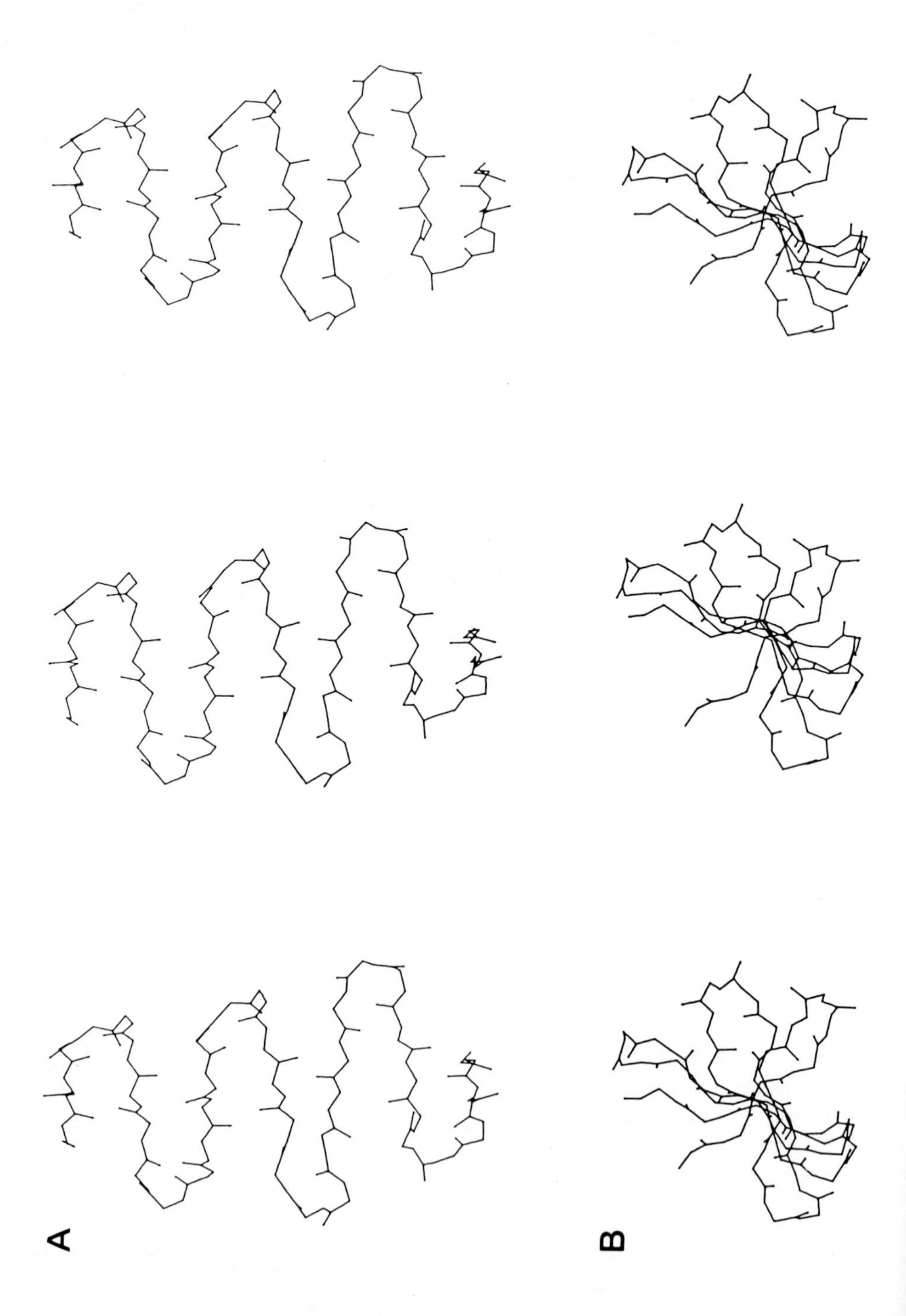
A
B

I. The type I′ turn differs from the type II turn in the phi and psi angles of only one residue (see Fig. 2 of Sibanda and Thornton 1985). Surprisingly, the type II turn is not observed to link adjacent antiparallel β-sheet strands in globular proteins.

A comparison of our theoretical search (Fig. 32) with Fig. 2 of Sibanda and Thornton (1985) clearly shows that our predicted structures do not differ substantially from those observed. Some of our type II turns can be considered as distorted I′ turns. Sibanda and Thornton (1985) suggest that the abundance of the type I′ turns is probably due to the fact that they have the correct twist to match the relative twist which is always observed between adjacent strands. Close study of our model verifies that this is indeed the case. Therefore, it appears that a further refinement of our proposed model is to adjust the phi and psi angles of only one residue, to belong to a type I′ turn rather than a type II turn. This modification leaves the remainder of the structure unaffected.

The model is further supported by the pattern of residues appearing in the β-turns. Sibanda and Thornton (1985) have found that the observed β-turns are strongly selective for amino acid type. For a type II′ turn they found that Gly predominates in the second position, whereas the third position is usually occupied by Ser, Thr, or a polar residue; for a type I′ turn, the second position is occupied mostly by Gly, Asp, or Asn and the third usually by Gly. Close study of our models (Figs. 30b, 31b) shows that for the type II′ turns the pattern is almost ideal: Gly is usually found in the second position, whereas in the third usually a Thr. For the type I′ (or II) turns the second position is occupied mostly by Gly, in agreement with the observed data, whereas, in the third position, charged (Glu) or polar (Asn) residues frequently appear.

5.1.6 Model Features

This model structure of the central domain, common to all chorion protein families and subfamilies, has the following characteristics:

1. It is highly conservative in each family and subfamily. Greater variability in sequence and perhaps in secondary structure (although this is not certain), is seen in the remainder of the molecules ("arms").
2. In the short four-residue β-strands relatively "small" residues (e.g., G) tend to alternate with "bulky" residues in several cases, e.g., VAVA, VSIG etc. This is reminiscent of the alternation of small (G) and bulky (A, S) residues to opposite sides of the β-sheet structure in silk fibroin (Marsh et al. 1955) and may be important in chorion for the packing of β-sheets to form higher order structure.
3. Both faces of the proposed β-sheets have a pronounced hydrophobic character, except for certain regions (e.g., in the strand VDFC), which might serve as sites

Fig. 33A, B. A skeletal model, obtained from an interactive graphics system, showing the characteristic β-pleated sheet fold of the central conservative domain of the A protein pc609 (main chain and carbonyl oxygens only). **A** View perpendicular to the "plane" of the β-sheet. β View perpendicular to the strands, parallel to the β-sheet "plane". The three-picture stereo system used in this figure enables readers with both normal and cross-over stereo vision to view the image. For normal vision select the *left* and *center images*, for cross-over vision use the *center* and *right images*. (Hamodrakas et al. 1988)

for specific recognition. The existence of charged or polar residues in these regions shows that they should be counterbalanced by complementary residues in neighboring β-sheets for the formation of hydrogen or salt bonds during morphogenesis of higher-order structure. It is well known (for a recent analysis, see Rashin and Honig 1984) that this is always the case whenever a charged or polar residue is found in the interior of water soluble globular proteins. A detailed analysis of protein-protein interactions which is well under way for silkmoth chorion proteins will provide further insights (Orfanidou C and Hamodrakas SJ, unpubl.). However, the hydrophobic character of both faces of the β-sheet structures clearly promotes favorable packing of protein molecules in the three-dimensional space along a radial direction of the eggshell.

4. Although both "edges" of the proposed β-sheet structure, i.e., the two central residues of the β-turns, consist mostly of Gly and polar residues, it is clearly seen from Figs. 30b and 31b that they show an uneven distribution of charges. Obviously, the right-hand side of the proposed sheets contains a number of charged residues (particularly Glu in the β branch and Glu, Arg in the α branch; Figs. 30b, 31b). The role of these charges has not yet been clarified, but they appear to be very important for the formation of higher order structure.
5. Certain residues occupy characteristic positions in this β-sheet structure: cysteines are often found in β-turns, in positions favorable to create disulfide bonds, which cross-link adjacent protein molecules during the late choriogenetic stages (Hamodrakas et al. 1984).

Two prolines in the A, B, and C proteins are always highly conserved and are always accompanied by large hydrophobic residues (I, V, L) in two β-strands which are distant by one strand. Their exact role is not yet fully understood. Prolines are not very common in β-strands and they do not favor β-sheet structure (Richardson 1981). Sometimes their presence is marked by the formation of β-bulges (Richardson 1981). Their appearance has seriously impaired our efforts to create satisfactory models of silkmoth chorion proteins. Perhaps they simply serve to increase the sheet twist, if the β-sheets are actually twisted.

Regarding the relation of the proposed structures (Fig. 33) with the helicoidal, higher-order structure of silkmoth chorion, a biological analog of a cholesteric liquid crystal, we can say that the (twisted? helical?) β-sheet ribbons of the central domains of chorion proteins are the basis for the morphogenesis of the helicoidal architecture (Hamodrakas 1984; Hamodrakas et al. 1986). We have already pointed out (cf. Sect. 2.2.1) that Rudall (1956) proposed some 35 years ago, and more recently Bouligand (1978a) as well, that a helicoidal structure can be formed from interactions of helical molecules.

5.2 Structure of the Flanking "Arms"

The quantitative measurement of secondary structure in silkmoth chorion proteins by analysis of the amide I band of the laser-Raman spectrum of chorion, suggests that chorion protein components consist of 60–70% antiparallel β-sheet and 30–40% β-turns, in a ratio 2:1 (cf. Sect. 4.3). These estimates, although in good agreement with the proposed model structure, cannot be attributed solely to the central conservative domains of chorion proteins since these domains account only for about 50% of total chorion mass. To fully account for the observed experimental percentages of secondary structure, the question remains, of course, what structure the variable chorion protein "arms" adopt. They also contain tandemly repetitive peptides evident from the sequences, different in nature from the peptides

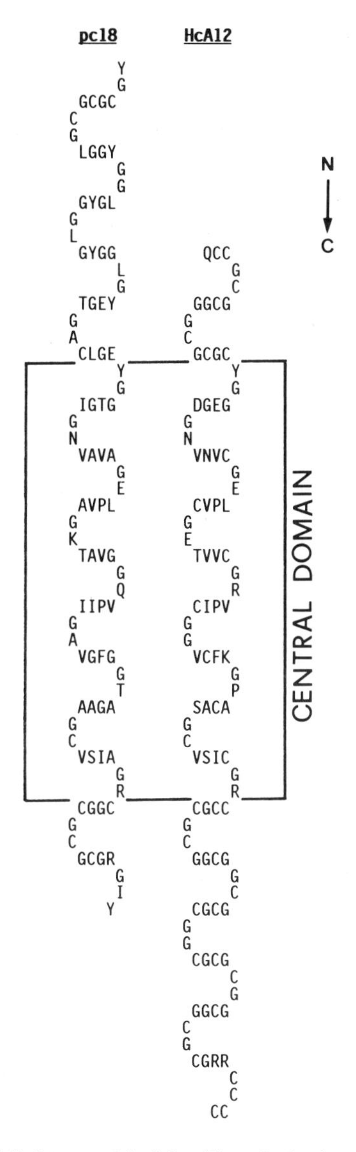

Fig. 34. Schematic, antiparallel β-sheet model of the silkmoth chorion A protein pc18 (*A. polyphemus*) and the HcA protein HcA12 (*B. mori*), assuming a uniform fold throughout their entire length.

of the central region and most probably adopting a β-sheet type of structure (cf. Sect. 4.1). The experimental percentages of protein secondary structure together with the uniform structure and packing of chorion fibrils evident from freeze-fracturing (Sect. 2.2.1) led us to attempt a folding of the protein "arms" similar to the folding of the central domain.

To our surprise, this folding can be done with no serious stereochemical constraints for all chorion proteins, with the possible exception, perhaps, of the "early" minor C proteins which are enriched in Pro in the "arms". Schematic sample folds for two proteins of the α branch are shown in Fig. 34 and a space-filling model of the amino-terminal arm of an A protein in Fig. 35. The resulting structures are attractive in several respects: (1) ensure uniform folding of the proteins throughout their entire length, (2) retain the characteristic features of the central domains: hydrophobicity of sheet surfaces, preferred patterns of residues in the β-turns, unequal charge distribution in the sheet "edges" and usually form "polar" β-sheets, having one "face" occupied mostly by Gly and the other by large hydrophobic residues (Tyr, Leu) which may promote efficient packing. The abundance of Gly in the "arms" implies high arm flexibility, which should be important during chorion formation. This aspect is further discussed by Papanicolaou et al. (1986). Our postulate is, however, that the protein "arms" may adopt a variety of conformations (with a prevalence of β-sheet structure) during choriogenesis and obtain their final uniform structure when "cementing" occurs, through disulfide bond formation at/or near ovulation.

Alternative model structures should not be excluded, however, for the protein arms, i.e., a model β-sheet of three-residue β-strands alternating with β-turns, taking into account the observed pentapeptide periodicity.

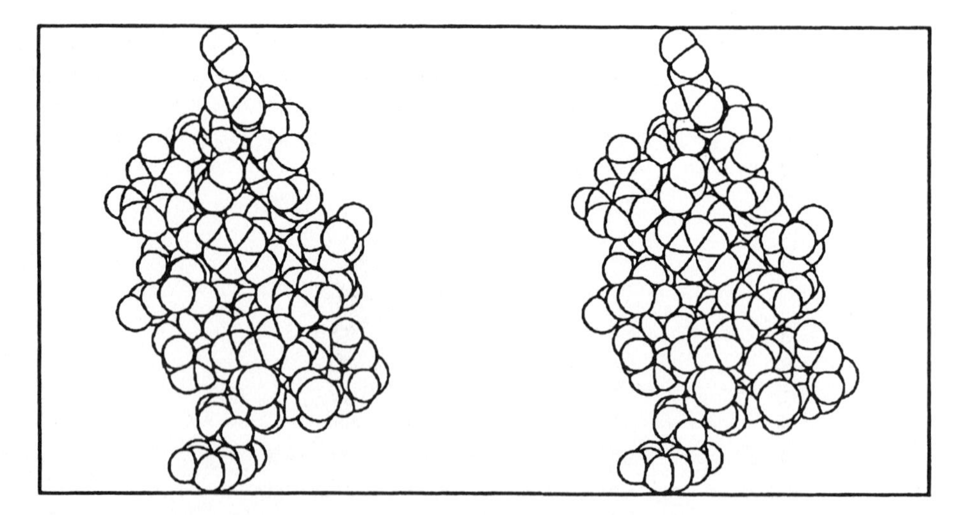

Fig. 35. A stereo (normal vision) space-filling model of the amino-terminal arm of the silkmoth chorion A protein pc18 (*A. polyphemus*). View perpendicular to the "plane" of the β-sheet

6 Models of Helicoidal Eggshell Protein Assembly

Formation of helicoidal eggshell architecture depends on the close packing of chorion proteins. The modes of packing of the detailed structural models presented above are currently under investigation to determine the rules of formation of the helicoidal structure (Orfanidou C and Hamodrakas SJ, in prep.); since β-sheet is the prevalent molecular conformation of individual proteins, these should be based on simple packing rules of β-sheets (Cohen et al. 1981; Chothia and Janin 1982). Twisted β-sheets in globular proteins usually pack with a small negative angle (~ 10–20°) between the sheet axes (or strands), or in an orthogonal fashion (angle close to 90°). Packing of the proposed β-sheets with a small negative angle would explain the observed anticlockwise (left-handed) sense of rotation of the helicoids.

Although it would clearly be premature to propose detailed models of protein-protein interactions due to the complexity of the system, some speculations may have heuristic value: to integrate available information and present a detailed model of three-dimensional chorion architecture, it is necessary to answer the fundamental question "how is a fibril formed?". The cross-sections of individual protein molecules (assuming they fold along their entire length as β-sheet ribbons) must be approximately 30 Å in diameter, taking into account the side chains of the turn residues. Measurements were made directly from molecular CPK space-filling and Kendrew skeletal models and on an interactive graphics screen. These values are in close agreement to the fibril diameter "seen" by freeze-fracturing and X-ray diffraction and are in favor of the attractive notion that one molecule corresponds to one fibril.

However, other key features drive us to assume that a basic structural unit in the bulk of the silkmoth chorion may be formed by a dimer of an A and a B protein instead of a single molecule (the fibril or a disulfide bonded dimer?): experimentally, it has been shown that (1) A and B protein pairs are produced in parallel and in equimolar amounts; their respective genes are clustered, divergently oriented and coordinately expressed (Jones and Kafatos 1980a), (2) disulfide bonded multimers (mostly dimers and trimers) may act as intermediates in the assembly process, showing preferred family linkage patterns (Regier and Wong 1988). Extensive analysis of the sequences and of models (Orfanidou C and Hamodrakas SJ, in prep.) suggests that all known, coordinately expressed A and B protein pairs (i.e., pc18-pc401, pc292-pc10a, AL11-BL11, AL12-BL12, etc; Regier and Kafatos 1985; Spoerel et al. 1989) may form dimers of the general form shown in Fig. 36.

These dimers which consist of an A and its complementary B molecule, each folded as a β-sheet ribbon throughout its entire length, and packed in an antiparallel fashion so that the "faces" of the β-sheet ribbons containing the side-chains of the invariant proline residues of their central domains (cf. Sect. 5.1) are "buried", combine the following significance features (Fig. 36):

1. Their cross-sections are of the same order of magnitude (30–40 Å) as those of individual molecules (~ 30 Å).

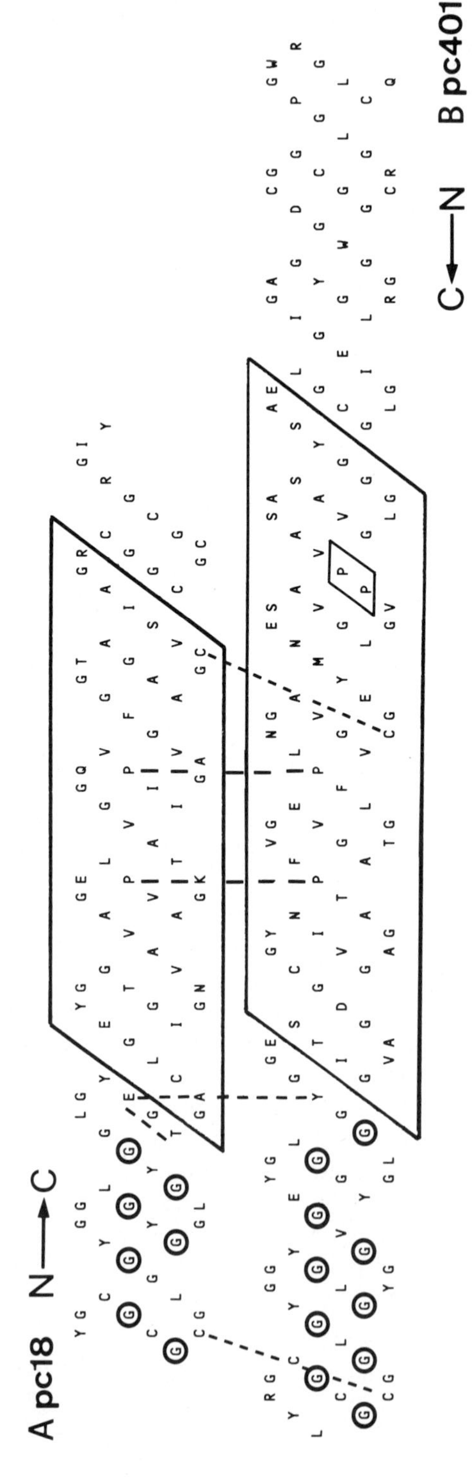

Fig. 36. Schematically, a packing model of an A–B silkmoth chorion protein dimer. The molecules pack in an antiparallel fashion and with the β-sheet faces containing the invariant proline residues of the central domain (*dotted lines*) "buried" in the hydrophobic core of the β-sheet sandwich. For details of the model features and specific interactions see text

2. Their central domains form a β-sheet sandwich, with the β-sheet surfaces exhibiting the most pronounced hydrophobic character buried in the interior.
3. The antiparallel fashion of packing ensures that the N-terminal arm of the A protein is packed against the distantly homologous (Regier and Kafatos 1985) carboxyl-terminal arm of the B, which has a similar length. Both contain tandem repeats of the pentapeptide LGYGG and its variants. The packing is such that the faces of the β-sheet ribbons containing mostly Gly are packed against each other, which ensures uniform packing. The beginning of the A central domain corresponds to the end of the B, whereas the end β-strand of the A central domain has as counterpart in the B, a β-strand containing two conserved, consecutive prolines in *A. polyphemus* sequences. This may be an "inflection point" in the conformation of the amino-terminal arm of the B's which has no "partner" from the A molecule. In *B. mori* sequences, the strain imposed by the two consecutive prolines is relieved since only one of them is conserved.
4. The four (two in each protein) invariant proline residues of the central domains pack with their side chains facing each other.
5. Buried polar or charged residues are fully counterbalanced by adjacent polar or of opposite charge residues, which is very important for the stability of the hydrophobic core. In the example shown (Fig. 36) a glutamyl residue just before the beginning of the central domain in Apc18 is counterbalanced by a Thr in the same strand and/or a "facing" Tyr in Bpc401.
6. Cysteines are found in favorable positions to form disulfide bonds, stabilizing the dimer and may also promote efficient cross-linking in three dimensions with other monomers or dimers.

These models were found intuitively. However, since there are eight possible ways of packing for each dimer (each molecule has two "faces" and the molecules may pack in a parallel or antiparallel fashion), efficient packing was checked in each case, by "sliding" surfaces on top of each other and scoring favorable interactions (gain in free energy by the burial of hydrophobic chains, counter-balancing of buried polar or charged residues, possible formation of disulfide bonds, etc.) and it was found that the models shown represent the most stable possible dimers. Furthermore, an additional check was performed utilizing an interactive graphics system.

A full account of these findings will be given elsewhere (Orfanidou C and Hamodrakas SJ, in prep.). Here, we propose, however, that these dimers may represent our starting point to fully understand the molecular interactions leading to helicoidal self-assembly. Refined biochemical work is needed along the lines suggested by Regier and Kafatos (1985) and also further modeling studies to define their possible roles in the formation of higher order structure and also verify their validity.

7 Solution Structure and Assembly of Helicoidal Eggshell Proteins

One of our future aims is to study the in vitro assembly process of chorion proteins in detail and determine the structural parameters of reconstituted units, making comparisons with native chorion.

As a first step, lepidopteran chorions were solubilized in the presence of denaturing and reducing agents (6 M urea, 0.4 M Tris-HCl, 1% 2-mercaptoethanol). Extracted chorion proteins were reassembled by a 12–18 h dialysis at room temperature against double-distilled water. Reassembled (polymerized) units were studied by negative staining (Fig. 37), dark field electron microscopy and circular dichroism (CD) spectroscopy (Fig. 38) to determine: (1) size and shape of reconstituted units (2) secondary structure of their constituent proteins.

Negative staining (Fig. 37) shows that chorion polypeptides assemble to form globular structures of varying diameters, 50–200 Å, which are cross-linked, at the low pH conditions of the experiment (pH ~3). Dark field microscopy reveals that similar in diameter globular structures result from polymerization of dissolved chorion components, even at near normal pH (~7) (data not shown). Nevertheless, the globules are not cross-linked in this case.

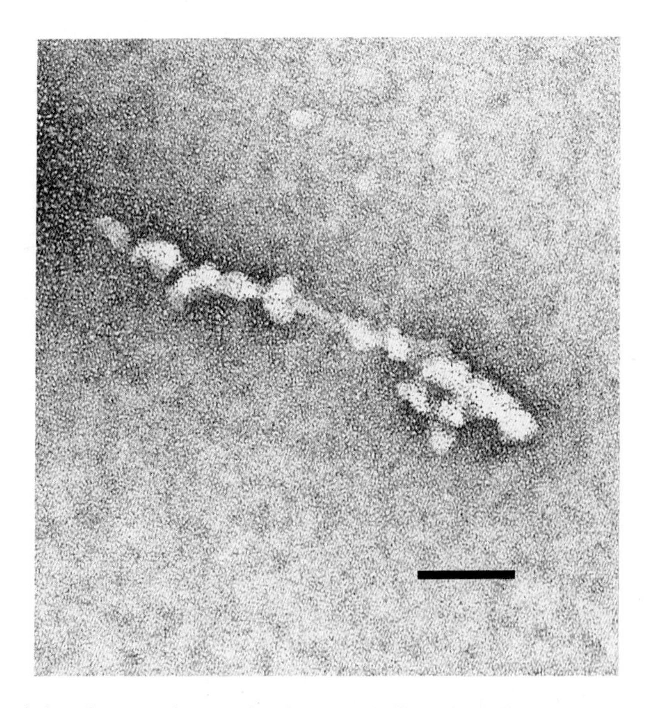

Fig. 37. Transmission electron micrographs of extracted silkmoth chorion proteins (6 M urea, 0.4 M Tris-HCl, 1% 2-mercaptoethanol), reassembled by dialysis at room temperature against double distilled water and contrasted by negative staining (1% uranyl acetate). Cross-linked globular structures of varying diameters (5–20 nm) are seen. *Bar* 100 *nm*

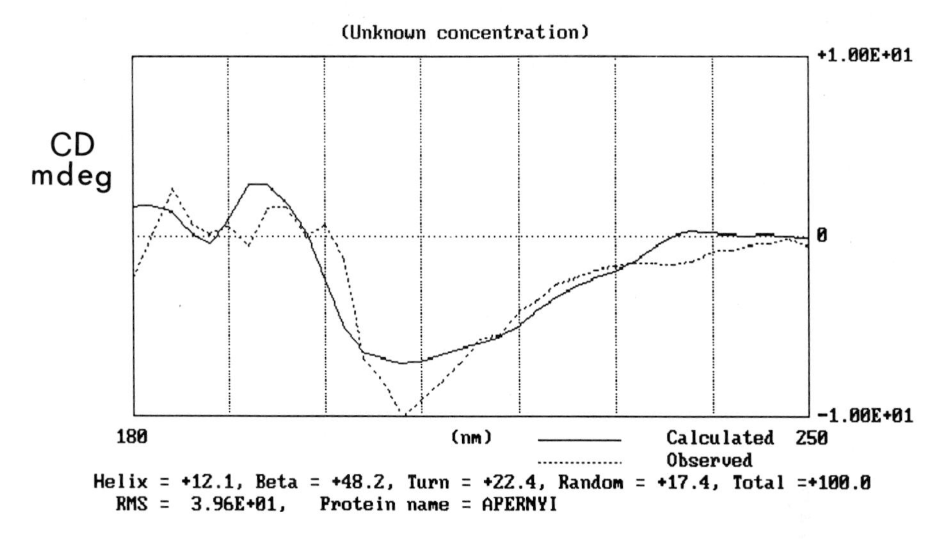

Fig. 38. Circular dichroism (*CD*) spectrum of a water solution containing reassembled units from extracted *A. pernyi* chorion proteins (details as in Fig. 37). The observed spectrum is shown by the *dotted line*. The *solid line* represents a (fitted) calculated spectrum by expressing the observed spectrum as a linear combination of reference spectra of four types of secondary structure: α-helix (α), β-sheet (β), β-turn (t), and coil (c; random). The linear coefficients representing secondary structure estimates were calculated by a linear least-squares method. The reference spectra of the four types of secondary structure were calculated from the CD spectra of four proteins with crystallographically known secondary structures: lysozyme (41%α, 16%β, 23%t, 20%c), ribonuclease A (23%α, 40%β, 13%t, 24% c), papain (28%α, 14%β, 17%t, 41%c), chymotrypsin A (9%α, 34%β, 34%t, 23%c)

The most surprising finding of these experiments, however, was that, in solution, the proteins which constitute these units, retain a very high percentage of β-sheet structure (usually more than 50%), in a ratio with β-turns of approximately 2:1, in many respects reminiscent of the secondary structure percentages obtained for the proteins of native chorions (Fig. 38; Sect. 4.3). It is, therefore, tempting to make the reasonable hypothesis that the prominent regular β-sheet structure of chorion proteins may persist in solution, dictating the aggregation and polymerization process in solution as well. A full account of these preliminary but instructive experiments will be given elsewhere (Hamodrakas SJ, Wellman SE, Case ST, and Ottensmeyer FP, in prep.).

8 Future Aims

Our molecular understanding of helicoidal proteinaceous eggshell structure is still quite primitive due to the complexity of the system and of its constituent proteins. However, many interesting questions will be experimentally and theoretically approachable in the near future.

Reconstitution studies are likely to furnish new insights into fibril structure and assembly (in space and time). Variation of several parameters in vitro will be attempted to simulate molecular events occurring in eggshell assembly.

Our theoretical studies of protein-protein interactions will continue, with particular emphasis in detecting specific sites of recognition between protein pairs and higher possible multimers. Additional helicoidal eggshell protein sequences are desirable, particularly from species other than the silkmoths.

The identification of the exact types of β-turns in eggshell proteins unequivocally will be a major breakthrough. Perhaps it will be possible to obtain partial answers to the problem by synthesizing peptides representative of segments of the central domain and the arms of chorion proteins and studying their structure utilizing biophysical methods. We have no guarantee, however, that these peptides will fold to a conformation similar to the one in vivo. In collaboration with Professor Steven T. Case, we have been able to synthesize such peptides and the study of their structure has just begun.

Finally, isolation of pure major protein components and study of their molecular structure may provide the necessary clues towards unraveling in molecular detail the secrets of helicoidal eggshell architecture.

9 Synopsis

I will not attempt to summarize this chapter, which is already a summary. I merely wish to point out that experimental and theoretical evidence to date clearly suggests that antiparallel β-pleated sheet dictates self-assembly in helicoidal proteinaceous eggshells. Molecular details of this process have started to become clear after the development of the specific, most probably correct, protein structural models in the case of the silkmoths, where amino acid information is available and with the help of several experimental techniques. However, for people seeking universal mechanisms the picture should still be far from complete. Several analogous systems should be studied before providing final answers.

Acknowledgments. I am indebted to all those whose publications are reviewed. I wish to thank Professor Lucas H. Margaritis, my Ph.D students Paraskevi Papadopoulou and Constance Orfanidou, and my son John for their invaluable help during the preparation of this manuscript.

References

Aebi U, Fowler WE, Rew P, Sun TT (1983) The fibrilar substructure of keratin filaments unraveled. J Cell Biol 97:1131–1143

Aggeli A, Hamodrakas SJ, Kanitopoulou K, Konsolaki M (1991) Tandemly repeating peptide motifs and their secondary structure in *Ceratitis capitata* eggshell proteins Ccs36 an Ccs38. Int J Biol Macromol 13:307–315

Anderson E (1967) The formation of the primary envelope during oocyte differentiation in teleosts. J Cell Biol 35:193–212

Anfinsen CB (1973) Principles that govern the folding of protein chains. Science 181:223–230
Barker WC, Hunt LT, George DG, Yeh LS, Chen HR, Blomquist MC, Seibel-Rose EI, Elzanowski A, Hong MK, Ferrick DA, Bair JK, Chen SL, Ledley RS (1986) Protein sequence database. National Biomedical Research Foundation, Georgetown University, Washington, DC
Bernstein FC, Koetzle TF, Williams GJB, Meyer EF Jr, Brice MD, Rodgers JR, Kennard O, Shimanouchi T, Tasoumi M (1977) The protein data bank: a computer-based archival file for macromolecular structures. J Mol Biol 112:535–542
Blau HM, Kafatos FC (1979) Morphogenesis of the silkmoth chorion: patterns of distribution and insolubilization of the structural proteins. Dev Biol 72:211–225
Blundell TL, Johnson LN (1976) Protein crystallography. Academic Press, New York
Bouligand Y (1972) Twisted fibrous arrangements in biological materials and cholesteric mesophases. Tissue Cell 4:189–217
Bouligand Y (1978a) Cholesteric order in biopolymers. Am Chem Soc Symp Ser 74:237–247
Bouligand Y (1978b) Liquid crystalline order in biological materials. In: Blumstein A (ed) Liquid crystalline order in polymers. Academic Press, New York, pp 261–297
Burke WD, Eickbush TH (1986) The silkmoth late chorion locus. I. Variation within two paired multigene families. J Mol Biol 190:343–356
Carey PR (1982) Biochemical applications of Raman and resonance Raman spectroscopies. Academic Press, New York
Chothia C, Janin J (1982) Orthogonal packing of β-pleated sheets in proteins. Biochemistry 21:3955–3965
Chou PY, Fasman GD (1977) β-turns in proteins. J Mol Biol 115:135–175
Chou PY, Fasman GD (1978) Prediction of the secondary structure of proteins from their amino acid sequence. Adv Enzymol 47:45–148
Cohen FE, Sternberg MJE, Taylor WR (1981) Analysis of the tertiary structure of protein β-sheet sandwiches. J Mol Biol 148:253–272
Crick FHC (1953) The packing of α-helices in simple coiled-coils. Acta Cryst 6:689–697
Davenport J, Lonning S, Kjorsvik E (1986) Some mechanical and morphological properties of the chorions of marine teleost eggs. J Fish Biol 29:289–301
Eickbush TH, Kafatos FC (1982) A walk in the chorion locus of *Bombyx mori*. Cell 29:633–643
Eickbush TH, Rodakis GC, Lekanidou R, Kafatos FC (1985) A complex set of early chorion DNA sequences from *Bombyx mori*. Dev Biol 112:368–376
Fehrenbach H, Dittrrich V, Zissler D (1987) Eggshell fine structure of three lepidopteran pests: *Cydia pomonella* (Tortricidae), *Heliothis virescens* and *Spodoptera littoralis* (Noctuidae). Int J Insect Morphol Embryol 16(3):201–219
Filshie BK, Rogers GE (1962) An electron microscope study of the fine structure of feather keratin. J Cell Biol 13:1–12
Filshie BK, Smith DS (1980) A proposed solution to a fine-structural puzzle: the organisation of gill cuticle in a crayfish (panulirus). Tissue Cell 12(1):209–226
Flugel H (1967) Licht- und elektronenmikroskopiche Untersuchungen an Oozyten und Eiern einiger Knochenfische. Z Zellforsch Mikrosk Anat 83:82–116
Fraser RDB, McRae TP (1959) Molecular organization in feather keratin. J Mol Biol 1:387–397
Fraser RDB, McRae TP (1973) Conformation in fibrous proteins. Academic Press, New York
Fraser RDB, McRae TP (1976) The molecular structure of feather keratin. In: Frith HJ, Calaby JH (eds) Proc 16th Int Ornithological Congress, Canberra. Australian Academy of Science, Canberra, pp 443–451
Friedel MG (1922) Les états mésomorphes de la matière. Ann Phys (Paris) 18:273–474
Frushour BJ, Koenig JL (1975) Raman Spectroscopy of proteins. In: Clark RJH, Hester RE (eds) Advances in infrared and Raman spectroscopy, vol I. Heyden, London, p 35
Furneaux PJS, Mackay AL (1972) Crystalline protein in the chorion of insect eggshells. J Ultrastruct Res 38:343–359
Geddes AJ, Parker KD, Atkins EDT, Beighton E (1968) Cross-β conformation in proteins. J Mol Biol 32:343–358
Giraud MM, Castanet J, Meunier FJ, Bouligand Y (1978) The fibrous structure of coelacanth scales: a twisted "plywood". Tissue Cell 10:671–686
Goldsmith MR, Kafatos FC (1984) Developmentally regulated genes in silkmoths. Annu Rev Genet 18:443–487
Green NM, Wrigley NG, Russel WC, Martin SR, McLachlan AD (1983) Evidence for a repeating cross-β sheet structure in the adenovirus fibre. EMBO J 2:1357–1365
Gregg K, Wilton SD, Parry DAD, Rogers GE (1984) A comparison of genomic coding sequences for feather and scale keratins: structural and evolutionary implications. EMBO J 3:175–181

Grierson JP, Neville AC (1981) Helicoidal architecture of fish eggshell. 13:819–830
Groot EP, Alderdice DF (1985) Fine structure of the external egg membrane of five species of Pacific salmon and steelhead trout. Can J Zool 63:552–566
Gubb D (1975) A direct visualisation of helicoidal architecture in *Carcinus maenas* and *Halocynthia papillosa* by scanning electron microscopy. Tissue Cell 7:19–32
Hagenmaier HE, Schmitz J, Fohles J (1976) Zum Vorkommen von Isopeptidbindungen in der Eihülle der Regenbogenforelle (*Salmo gairdneri* Rich). Hoppe-Seyler's Z Physiol Chem 357:1435–1438
Hamodrakas SJ (1984) Twisted β-pleated sheet: the molecular conformation which possibly dictates the formation of the helicoidal architecture of several proteinaceous eggshells. Int J Biol Macromol 6:51–53
Hamodrakas SJ (1988) A protein secondary structure prediction scheme for the IBM PC and compatibles. CABIOS 4:473–477
Hamodrakas SJ, Kafatos FC (1984) Structural implications of primary sequences from a family of Balbiani ring-encoded proteins in *Chironomus*. J Mol Evol 20:296–303
Hamodrakas SJ, Jones CW, Kafatos FC (1982a) Secondary structure predictions for silkmoth chorion proteins. Biochim Biophys Acta 700:42–51
Hamodrakas SJ, Asher SA, Mazur GD, Regier JC, Kafatos FC (1982b) Laser-Raman studies of protein conformation in the silkmoth chorion. Biochim Biophys Acta 703:216–222
Hamodrakas SJ, Paulson JR, Rodakis GC, Kafatos FC (1983) X-ray diffraction studies of a silkmoth chorion. Int J Biol Macromol 5:149–153
Hamodrakas SJ, Kamitsos EI, Papanicolaou A (1984) Laser-Raman spectroscopic studies of the eggshell (chorion) of *Bombyx mori*. Int J Biol Macromol 6: 333–336
Hamodrakas SJ, Etmektzoglou T, Kafatos FC (1985) Amino acid periodicities and their structural implications for the evolutionary conservative central domain of some silkmoth chorion proteins. J Mol Biol 186:583–589
Hamodrakas SJ, Margaritis LH, Papasideri I and Fowler A (1986) Fine structure of the silkmoth *Antheraea polyphemus* chorion as revealed by X-ray diffraction and freeze fracturing. Int J Biol Macromol 8:237–242
Hamodrakas SJ, Kamitsos EI, Papadopoulou PG (1987) Laser-Raman and infrared spectroscopic studies of protein conformation in the eggshell of the fish *Salmo gairdneri*. Biochim Biophys Acta 913:163–169
Hamodrakas SJ, Bosshard HE, Carlson CN (1988) Structural models of the evolutionarily conservative central domain of silk-moth chorion proteins. Prot Eng 2:201–207
Hamodrakas SJ, Batrinou A, Christoforatou T (1989) Structural and functional features of *Drosophila* chorion proteins s36 and s38 from analysis of primary structure and infrared spectroscopy. Int J Biol Macromol 11:307–313
Hinton H (1981) Biology of insect eggs. Pergamon Press, Oxford
Hojrup P, Andersen SO, Roepstorff P (1986) Isolation, characterization and N-terminal sequence studies of cuticular proteins from the migratory locust *Locusta migratoria*. Eur J Biochem 154:153–159
Hurley DA, Fischer KC (1966) The structure and development of the external membrane in young eggs of the brook trout, *Salvelinus fontinallys* (Mitschill). Can J Zool 44:173–190
Iatrou K, Tsitilou SG, Kafatos FC (1984) DNA sequence transfer between two high-cysteine chorion gene families in *Bombyx mori*. Proc Natl Acad Sci USA 81:4452–4456
Jones CW, Kafatos FC (1980a) Coordinately expressed members of two chorion multi-gene families are clustered, alternating and divergently oriented. Nature 284:635–638
Jones CW, Kafatos FC (1980b) Structure, organization and evolution of developmentally regulated chorion genes in a silkmoth. Cell 22:855–867
Jones CW, Kafatos FC (1982) Accepted mutations in a gene family: evolutionary diversification of duplicated DNA. J Mol Evol 19:87–103
Kafatos FC, Regier JC, Mazur GD, Nadel MR, Blau HM, Petri WH, Wyman AR, Gelinas RE, Moore PB, Paul M, Efstratiadis A, Vournakis JN, Goldsmith MR, Hunsley JR, Baker B, Nardi J, Koehler M (1977) The eggshell of insects: differentiation-specific proteins and the control of their synthesis and accumulation during development. In: Beerman W (ed) Results and problems in cell differentiation, vol 8. Springer, Berlin Heidelberg New York, pp 45–145
Kakudo M, Kasai N (1972) X-ray diffraction by polymers. Elsevier, Amsterdam
Kawasaki H, Sato H, Suzuki M (1971) Structural proteins in the silkworm eggshells. Insect Biochem 1:130–148
King RC, Aggarwal SK (1965) Oogenesis in *Hyalophora cecropia*. Growth 29:17–83
Kobayashi W (1982) The fine structure and amino acid composition of the envelope of the chum salmon egg. J Fac Sci Hokkaido Univ Ser 6 23:1–12

Lecanidou R, Rodakis GC, Eickbush TH, Kafatos FC (1986) Evolution of the silkmoth chorion gene superfamily: gene families CA and CB. Proc Natl Acad Sci USA 83:6514–6518
Lipman DJ, Pearson WR (1985) Rapid and sensitive protein similarity searches. Science 227:1435–1441
Livolant F, Bouligand Y (1989) Freeze-fractures in cholesteric mesophases of polymers. Mol Cryst Liq Cryst 166:91–100
Lonning S, Kjorsvik E, Davenport J (1984) The hardening process of the egg chorion of the cod, *Gadus morhua* L., and lumpsucker, *Cyclopterus lumpus* L. J Fish Biol 24:505–522
Lotz B, Gouthier-Vassal A, Brack A, Magoshi J (1982) Twisted single crystals of *Bombyx mori* silk fibroin and related model polypeptides with β-structure. J Mol Biol 156:345–357
Margaritis LH (1985) Structure and physiology of the eggshell. In: Gilbert LI, Kerkut GA (eds) Comprehensive insect biochemistry, physiology and pharmacology, vol I. Pergamon, Oxford, pp 153–230
Marsh RE, Corey RB, Pauling L (1955) The structure of silk fibroin. Biochim Biophys Acta 16:1–34
Mazur GD, Regier JC, Kafatos FC (1980) The silkmoth chorion: Morphogenesis of surface structures and its relation to synthesis of specific proteins. Dev Biol 76:305–321
Mazur GD, Regier JC, Kafatos FC (1982) Order and defects in the silkmoth chorion, a biological analogue of a cholesteric liquid crystal. In: Akai H, King RC (eds) Insect ultrastructure, vol I. Plenum, New York, pp 150–183
McLachlan AD (1977) Analysis of periodic patterns in amino acid sequences: collagen. Biopolymers 16:1271–1297
McLachlan AD, Stewart M (1976) The 14-fold periodicity in α-tropomyosin and the interaction with actin. J Mol Biol 103:271–298
Neville AC (1975) Biology of the arthropod cutide. Springer, Berlin Heidelberg New York
Neville AC (1981) Cholesteric proteins. Mol Cryst Liq Cryst 76:279–286
Neville AC (1986) The physics of helicoids: multidirectional "plywood" structures in biological systems. Phys Bull 37:74–76
Ohzu E, Kusa M (1981) Amino acid composition of the egg chorion of rainbow trout. Annot Zool Jpn 54:241–244
Orcutt BC, George DG, Dayhoff MO (1983) Protein and nucleic acid sequence database systems. Annu Rev Biophys Bioeng 12:419–441
Papanicolau AM, Margaritis LH, Hamodrakas SJ (1986) Ultrastructural analysis of chorion formation in the silkmoth *Bombyx mori*. Can J Zool 64:1158–1173
Parker FS (1971) Applications of infrared spectroscopy in biochemistry, biology and medicine. Plenum, New York
Parry DAD (1979) Determination of structural information from the amino acid sequences of fibrous proteins. In: Parry DAD, Creamer LK (eds) Fibrous proteins: scientific, industrial and medical aspects, vol I. Academic Press, London, pp 393–427
Parry DAD, Fraser RDB, McRae TP (1979) Repeating patterns of amino acid residues in the sequences of some high-sulphur proteins from α-keratin. Int J Biol Macromol 1:17–22
Pau RN (1984) Cloning of cDNA for a juvenile hormone-regulated oothecin mRNA. Biochim Biophys Acta 782:422–428
Rashin AA, Honig B (1984) On the environment of ionizable groups in globular proteins. J Mol Biol 173:515–521
Regier JC (1986) Evolution and higher-order structure of architectural proteins in silkmoth chorion. EMBO J 5:1981–1989
Regier JC, Kafatos FC (1985) Molecular aspects of chorion formation. In: Gilbert LI, Kerkut GA (eds) Comprehensive insect biochemistry, physiology and pharmacology, vol I, Pergamon, Oxford pp 113–151
Regier JC, Vlahos NS (1988) Heterochrony and the introduction of novel modes of morphogenesis during the evolution of moth choriogenesis. J Mol Evol 28:19–31
Regier JC, Wong JR (1988) Assembly of silkmoth proteins: in vivo patterns of disulphide bond formation. Insect Biochem 18:471–482
Regier JC, Kafatos FC, Goodfliesh R, Hood L (1978a) Silkmoth chorion proteins: sequence analysis of the products of a multigene family. Proc Natl Acad Sci USA 75:390–394
Regier JC, Kafatos FC, Kramer KJ, Heinrikson RL, Keim PS (1978b) Silkmoth chorion proteins: their diversity, amino acid composition and the NH_2-terminal sequence of one component. J Biol Chem 253:1305–1314
Regier JC, Mazur GD, Kafatos FC (1980) The silkmoth chorion: morphological and biochemical characterization of four surface regions. Dev Biol 76:286–304
Regier JD, Mazur GD, Kafatos FC, Paul M (1982) Morphogenesis of silkmoth chorion: initial framework formation and its relation to synthesis of specific proteins. Dev Biol 92:159–174

Regier JC, Kafatos FC, Hamodrakas SJ (1983) Silkmoth chorion multigene families constitute a superfamily: comparison of C and B family sequences. Proc Natl Acad Sci USA 80:1043–1047
Richardson JS (1981) The anatomy and taxonomy of protein structure. Adv Prot Chem 34:167–339
Rill RL, Livolant F, Aldrich HC, Davidson MW (1989) Electron microscopy of liquid crystalline DNA: direct evidence for cholesteric-like organisation of DNA in dinoflagellate chromosomes. Chromosoma (Berl) 98:280–286
Rodakis GC, Kafatos FC (1982) Origin of evolutionary novelty in proteins: how a high-cysteine chorion protein has evolved. Proc Natl Acad Sci USA 79:3551–3555
Rodakis GC, Moschonas NK, Kafatos FC (1982) Evolution of a multigene family of chorion proteins in silkmoths. Mol Cell Biol 2:554–563
Rodakis GC, Lekanidou R, Eickbush TH (1984) Diversity in a chorion multigene family created by tandem duplications and a putative gene conversion event. J Mol Evol 20:265–273
Rudall KM (1956) Protein ribbons and sheets. In: Lectures on the scientific basis of medicine 5. Athlone Press, London, pp 217–230
Schulz GE, Schirmer RH (1978) Principles of protein structure. Springer, New York Heidelberg Berlin
Siamwiza MN, Lord RC, Chen MC, Takamatsu T, Harada I, Matsuura H, Shimanouchi T (1975) Interpretation of the doublet at 850 and 830 cm^{-1} in the Raman spectra of tyrosyl residues in proteins and certain model compounds. Biochemistry 14:4870–4876
Sibanda BL, Thornton JM (1985) β-hairpin families in globular proteins. Nature 316:170–174
Smith DS, Telfer WH, Neville AC (1971) Fine structure of the chorion of a moth, *Hyalophora cecropia.* Tissue Cell 3:477–498
Spiro TG, Gaber BP (1977) Laser-Raman scattering as a probe of protein structure. Annu Rev Biochem 46:553–572
Spoerel NA, Nguyen HT, Eickbush TH, Kafatos FC (1989) Gene evolution and regulation in the chorion complex of *Bombyx mori*: hybridization and sequence analysis of multiple developmentally middle A/B chorion gene pairs. J Mol Biol 209:1–19
Squire JM, Vibert PJ (1987) Fibrous protein structure. Academic Press, London
Stewart M (1977) The structure of chicken scale keratin. J Ultrastruct Res 60:27–33
Sugeta H, Go A, Miyazawa T (1972) S–S and C–S stretching vibrations and molecular conformations of dialkyl disulphides and cystine. Chem Lett 1:83–86
Taylor WR (1987) Protein structure prediction In: Bishop MJ, Rawlins CJ (eds) Nucleic acid and protein sequence analysis: a practical approach. IRL Press, Oxford, pp 285–322
Telfer WH, Smith DS (1970) Aspects of egg formation. Symp R Entomol Soc Lond 5:165–185
Tesoriero JV (1977) Formation of the chorion (zona pellucida) in the teleost *Oryzias latipes*.I. Morphology of early oogenesis. J Ultrastruct Res 59:282–291
Vanderlei R, Chaudhri M, Knight M, Meadows H, Chambers A, Taylor W, Kelly C, Simpson AJG (1989) Predicted structure of a major *Schistosoma mansoni* eggshell protein. Mol Biochem Parasitol 32:7–14
Walker ID, Bridgen J (1976) The keratin chains of avian scale tissues. Eur J Biochem 67:283–293
Williams RW, Dunker AK (1981) Determination of the secondary structure of proteins from the amide I band of the laser-Raman spectrum. J Mol Biol 152:783–813
Wourms JP (1976) Annual fish oogenesis. I. Differentiation of the mature oocyte and formation of the primary envelope. Dev Biol 50:338–354
Xu M, Lewis RV (1990) Structure of a protein superfiber: spider dragline silk. Proc Natl Acad Sci USA 87:7120–7124
Young EG, Inman WR (1938) The protein casing of salmon eggs. J Biol Chem 124:189–193
Yu NT (1977) Raman spectroscopy: a conformational probe in biochemistry. CRC Crit Rev Biochem 4:229–280

Chapter 7

Secretory Proteins of *Chironomus* Salivary Glands: Structural Motifs and Assembly Characteristics of a Novel Biopolymer[1]

Steven T. Case[2] and Lars Wieslander[3]

1 Introduction

Aquatic larvae of the midge, *Chironomus* (Diptera), historically have been known for the giant polytene chromosomes found in the nuclei of their salivary glands. More than a century ago, Balbiani (1881) described large puffs on these chromosomes. These puffs, referred to as Balbiani rings (BRs), were recognized by Beermann (1952, 1961) as loci of tissue-specific gene expression and are now known to be sites of intensive transcription (Pelling 1964; Case and Daneholt 1977; Grossbach 1977). Cytogenetic studies by Beermann (1961) and Grossbach (1969) led to the proposal that BRs contain genes encoding secretory proteins, the major product of the salivary gland. Secretory proteins assemble in vivo into insoluble silken threads that larvae use for feeding and construction of housing and pupation tubes.

The classic relationship between salivary gland BRs and secretory proteins has been reviewed (Beermann 1973; Case and Daneholt 1977; Grossbach 1977). During the past decade, data obtained by molecular cloning have confirmed that BRs contain genes for secretory proteins. Moreover, significant progress in understanding the primary structure of BR genes and their encoded secretory proteins have revealed two new and significant themes. Firstly, secretory proteins are encoded by a multigene family that may have emerged by tandem duplication, homogenization, translocation, and divergence of small ancestral sequences (Pustell et al. 1984; Wieslander et al. 1984; Dreesen et al. 1985; Bäumlein et al. 1986; Höög et al. 1986, 1988; Grond et al. 1987; Saiga et al. 1987; Dignam et al. 1989; Galli et al. 1990; Paulsson et al. 1990, 1992a, b; Galli and Wieslander 1992). Secondly, secretory proteins have unique structures and, when assembled into silk, become components of a novel and dynamic composite biopolymer. The latter theme will be the focus of this chapter.

---

[1] The authors wish to dedicate this chapter to Professor Bertil Daneholt who, earlier in our careers, provided a stimulating atmosphere, freedom to inquire and encouragement to pursue experiments that brought us to our current understanding of the secretory proteins and their genes.

[2] Department of Biochemistry, The University of Mississippi Medical Center, Jackson, MS 39216-4505 USA.

[3] Department of Molecular Genetics, Medical Nobel Institutet, Karolinska Institute, Box 60400, S-104 01 Stockholm, Sweden

Results and Problems in Cell Differentiation 19
Biopolymers
Case, S. T. (Ed.)

To fully appreciate these secretory proteins and their biopolymer, it is essential to relate their structure, regulated synthesis, and assembly to what is known about the biology of *Chironomus* and the morphology and function of larval salivary glands.

2 The Biology of *Chironomus*

2.1 Taxonomy and Habitat

Microphagus macroinvertebrates constitute a vital pathway for energy flow in aquatic environments (Anderson and Sedell 1979; Wallace 1980; McCafferty 1983). After consuming small plants, animals, and detritus, they themselves become a prime source of nutrients for larger species including vertebrates (Walshe 1951).

The Chironomidae (for taxonomic classification, see Table 1) are the most diverse family of insects in selection of their aquatic habitats (Oliver 1971; Neumann 1976; McCafferty 1983). Their evolution most likely began in a stable primitive habitat in the upper portion of mountain streams that ran through temperate forests (Oliver 1971). Subfamilies eventually became distributed within geographical ranges specified by the availability of water ecologically suitable for their larvae. The subfamily Chironominae is composed of genera each adapted to living in different types of water; lentic or lotic, cold or warm, oxygen-rich or oxygen-poor. They are the most flourishing fresh-water macroinvertebrates, distributed worldwide on all continents and Arctic lands, and typically constitute the dominant species within their habitat (McCafferty 1983).

Since the habitat of *Chironomus* is reasonably easy to duplicate, several species have been successfully cultivated in the laboratory (Sadler 1935; Grossbach 1977; Case and Daneholt 1978; Kiknadze et al. 1990). Being generally microphagous,

Table 1. The taxonomic classification of *Chironomus*[a]

Phylum, Arthropoda;
Class, Insecta;
Subclass, Pterygota (wings primarily present in adults);
Division, Holometabola (Endoterygota, larvae with internal wing buds and special larval eyes, metamorphosis complete);
Order, Diptera (true flies, mosquitoes, midges, gnats);
Infraorder, Culicomorpha;
Superfamily, Chironomoidea,
Family, Chironomidae (previously Tendipedidae)
Subfamily, Chironominae; Tribe, Chironomini;
Genus; *Chironomus*
(e.g., *C. thummi*, *C. plumosus*);
Subgenus: *Camptochironomus*
(*C. tentans*, *C. pallidivittatus*).

[a] Information for this table was compiled from several sources (Hickman 1967; Oliver 1971; Grossbach 1977; McCafferty 1983; Wood and Borkent 1989).

they will eat almost any available alga, protozoan, diatom, smaller invertebrate, or detritus.

2.2 Life Cycle

There are four basic stages in the life cycle of *Chironomus*: egg, larva, pupa, and adult (Fig. 1). The first three stages are aquatic whereas the adult stage is nonaquatic. The time required for growth and development varies from 4 to 8

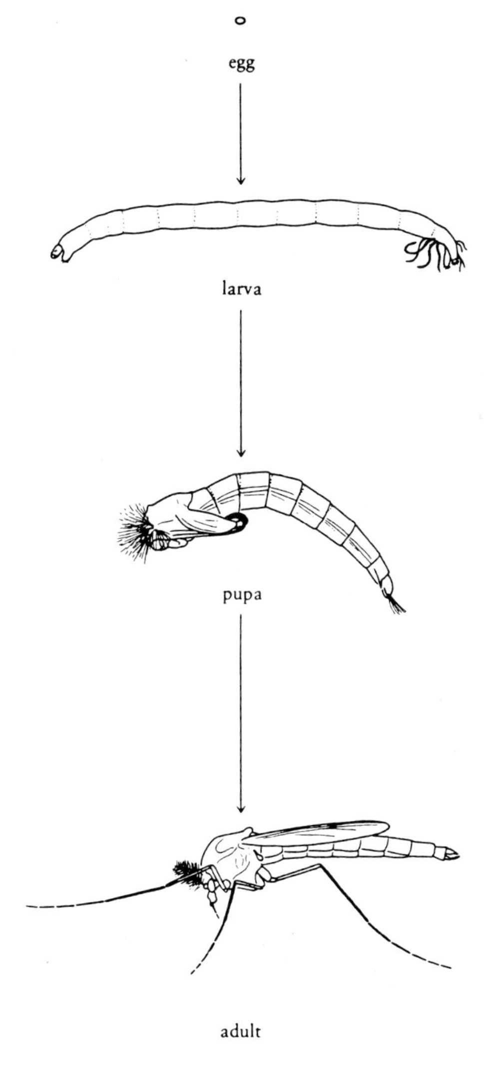

Fig. 1. Illustrations of different stages in the life cycle of *Chironomus*. (McCafferty 1983)

weeks depending on both the species and environmental factors such as temperature, photoperiod, availability of food, and population density. Below is a simplified summary of salient features of each stage common to most species. Detailed descriptions for specific species have been published elsewhere [*Chironomus tentans*, (Sadler 1935); *Chironomus pallidivittatus*, (Beermann 1955) *Chironomus plumosus*, (Hilsenhoff 1966); *Chironomus thummi*, (Kiknadze et al. 1990)].

2.2.1 Egg

The size of eggs varies among species (Sadler 1935; Hilsenhoff 1966; Kiknadze et al. 1990); however, all are oval in shape, ranging from 0.2–0.5 mm in length and 0.1–0.2 mm in diameter. A single female will lay between 300 to 3000 eggs at one time in a mass that is approximately 2–4 mm in diameter and 7 mm long. Immediately upon contact with water, it swells to as much as 7 mm in diameter and 25 mm in length. At this point one can observe the eggs deposited within a transparent gelatinous ribbon wound around two nearly invisible longitudinal cords. Fertilized eggs undergo embryogenesis (Sadler 1935) which can last from a few days to 2 weeks, depending upon the temperature of the water. At temperatures between 20–24 °C, eggs typically hatch within 3 to 4 days.

2.2.2 Larva

The larval stage of development is divided into four instars demarcated by molting of the larval cuticle. The duration of each instar and the size of larvae vary with species and growth conditions. For example, as the environmental temperature increases, development proceeds more rapidly, resulting in mature larvae that are generally smaller than those which developed at lower temperatures. One quick and reliable way to distinguish between instars within a species raised under the

Table 2. A comparison of head diameter and larval length during four instars of larval development in three species of *Chironomus*

Instar	1st	2nd	3rd	4th
Duration (reported range in days)	3–9	3–8	6–10	5–21
C. thummi[a]				
Length of Larva (mm)	0.88	1.73	3.3	6–12
Head capsule diameter (mm)	0.104	0.20–0.25	0.33–0.38	0.48–0.68
C. tentans[b]				
Length of Larva (mm)	0.68–0.82	4–5	10–12	14–18
Head capsule diameter (mm)	< 0.102	0.21–0.22	0.37–0.47	0.71–0.74
C. plumosus[c]				
Length of Larva (mm)	1.4–2.5	2.3–5.5	5.3–12	18–28
Head capsule diameter (mm)	0.12–0.15	0.22–0.28	0.44–0.55	0.88–1.01

[a] Kiknadze et al. (1990).
[b] Sadler (1935).
[c] Hilsenhoff (1966).

same conditions is the diameter of the head capsule which, while constant during any one instar, increases geometrically at each larval/larval molt (Table 2).

2.2.2.1 First instar

Upon hatching, larvae spend the first day within the gelatinous egg mass, becoming free-swimming on the 2nd and 3rd days. These first instar larvae are generally transparent and have four anal gills (for anatomical diagrams, see Miall and Hammond 1900). They exhibit a strong positive phototropism unless they are able to burrow in mud or silt. With the possible exception of *C. plumosus* (Hilsenhoff 1966), first instar larvae are capable of constructing tubes fabricated from viscous threads of secretion which are expelled from their mouths and bond together any available particulate matter (such as mud and detritus). The construction and function of these tubes will be discussed below (Sect. 3.4)

2.2.2.2 Second Instar

Larvae begin to turn pink to dark red coincident with the appearance of three pairs of blood gills located in the second-to-last and penultimate posterior segments. This color change gives rise to the common name for *Chironomus* larvae, bloodworms. Growth is most rapid during this instar.

2.2.2.3 Third Instar

Larvae become bright red and avoid light. Most time is spent in larval tubes, with larva only coming out to search for food when the local supply is scarce.

2.2.2.4 Fourth Instar

While this larval stage is most used for cytological and biochemical studies, development during this instar is naturally asynchronous and particularly susceptible to alteration by environmental factors. For example, *C. tentans* larvae can undergo subitaneous (uninterrupted) development, in which this stage can last from 1 to 3 weeks. Alternatively, they can be induced into a nonobligatory and reversible form of diapause known as oligopause (Ineichen et al. 1979). Many species overwinter as fourth instar larvae (Hilsenhoff 1966; Oliver 1971).

While the diameter of the head capsule can identify larvae within the fourth instar, natural asynchrony makes it impossible to further classify larvae simply based on a time-scale or larval length. However, the precise physiological stage of development can be ascertained by means of morphological criteria for developing imaginal disks. For example, the fourth instar of *C. thummi* (Wulker and Gotz 1968; Kiknadze et al. 1990) and *C. tentans* (Ineichen et al. 1983) can be divided into larval and prepupal stages or classified further into nine or more stages and substages. Prepupae outwardly appear like larvae; however, internally, epidermal cells switch to synthesis of pupal cuticle concurrent with pronounced growth of imaginal disks from which adult tissues will arise. Partial apolysis of imaginal epidermis from the prospective pupal cuticle takes place prior to the observed larval/pupal ecdysis.

2.2.3 Pupa

Pupation takes place inside a modified housing tube or newly constructed pupation tube (see Sect. 3.4). This stage is highly temperature-dependent (Hilsenhoff 1966) and lasts from 2 to 10 days. After spending the 1st day within the pupation tube, a pupa emerges into the surrounding water. In the pupa, larval organs (such as salivary glands) finish undergoing lysis and formation of adult organs from imaginal disks is completed. Finally, the pupa enters an active swimming phase during which it rises to the surface of the water. The pupal integument splits along the mediodorsal portion of the thorax and within seconds, an adult fly emerges.

2.2.4 Adult

The adult fly, commonly known as a midge, is the only nonaquatic stage of this organism. It lives for 2 to 11 days, during which time mating occurs. The most common mating behavior involves a female midge flying into a swarm of males and participating in copulation in flight. However, some species, such as *C. tentans* and *C. pallidivittatus* can, at least in captivity, undergo pair-wise mating (Beermann 1955). Oviposition of fertilized eggs generally occurs within 2 to 3 days.

3 Structure and Function of Larval Salivary Glands

3.1 Salivary Gland Morphology

Each larva has a pair of salivary glands; flat, sac-like bodies, generally spanning the 2nd and 3rd thoracic segments (Miall and Hammond 1900). First instar larvae already have functional salivary glands. While these increase in size during subsequent instars, their morphological appearance will remain essentially unaltered (for example, see Fig. 1 in Grossbach 1977). This is due to the fact that although the dimensions of the gland increase geometrically during larval growth, there are no concurrent cell divisions. By the fourth instar the salivary gland is approximately 1–2 mm in length. The perimeter of the gland consists of a single layer of 30–40 secretory cells with huge nuclei readily visible in the light microscope (Fig. 2). The flat upper and lower surfaces of the gland are each covered by one large, thin epithelial cell whose nucleus is difficult to see. The cavity between the epithelial cells and peripheral secretory cells forms the glandular lumen. Contents of the lumen exit via a secretory duct located near the anterior portion of the gland. The intima of the duct is lined by spirals of chitin surrounded by a thin layer of flat epithelial cells (Churney 1940). The duct is nearly as long as the gland. The ducts from the two glands join in a short common secretory canal which terminates at the floor of a larva's mouth near the posterior tip of the hypopharynx.

The distribution of secretory cells is asymmetric resulting in several distinct lobes (Fig. 2): posterior, main, lateral (or side), and two anterior lobes. One anterior

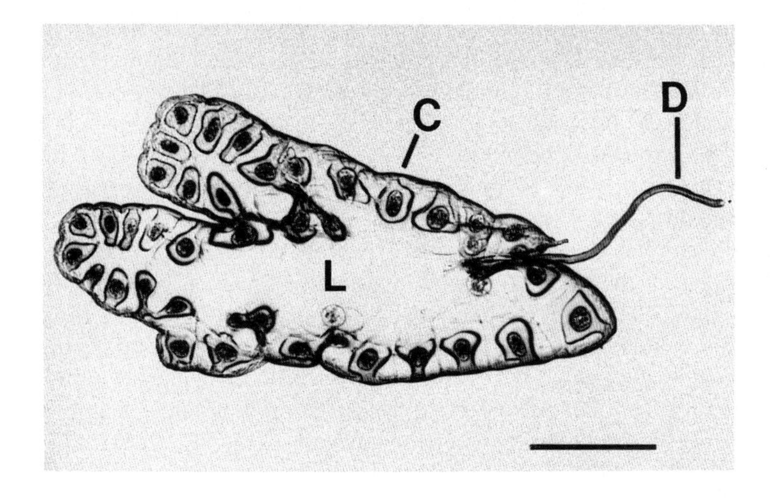

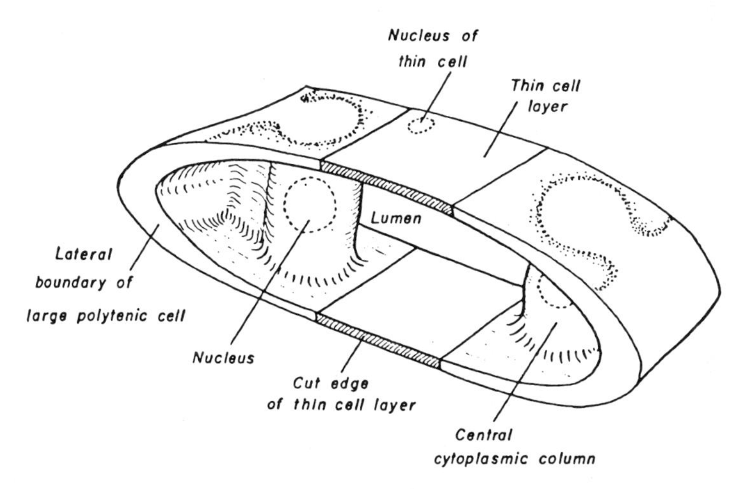

Fig. 2. *Upper panel* a salivary gland from a fourth instar larva of *C. tentans*. Secretory cells (*C*), each with a giant nucleus, surround the central lumen (*L*) where secretory proteins accumulate. Secretory proteins exit the lumen via the salivary duct (*D*) which emanates between the anterior lobes. *Bar* 200 μm. *Lower panel* a diagramatic three-dimensional representation of a section cut through the center of a gland. (Kloetzel and Laufer 1969)

lobe is broad and curved while the other tends to be thin and elongated. The latter is referred to as the "special lobe" and its constituent secretory cells are referred to as "special cells." What makes this lobe "special" is that it is morphologically and biochemically distinct from all other lobes. For example, in *C. thummi* and *C. pallidivittatus*, this lobe contains an abundance of secretory granules (Beermann 1961) and at least one additional secretory protein that is absent from the other lobes (Grossbach 1977; Kolesnikov et al. 1981). While the corresponding lobe in *C. tentans* exhibits neither the granules nor an extra protein, secretion in this lobe does have a unique appearance (Grossbach 1977).

The ultrastructure of the saddle-shaped secretory cells is distinctive. Nuclei contain giant polytene chromosomes; numerous rounds of endoreplication result in chromosomes with 4000 to 16000 times the haploid DNA content (Beermann 1952; Daneholt and Edström 1967). These chromosomes contain several distinctive features including BRs (Beermann 1973), giant salivary gland-specific puffs with actively transcribed genes encoding secretory proteins. The cytoplasm has three characteristic zones (Kloetzel and Laufer 1969; Thyberg et al. 1982). The basal cytoplasm (near the gland's periphery) is densely packed with mitochondria, the medial cytoplasm, which occupies the greatest volume, with rough endoplasmic reticulum and Golgi bodies. The apical cytoplasm contains numerous secretory granules and coated pits along the brush border facing the glandular lumen. These features are typical of eukaryotic cells actively synthesizing and secreting proteins. These cytoplasmic zones are visible during intermolt periods but undergo substantial changes during molting when the synthesis of secretory proteins is reduced (Kloetzel and Laufer 1970; Kiknadze et al. 1990).

3.2 The Secretory Protein Family

Secretory proteins are the major product of *Chironomus* salivary glands (Grossbach 1977; Rydlander and Edström 1980). These proteins are synthesized in the secretory cells, secreted into and temporarily stored in the lumen of the gland. Grossbach (1969) first recognized two important properties of these proteins; some were extraordinarily large and their overall amino acid composition was simple.

Salivary gland secretory proteins are actually quite diverse in size. When the lumenal contents of a salivary gland are denatured, reduced, alkylated, and separated by electrophoresis in concave exponential gradient polyacrylamide gels containing sodium dodecyl sulfate (Kao and Case 1985), the most abundant members of the family can be visualized simultaneously (Fig. 3). Their apparent size distribution covers more than two orders of magnitude.

From the outset it should be made clear that while we refer to these proteins by their apparent molecular weight (i.e., a secretory protein with an apparent molecular weight of 185 kDa is referred to as sp185), these values are probably imprecise. The unusual size and aberrant behavior of these proteins under certain electrophoretic conditions (Case ST, unpubl. data) indicate that accurate molecular weights will not be available until each protein can be purified biochemically and characterized by biophysical techniques.

The three best characterized species of *Chironomus* exhibit similar patterns of large, intermediate, and small size classes of secretory proteins. The large size class contains the spIs with an approximate mass of 1000 kDa (Rydlander and Edström 1980; Hertner et al. 1980). In *Camptochironomus* (*C. tentans* and *C. pallidivittatus*), four spIs in all (spIa, spIb, spIc, and spId) have been described (Rydlander and Edström 1980; Edström et al. 1980; Kao and Case 1985). At least three spIs have been reported for *C. thummi* (Serfling et al. 1983; Cortes et al. 1989; Diez et al. 1990). While spIs are related in size and structure (discussed below in detail), they

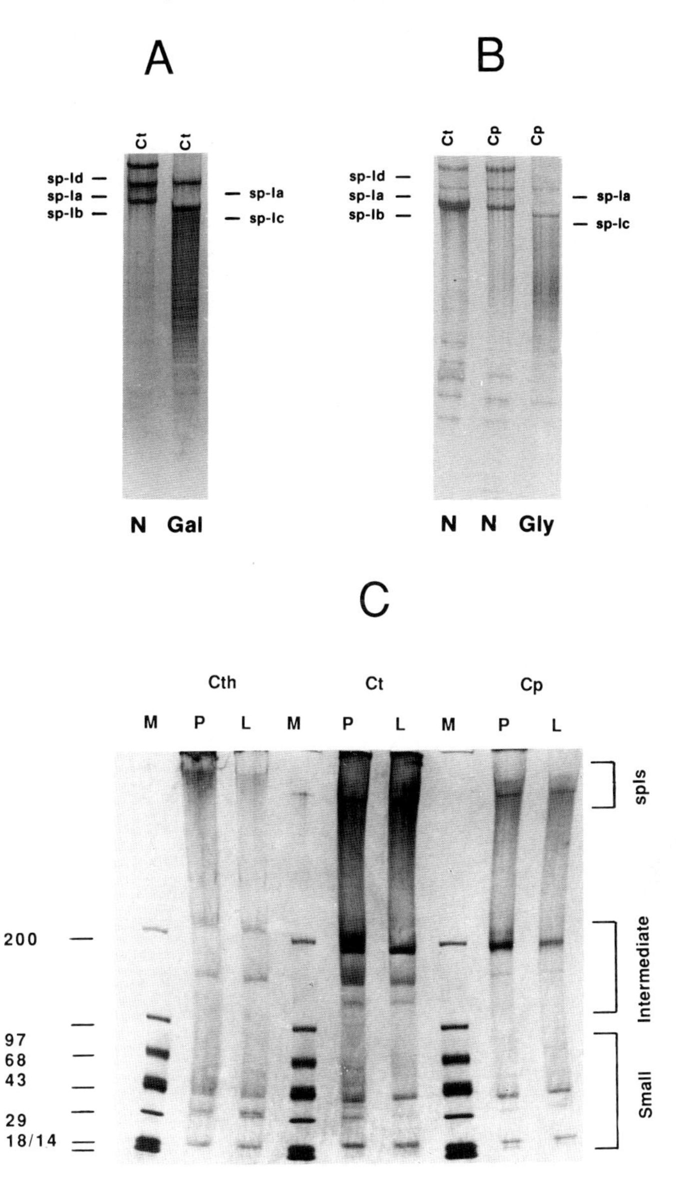

Fig. 3A–C. Electrophoretic comparison of secretory proteins extracted from salivary glands of three species of *Chironomus*. Salivary glands dissected from fourth-instar larvae were fixed in 70% ethanol and cells removed under a dissecting microscope. The remaining plug of secretion was denatured in 6 M guanidine hydrochloride, reduced with dithiothreitol, alkylated with iodoacetamide, and prepared for electrophoresis on a 3–20% concave exponential gradient of polyacrylamide as described (Kao and Case 1985). **A, B** Single-gland extracts were stained with silver nitrate to enhance identification of spIs in samples obtained from *C. tentans* (*Ct*) and *C. pallidivittatus* (*Cp*) larvae kept in normal (*N*), galactose (*Gal*)-, or glycerol (*Gly*)-containing culture medium. **C** Multiple-gland extracts stained with Coomassie Blue to enhance comparison of intermediate and small size secretory proteins from larvae (*L*) and prepupae (*P*) of *C. tentans* (*Ct*), *C. pallidivittatus* (*Cp*), and *C. thummi* (*Cth*). Numbers on the *left* indicate the sizes (in kDa) and positions of molecular weight markers (*M*)

are encoded by different genes. The intermediate size class contains secretory proteins from about 100 to 200 kDa. In *C. tentans*, for example, the members of this class include sp195, sp185, sp140, and sp115. Finally, there is the small size class (100 kDa or less) which includes sp65, sp55, sp40, sp39, sp38, and sp18. By varying electrophoretic conditions, it is possible to resolve additional members of both the intermediate and small size classes.

The special lobe cells and secretion in *C. pallidivittatus* (Beermann 1952; Grossbach 1969) and *C. thummi* (Kolesnikov et al. 1981) have an additional secretory protein not found in the other lobes of the salivary gland. In *C. thummi*, this secretory protein has an apparent molecular weight of 160 kDa and is glycosylated (Kolesnikov et al. 1981).

A detailed discussion of the structure of secretory proteins will be presented in Section 4.

3.3 Formation and Extrusion of Silk

Upon demand, the mixture of secretory proteins, also referred to as secretion, exits the lumen of each gland via its secretory duct (Fig. 4), passes through a short common secretory canal and empties through the floor of the mouth near the hypopharynx. The anatomy of the salivary apparatus and process of delivering secretion from the gland to the mouth has been described in detail (Churney 1940). The most anterior portion of the salivary apparatus consists of a pair of abductor muscles which attach to a chitinous band at the posterior tip of the hypopharynx. Also attached to this chitinous band are two broad elastic bands which, at the other end, insert within and block the aperture of the common canal. The flow of secretion coincides with the displacement of the common canal and is comparable to a modified air-pump in which the barrel moves about a stationary piston. The common canal forms the barrel which is movable by retractor muscles attached to its dorsal surface. The elastic bands act as the piston by virtue of the cone-like plug they form within the canal. Active movement is in the posterior direction; contraction of the retractor muscles stretches and separates the two elastic bands, opening the aperture of the canal and sucking secretion into the mouth. Passive movement is in the anterior direction; the retractor muscles relax and the elastic bands broaden and reinsert themselves within the aperture to plug it.

When secretion enters the larva's mouth, it appears as an insoluble silk-like thread which, for simplicity, we shall refer to as silk. Although it has not yet been proven biochemically, it has long been assumed that *Chironomus* silk is a mixture of salivary gland secretory proteins assembled and transformed into an insoluble silk-like thread. Secretory protein assembly will be discussed below (Sect. 7).

3.4 Uses of Silk and Formation of Tubes

Chironomus belongs to a diverse family of aquatic insects that spin silk for at least three physiologically important roles; construction of tubes for protective housing, assistance in feeding, and modification of tubes for pupation.

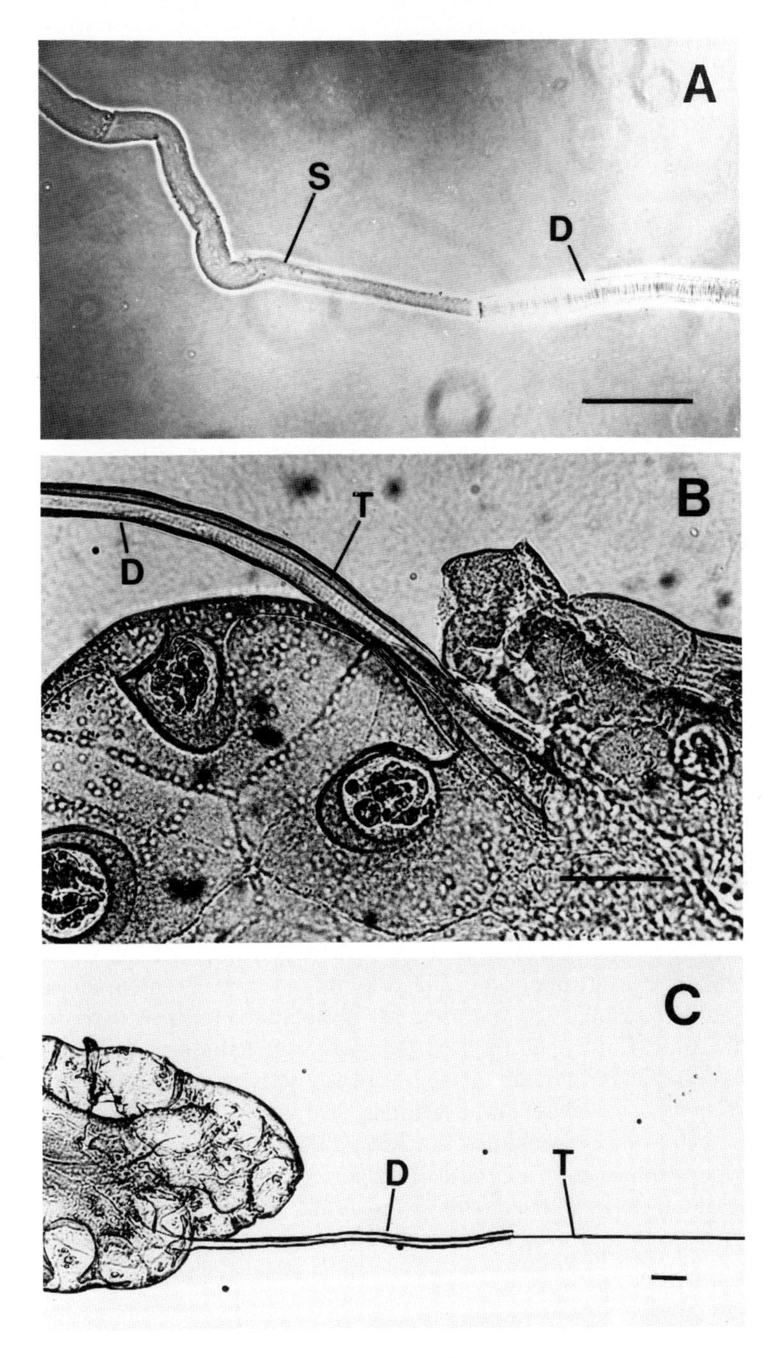

Fig. 4A–C. Secretion in the salivary duct. *C. tentans* salivary glands were dissected and observed in 0.9% NaCl with phase contrast optics (**A**) or dehydrated with 70% ethanol and observed with bright-field illumination (**B**, **C**). Under aqueous conditions, secretion (*S*) is viscous and fills the duct (*D*). When dehydrated, secretion shrinks away from the wall of the duct (**B**) and becomes an elastic thread (*T*). *Bars* 100 μm

3.4.1 Tubes for Protective Housing

Larvae in all instars use silk to build housing tubes (Sadler 1935; Darby 1962) which protect them from detection by predators. In soft lake bottoms, housing tubes consist of borrows dug in the mud with walls reinforced by silk. On hard surfaces, such as laboratory cultivation tubs, larvae weave horizontal tubes comprised of any available particulate matter that can adhere to the silk (Hodkinson and Williams 1980). Housing tubes can be individual or clustered and vary in size and shape depending upon the genus (Walshe 1951; Darby 1962; Scott 1967; Oliver 1971), stage of development (Hilsenhoff 1966; Scott 1967), and changes in environmental conditions including sediment (McLachlan and Cantrell 1976; Hodkinson and Williams 1980) and population density (McLachlan 1977). They are typically longer than their residents (Sadler 1935; Dreesen et al. 1988) and are constantly being extended to accommodate the size of growing larvae (Walshe 1951). Larvae rarely leave their tubes unless food becomes locally unavailable or their tube is destroyed (Sadler 1935). In the latter case, larvae will rapidly rebuild a tube or occupy one that is uninhabited (Sadler 1935; Scott 1967).

3.4.2 Assistance with Feeding

Larvae use silk indirectly and directly for feeding, including the adaptation of housing tubes to trap available sources of food (McLachlan 1977; Hodkinson and Williams 1980; Walentowicz and McLachlan 1980). For example, fine particles of food enter housing tubes in water currents and adhere to silk in the walls. In standing water, currents may be induced by sinuous undulations of the resident larva (Sadler 1935; Darby 1962). Passive feeding amounts to scraping adhering food particles off the tube walls (Sadler 1935; Kiknadze et al. 1990). Silk is actively used in filter feeding. Tubes used for filter feeding can be elaborate in design and indicative of the genus that built them (Walshe 1951; Darby 1962; Scott 1967). Three distinctive types of filter feeding have been described for Chironominae (Walshe 1951). For example, *C. plumosus* larvae spin a catch-net across the diameter of the housing tube and use body undulations to draw currents of water through the tube. Entrapped food particles are ingested along with the catch-net which is subsequently reconstructed. Another active use of silk is feeding by foraging. *Camptochironomus* larvae routinely feed this way when cultivated on hard substrates such as laboratory tubs. Anchoring the caudal end of their body to the housing tube with hooks at the end of their posterior prolegs, they feed in a circular area as far out as they can reach. Silk is spread on the anterior forelegs and used to gather up the food and drag it back into the housing tube (Walshe 1951; Oliver 1971; Kiknadze et al. 1990).

3.4.3 Tubes for Pupation

Larvae pupate in tubes (Sadler 1935; Darby 1962; Hilsenhoff 1966; Scott 1967) and silk is apparently vital for pupation. The prepupae of some species build new tubes especially for pupation, whereas other species modify existing housing tubes; pupation tubes are typically shorter and more constricted at the ends (Darby 1962;

Scott 1967). The walls of pupation tubes are also more resistant to tearing (Dreesen et al. 1988). In either case, prepupae apparently attach the larval cuticle to the inner walls of the pupation tube to anchor it while the pupa emerges from within. Prepupae usually fail to molt, and die if they are removed and restrained from reentering or constructing a tube (Dreesen TD and Case ST, unpubl. observations).

4 Structure of Secretory Proteins

During the past decade, molecular cloning has revealed much about the primary structure of secretory proteins and the location and expression of their genes. The species whose genes have been most extensively studied is *C. tentans* (Table 3). This section will summarize what is currently known about the structures of these proteins and how this may relate to their ability to assemble into insoluble threads of silk.

4.1 The spI Family

Grossbach (1969) originally described "fraction I" of salivary gland secretion as a protein with an unusually high molecular weight. Fraction I is actually a family (referred to as the "spI family") of several secretory proteins with apparent molecular weights of about 1000 kDa (Rydlander and Edström 1980; Hertner et al. 1980). Four members of this family (spIa, spIb, spIc, and spId) have been identified in *Camptochironomus* (Rydlander and Edström 1980; Edström et al. 1980; Hertner et al. 1983; Kao and Case 1985), whereas three members have thus far been identified in *C. thummi* (Serfling et al. 1983; Cortes et al. 1989; Diez et al. 1990). In spite of their size, data exist to substantiate the fact that these are monomeric proteins. The spI mRNAs, historically referred to collectively as 75S RNA (Daneholt 1972), are approximately 37000 nucleotides long (Case and Daneholt 1978) and can be recovered from larval salivary gland secretory cell polysomes (Daneholt et al. 1977; Wieslander and Daneholt 1977). This RNA can direct in vitro translation of polypeptides with immunological and biochemical properties common to spIs (Hardy and Pelling 1980; Rydlander et al. 1980; Weber et al. 1983). In *Camptochironomus*, four genes have been identified whose expression in vivo is correlated with the steady-state levels of individual mRNAs and spIs (Edström et al. 1980; Rydlander et al. 1980; Galler et al. 1984, 1985; Kao and Case 1985; Case 1986). In addition, incorporation of gene-specific encoded amino acids is in agreement with these correlations (Botella et al. 1988). The gene: protein assignments derived from these studies are indicated in Table 3.

Besides their size, spIs are related structurally. During the 1980s, five laboratories worldwide concurrently used genomic and cDNA cloning to study the spI genes in three species of *Chironomus* (*C. tentans*, *C. pallidivittatus*, and *C. thummi*) (for reviews, see Pustell et al. 1984; Wieslander et al. 1984; Grond et al. 1987) and a

Table 3. Summary of known secretory protein structures, gene locations, and expression in *C. tentans*

Secretory protein	Apparent molecular weight (kDa)	Major structural element	Chromosome	Locus	Expression during fourth instar	Reference
spIa	~ 980	82-residue tandem repeats; $[C + SR]_n$	IV	BR1	Throughout	[a]
spIb	~ 850	50–90 residue tandem repeats; $[C + SR]_n$	IV	BR2	Throughout	[a]
spIc	~ 750	79-residue tandem repeats; $[C + SR]_n$	III	BR6	Inducible by galactose throughout	[a]
spId	~ 1100	74-residue tandem repeats; $[C + SR]_n$	IV	BR2	Throughout	[a]
sp195	195	25-residue simple tandem repeats	IV	BR1	Prepupal stages	[b]
sp185	185	Cys-X-Cys-X-Cys every 22–26 residues	IV	BR3	Throughout	[c]
sp140	140	14-residue simple tandem repeats	I	17-B	Prepupal stages	[d]
sp115	115	14-residue simple tandem repeat	I	17-B	Prepupal stages	[d]
sp40	40	244-residue prototype unit	II	5	Only during larval stages	[e]
sp38	38	253-residue prototype unit	II	5	Prepupal stages	[e]

[a] Edström et al. (1980, 1982); Hertner et al. (1980); Rydlander and Edström (1980); Sumegi et al. (1982); Wieslander et al. (1982); Case and Byers (1983); Case et al. (1983); Wieslander and Lendahl (1983); Galler and Edström (1984); Höög and Wieslander (1984); Lendahl and Wieslander (1984, 1987); Galler et al. (1985); Kao and Case (1985, 1986); Case (1986); Dreesen et al. (1988).
[b] Dreesen et al. (1985, 1988); Dreesen and Case (1987).
[c] Dignam and Case (1990); Paulsson et al. (1990).
[d] Dignam et al. (1989); Galli et al. (1990).
[e] Galli and Wieslander (1992).

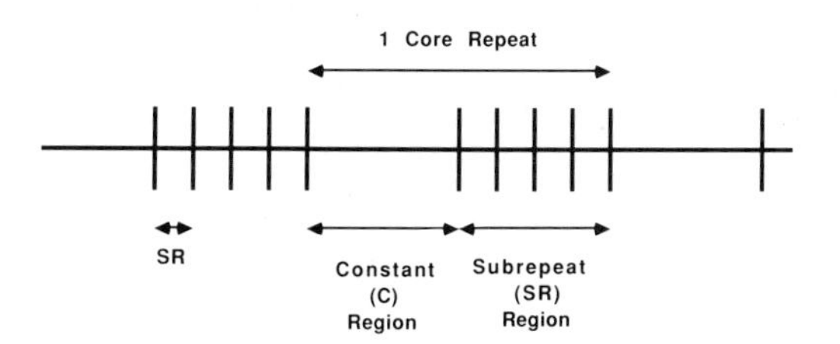

Fig. 5. A diagrammatic representation of the composite structure of tandemly arranged core repeats in spIs

fascinating story began to unravel. The primary structure of spIs is characteristic of fibrous proteins and consists mainly of tandem copies of distinctive "core repeats" (Fig. 5) dominated by hydrophilic amino acids. This predominant pattern of repeats, substantiated by tryptic peptide mapping (Hertner et al. 1980, 1983; Rydlander et al. 1980; Rydlander 1984; Kao and Case 1985), explains why such large and potentially complex proteins have such a simple amino acid composition (Grossbach 1969; Hertner et al. 1983; Rydlander 1984). Each core repeat contains 60–90 amino acids and has two domains: a constant (C) region and a subrepeat (SR) region (Fig. 5, Table 3). The C region of each core repeat is 30 residues long and has 4 Cys, 1 Met, and 1 Phe, all of which are invariant. SR regions may be regular or variable in length due to the size and number of shorter repeated amino acid sequence elements. This can be illustrated by an analysis of BR2.1 and BR2.2 repeats (Höög and Wieslander 1984) which encode spId and spIb, respectively (Case 1986). Core repeats in spId are about 74 residues in length and contain relatively uniform SR regions that consist of seven copies of a six-residue repeat. In contrast, the core repeat of spIb varies between 50 to 90 residues in length due to variable lengths of SR regions that consist of between two and six copies of a ten-residue repeat. Core repeats can be classified according to amino acid sequence elements within their SR region (Grond et al. 1987). In all instances, elements within SR regions contain a tripeptide motif: ⊕ Pro ⊖ (basic residue-Pro-acidic residue). The basic residues are typically Arg or Lys, and the acidic residues are Glu, phospho-Ser, or phospho-Thr. In summary, the primary structure of spIs consists primarily of an unusual arrangement of alternating domains represented by C and SR regions.

To gain some idea about the secondary structure of spIs, Hamodrakas and Kafatos (1984) used the known primary sequences of core repeats and a combination of six different methods to predict their protein secondary structure. Their results predict that the C region consists of 18–20% α-helix and 7% β-turn. Although the highly repetitive nature of the SR region was not amenable to predictive methods, these authors suggest that it would have some collagen-like structure.

An alternate approach to studying the higher order structure of spIs has been to synthesize peptides corresponding to model C and SR regions. While this approach

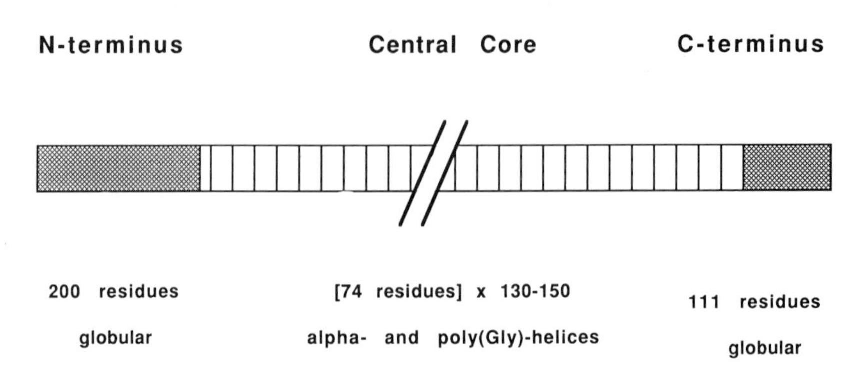

Fig. 6. A diagrammatic representation of the overall structure of spId

is limited to studying isolated domains, it provides biochemically pure substrates (lacking the heterogeneity found between core repeats) suitable for biophysical studies such as circular dichroism and Fourier transform infrared spectroscopy (Case et al. 1991; Wellman et al. 1992). The results of such studies indicate that the C domain of spIa is, in fact, primarily α-helical with as many as two β-turns. In marked contrast, an spIa SR domain appears to be a more extended, poly(Gly)II-type helix interrupted by two or more β-turns. Contrary to the circular dichroism spectrum, the infrared spectrum lacks evidence for the presence of β-sheet. Since infrared spectroscopy is more accurate than circular dichroism for detecting β-sheet, it appears unlikely that spIs contain this structure. The elucidation of these structures has made significant contributions to our understanding of the assembly of spIs into fibers (see Sect. 7.2). Moreover, it indicates that spIs contain a novel pattern of alternating helical structures with contrasting physical properties such as elasticity, flexibility and tensile strength.

It has been proposed (Hamodrakas and Kafatos 1984; Lendahl and Wieslander 1984; Wieslander et al. 1984) that core repeats may facilitate fiber formation by providing multiple in-phase alignments of juxtaposed SR regions on parallel spI molecules so that intermolecular electrostatic interactions can occur. Since C regions would also be aligned, subsequent formation of intermolecular disulfide bonds would provide covalent linkage of adjacent spIs. The central role of spI core repeats in fiber assembly will be discussed below (Sect. 7.2).

The overall structure of spIs was recently deduced from the complete nucleotide sequence organization of the protein-coding portions of their genes (Wieslander and Paulsson 1992; Paulsson et al. 1992a, b). All four spIs share a common architecture: a huge central domain of 130–150 core repeats flanked by two globular domains (Fig. 6). The amino-termini are about 200 residues long and each presumably forms a globular domain with several patches of negatively and positively charged residues on its surface. The carboxyl-termini of spIs are also unique and include an 111-amino acid sequence containing two stretches of hydrophobic residues that are more conserved among spIs than even the core repeats (Bäumlein et al. 1986; Höög et al. 1986, 1988; Saiga et al. 1987). If these terminal structures function in fiber assembly, then they presumably do so in a

```
                10                  20                  30                      40                  50                  60                  70
                 *                   *                   *                       *                   *                   *                   *
TAT CCA GCA AAC AAA AAA TGG AAC GAA AAC ACG |TGT |AGC CTG GAA |TGC |AAA ACT GAA AAG CCT AAA CCT GAT GAT
Tyr Pro Ala Asn Lys Lys Trp Asn Glu Asn Thr |Cys |Ser Leu Glu |Cys |Lys Thr Glu Lys Pro Lys Pro Asp Asp
                                                                                                ___________

         80                  90                 100             110                 120
          *                   *                   *               *                   *
ATT CCA GCA AAC AAA AAA TGG AAC GAA AAC ACG |TGT |TGC CTG GAA |
Ile Pro Ala Asn Lys Lys Trp Asn Glu Asn Thr |Cys |Cys Leu Glu |
                                                                                 130                 140                 150
                                                                                  *                   *                   *
                                                               |TGC |AAA ACT GAA AAG CCT AAA CCT GAT GAT
                                                               |Cys |Lys Thr Glu Lys Pro Lys Pro Asp Asp
                                                                                                ___________

               160                 170                 180                     190                 200                 210                 220
                *                   *                   *                       *                   *                   *                   *
ATT CCA GCA AAC AAA AAA TGG AAC GAA AAC ACG |TGT |AGC CTG GAA |TGC |AAA ACT GAA AAG CCT AAA CCT CAA GAT
Ile Pro Ala Asn Lys Lys Trp Asn Glu Asn Thr |Cys |Ser Leu Glu |Cys |Lys Thr Glu Lys Pro Lys Pro Gln Asp

         230                 240                 250            260                 270                 280                 290
          *                   *                   *               *                   *                   *                   *
ATT CCA GCA AAC AAA AAA TGG AAC GAA AAC ACG |TGT |AGC CTA GAA |TGC |AAA ACT GAA AAG CCT AAA CCT GAT
Ile Pro Ala Asn Lys Lys Trp Asn Glu Asn Thr |Cys |Ser Leu Glu |Cys |Lys Thr Glu Lys Pro Lys Pro Asp
                                                                                                ___________
```

Fig. 7. Partial cDNA and encoded amino acid sequences for sp195. The sequences of two partial cDNA clones [nucleotides 1–120 (Dreesen et al. 1985) and 121–297 (Hill VL and Case ST, unpubl.)] are aligned to illustrate the 25-residue repeat found in this secretory protein. Blocks are placed around conserved Cys residues and copies of the ⊕ Pro ⊖ motif are *underlined*

manner distinct from core repeats. Alternatively, it has been suggested that rather than participate in fiber assembly, the carboxyl-domain of spIb may be involved in regulating the expression of its gene (Botella et al. 1988; Silva et al. 1990).

4.2 The Structure of sp195

The view that the intermediate size class of secretory proteins contains repeated amino acid sequences and structural motifs related to spIs originated with the report of cDNA encoding sp195. A clone with a small insert was isolated from a partial cDNA library of salivary gland RNA and shown to originate from a 6.5-kb poly $(A)^+$ RNA transcribed from a gene in BR1 (Dreesen et al. 1985). Antibodies against synthetic peptides encoded by this cDNA were ultimately shown to react with sp195 [originally referred to as sp180, (Dreesen and Case 1987)]. Limited cDNA sequences (Fig. 7) combined with genomic Southern blotting data (Dreesen et al. 1985) suggest that sp195 is composed of about 50 tandemly arranged copies of a 25-residue repeat. An extrapolation of the sequence data obtained for three repeats suggests that sp195 is hydrophilic and shares features common to spIs: repeats contain two conserved Cys and one copy of the ⊕ Pro ⊖ tripeptide motif found in the SR regions of spIs.

4.3 The Structure of sp185

Antibodies against cDNA-encoded synthetic peptides have been used to demonstrate that BR3 contains a gene encoding mRNA for sp185 (Dignam and Case 1990; Paulsson et al. 1990). The complete amino acid sequence (deduced from the full-length 5500-nucleotide cDNA sequence) and gene structure is known (Paulsson et al. 1990). The predominant pattern of sp185 is 50–60 tandem copies of highly diverged repeats of hydrophilic amino acids. The most striking feature is that, similar to the C region of spIs, it contains a remarkable motif of conserved Cys residues: Cys-X-Cys-Y-Cys approximately every 22–26 residues. These divergent repeats also contain sporadic copies of the tripeptide motif, ⊕ Pro ⊖ motif.

4.4 The Structure of sp140 and sp115

Hybridization and antipeptide antibody probes have been used to demonstrate that region 17-B on polytene chromosome I contains a gene encoding a 3.6-kb mRNA for sp140 (Dignam et al. 1989; Galli et al. 1990). This marked the discovery of the first non-BR-associated secretory protein-encoding gene and extended the distribution of this multigene family to three chromosomes. The complete structure of this gene is known (Galli et al. 1990). The central portion of sp140 contains 65 copies of a 14-residue repeat that is Lys- and Gly-rich, composed of hydrophilic amino acids, with most containing the ⊕ Pro ⊖ motif or related derivatives. In fact, the pattern is strikingly similar to subrepeats seen in spIs. The amino terminus consists of a 60-residue nonrepetitive region and the carboxyl terminus, a 31-residue nonrepetitive region. The protein contains seven Cys: one each in repeats one, four, and seven, and four at the carboxyl terminus.

Antibodies exist for two synthetic peptides derived from two slightly different sp140 repeats (Dignam et al. 1989; Galli et al. 1990). One reacts exclusively with sp140, while the other reacts with both sp140 and sp115, suggesting that these proteins are encoded by two similar genes. Thus, similar to the large size class, the intermediate size class of secretory proteins may contain a subfamily of structurally related proteins.

4.5 The Structure of sp40 and sp38

The complete cDNA and genomic DNA sequences are known for a gene encoding one of the small size class secretory proteins, sp40 (Galli and Wieslander 1992). Two nearly identical copies of the gene are located in region 5 of chromosome II, the only chromosome of the four still lacking a secretory protein-encoding gene. The central portion of sp40 contains six Cys residues whose spacing is similar to C regions in spIs. There are also several short tandem repeats containing Pro and hydrophilic residues. Antibodies against sp40 also react with two additional closely migrating secretory proteins, sp38 and sp39. It is likely that two of these proteins are encoded by the cloned pair of tandem genes. The third protein may be encoded by an additional gene or result from post-translational modification of one of the other proteins.

The isolation of a tandem pair of nearly identical genes and the immunological cross-reaction of anti-sp40 antibodies with sp38, sp39, and sp40 further support the existence of structurally related subfamilies of secretory proteins.

4.6 Overall Primary Structures

The prominent structural features of secretory proteins are tandemly repeated sequences of hydrophilic amino acids and, in most instances, conserved Cys residues. Repeats may be composite (spIs and sp40) or simple (sp195, sp185, sp140, and sp115). Whereas the number of repeats per protein molecule differs only threefold (between 50 to 150 tandem copies), the length of these repeating units varies considerably (14 to 90 residues) and generally decreases as the size of the secretory protein decreases (Table 3). The smallest protein studied thus far (sp40) represents a prototype unit related to the core repeats in spIs (Sect. 5). The significance of this pattern, if any, is currently unknown. All secretory proteins studied thus far also posses either a conserved pattern of Cys residues or ⊕ Pro ⊖ motifs, or both. These motifs may play an essential role in secretory protein assembly (Sect. 7).

5 Evolutionary Relationship of Genes Encoding Secretory Proteins

5.1 The Genes Encoding spIs

Upon first recognition of the tandem repetitive sequence organization of an spI-encoding gene, it was apparent that it must have evolved by repetitive duplication of smaller sequences (Wobus et al. 1980; Sumegi et al. 1982). Furthermore, the hierarchic repeat structure suggested that such duplications had taken place at different levels, i.e., at the level of core repeats and the level of subrepeats within SR regions of core repeats. In time, it became obvious that the genes encoding spIa, spIb, spIc, and spId must be closely related to each other, with the gene coding for spIc being most divergent (Lendahl and Wieslander 1984). These relationships are supported by the similarity in the nonrepetitive amino- and carboxyl-termini of these proteins and the intron-exon organization of the genes (Höög et al. 1988; Paulsson et al. 1992b).

Subsequent comparative analyses of spI genes in two sibling species of *Camptochironomus* emphasized that the core repeats within an spI-encoding gene may change during a short evolutionary time (Lendahl et al. 1987). Core repeats are dynamic and one type may be largely replaced by a variant. Therefore, it is likely that the four genes coding for spIs have evolved by translocative duplications, placing them at different chromosomal loci, followed by concerted divergence of the repeats within each gene (Pustell et al. 1984; Höög et al. 1988). Interestingly, while large-scale changes in repeats have taken place, the overall length of each gene has been conserved, suggesting that the lengths of spIs are important.

5.2 The Relationship Between Genes Encoding spIs, Intermediate, and Small Secretory Proteins

Two features have emerged from the structure of genes encoding intermediate and small secretory proteins. Firstly, the genes coding for sp195, sp185, sp140, sp115 and sp40 share structural units common to genes encoding spIs. All genes either contain multiple repeats (genes encoding sp195, sp185, sp140, and sp115) or represent a prototype unit (sp40) for core repeats. In the sp185 encoding gene, repeats contain conserved Cys codons similar to the C region of spI core repeats, whereas in the sp115 and sp140 encoding genes, the repeats contain hydrophilic sequences with the $\oplus$ Pro $\ominus$ motif, similar to SR regions of spI core repeats. In the sp40 and sp195 genes, the repeats contain both conserved Cys codons and hydrophilic, Pro-rich regions. Based upon these structural similarities, it is likely that the genes studied thus far originated from a common ancestor. Secondly, the intermediate and small size classes of secretory proteins may contain structurally related subfamilies, such as genes encoding sp140 and sp115, and those encoding sp40, sp39, and sp38.

5.3 Evolutionary History of the Multigene Family

Based on the structural properties of the secretory protein encoding genes, it is likely that all genes encoding secretory proteins may have evolved from a common ancestor (Fig. 8). It is presumed that such an ancestor coded for a small protein with components shared by the family members of today: at least one Cys codon and hydrophilic sequences containing the ⊕ Pro ⊖ motif. Sequence duplications have given rise to separate gene copies in which reduplications amplified parts or all of the ancestor sequences, resulting in repeats composed of either one or both of the original components. Reduplications have occurred at different times in the history of this gene family and along the various branches of the gene family tree to produce subfamilies. In individual genes, particularly in those composed of many repeats, dynamic processes known to operate on repetitive sequences have continuously redesigned their structures resulting in gene-sepecific repeats. During these remodelings, certain fundamental properties have been conserved such as Cys codons and hydrophilic regions containing the ⊕ Pro ⊖ motif. The regulation of the expression of individual genes must have also evolved in parallel with the growing gene family. The differences in transcriptional regulation between gene family members have resulted in individual differences in transcription during ontogeny and in response to external factors. The evolution of coding and

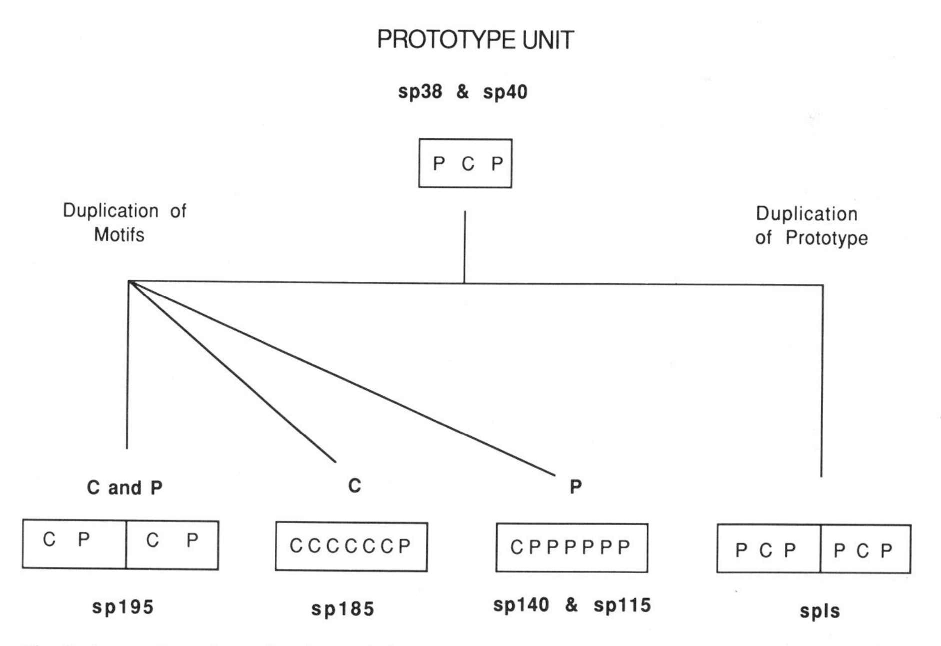

Fig. 8. A tentative scheme for the evolution of the multigene family encoding secretory proteins in *C. tentans*. All secretory proteins share the same structural features; regions with Cys codons (*C*) and regions with the ⊕ Pro ⊖ motif (*P*). The organization of C and P in secretory proteins today may have originated from a short prototype unit similar to sp38 and sp40

regulatory elements has resulted in a gene family collectively capable of producing a larval tube serving versatile functions under variable environmental conditions such as those described in this chapter.

6 Regulated Expression of Genes Encoding Secretory Proteins

Since secretory proteins form silk which is required for the construction of tubes throughout all larval instars, one might predict that larvae constantly express these genes. In fact, this is not the case.

6.1 Genes Encoding spIs

6.1.1 Expression During Ontogeny

As expected, expression of genes encoding spIs is restricted to salivary glands and occurs throughout larval stages of development (Lendahl and Wieslander 1987). However, a comparison of the steady-state level of spI mRNAs at each instar reveals a developmental pattern in which the relative level of each mRNA is highest during the second instar, the time when larval growth is most rapid (Lendahl and Wieslander 1987). Synthesis of spIs is not coupled; widespread variations occur in the relative levels of each spI and its mRNA during the fourth instar (Case 1986; Kao and Case 1986; Dreesen et al. 1988). Whereas the expression of spI-encoding genes during larval instars is regulated primarily at the level of transcription, discrepancies between transcription rate, steady-state level of mRNA and glandular content of protein during the fourth instar suggest that translational control may also be involved at this stage of development (Dreesen et al. 1988).

6.1.2 Response to Environmental Change

The synthesis of spIs responds to changes in environment. The best-studied example involves a chemical change; the addition of galactose or glycerol to larval culture medium. In *Camptochironomus*, this phenomenon has been examined by chromosome puffing (Beermann 1973), measurements of transcription (Nelson and Daneholt 1981), and steady-state levels of mRNA and/or protein (Edström et al. 1980; Galler et al. 1985; Kao and Case 1985, 1986; Lendahl and Wieslander 1987). Addition of these compounds leads to two antagonistic events. On one hand, a puff, BR6, is induced on chromosome III along with an increase in transcription of spIc mRNA (Table 3). The cytoplasmic concentration of this mRNA rapidly increases along with the synthesis and accumulation of spIc. Concurrently, the size of BR2 decreases rapidly along with cessation of initiation of transcription of the spI genes contained therein: spId mRNA decays more rapidly than spIb mRNA (Galler et al. 1985; Case 1986). The differential rate of decay of these mRNAs parallels the disappearance of their encoded proteins from salivary glands. Galactose and glycerol can exert a biphasic effect on spIa mRNA in that its level initially rises then falls (Case 1986; Lendahl and Wieslander 1987). Thus, the glandular concentration of spIa may appear to increase, decrease, or be stable depending upon when samples are obtained. The spIs in *C. thummi* exhibit similar responses to galactose (Cortes et al. 1989; Diez et al. 1990).

A number of other compounds such as pilocarpine (Meyer et al. 1983), dimethylsulfoxide (Sass 1981), and sugars other than galactose (Beermann 1973) modify the expression of genes in BRs; however, their effects on the synthesis of specific secretory proteins have not been examined.

6.2 Genes Encoding sp195, sp185, sp140, sp40, and sp38

The developmental expression of genes encoding intermediate and small size proteins has been examined only during stages within the fourth instar. Whereas sp185 is synthesized throughout the fourth instar (Dignam and Case 1990), synthesis of sp195, sp140, sp40, and sp38 differs. Expression of the gene encoding sp40 goes from a high level during the first half of the fourth instar to very low levels at stages 8–10, the prepupal stages (Galli and Wieslander 1992). In contrast, the expression of genes encoding sp195 (Dreesen et al. 1988), sp140 (Dignam et al. 1989), and sp38 (Galli and Wieslander 1992), goes from nearly undetectable levels at stages 3–4 to maximal levels at the prepupal stages. In all instances, the increased level of protein is due to a parallel increase in rate of transcription and steady-state level of mRNA. At maximal levels, these mRNAs reach concentrations comparable to spI mRNAs, approximately $2–10 \times 10^6$ molecules/cell (Dreesen et al. 1988; Dignam et al. 1989; Dignam and Case 1990).

6.3 Why Does the Expression of These Genes Change?

It has been proposed (Dreesen et al. 1988) that genes encoding secretory proteins may be divided into two expression classes: a larval class and prepupal class. Larval class genes may be expressed during all instars and encode secretory proteins involved in formation of larval silk used for housing and feeding (Fig. 9). Prepupal class genes encode additional proteins essential for formation of the prepupal silk required for ecdysis. An essential component of this hypothesis is that larval silk is dynamic; subtle changes in composition may lead to alterations in mechanical properties required for specialized functions.

There is additional support for the notion that silk is dynamic. When individual animals are examined, salivary glands contain any one or combination of two or more spIs; however, an spI is never lacking (Kao and Case 1986). The motifs shared by spIs suggest that they may be structural homologs capable of forming homo- and heteropolymers. Yet, sequence variations among spIs may result in polymers with differences in structure and mechanical properties. Such differences may be

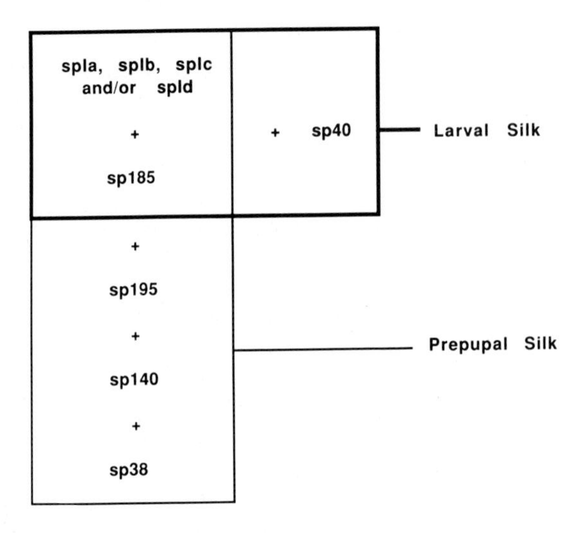

Fig. 9. The differential contribution of secretory proteins to *C. tentans* larval and prepupal silk. This illustration summarizes data obtained from Dreesen et al. (1988), Dignam et al. (1989), and Galli and Wieslander (1992)

vital for larvae to construct and modify tubes in response to changes in habitat such as sediment and population density and changes in feeding habits due to the availability and particle size of food (Sect. 3.4). Hence, an spI may be essential to the formation of silk (Wellman and Case 1989) but spIs may be specialized (Lendahl and Wieslander 1987).

Although sp185 has not been studied equally thoroughly, every individual examined contained this protein (Dignam SS, Brumley LL and Case ST, unpubl. data). We therefore place its gene in the larval class and suggest that, whatever its role in silk assembly might be, it functions independently from variations in the content of spIs.

The gene encoding sp40 is the only gene described thus far which is expressed only during the first half of the fourth instar. In contrast, genes encoding sp195, sp140 and sp38 belong to the prepupal class whose expression is limited to, or most prominant at, stages 8–10 of the fourth instar (the prepupal stages). The synthesis of these proteins coincides with noticeable changes in the microscopic appearance of silk (Fig. 10) and the construction of pupation tubes which are more resistant to

Fig. 10a–d. Microscopic comparison of larval and prepupal silk. Silk spun by stage 7 (**a** and **b**) and stage 10 (**c** and **d**) larvae was gently (**a** and **c**) then extensively (**b** and **d**) teased apart and reacted with an affinity purified rabbit antipeptide antibody specific for sp195 (Dreesen and Case 1987). The second antibody was a fluorescein-conjugated goat anti-rabbit antibody. Panels **a**–**d** and **a′**–**d′** show corresponding fields photographed under bright-field and epifluoresence conditions, respectively. Larvae spun thick fibers (*tf*) that usually occurred as loose masses of thick fibers (*mtf*) which sometimes unraveled (*arrow* in **b**) to reveal that they were comprised of thinner fibers (*f*). Prepupae spun bundles of thick fibers (*btf*) that were difficult to unravel and often contained nonfibrous (*nf*) component. Antibody to sp195 is localized to nonfibrous material by immunofluoresence. *Bars* **a** and **d** 250 μm; **b** and **c** 100 μm. (Dreesen et al. 1988)

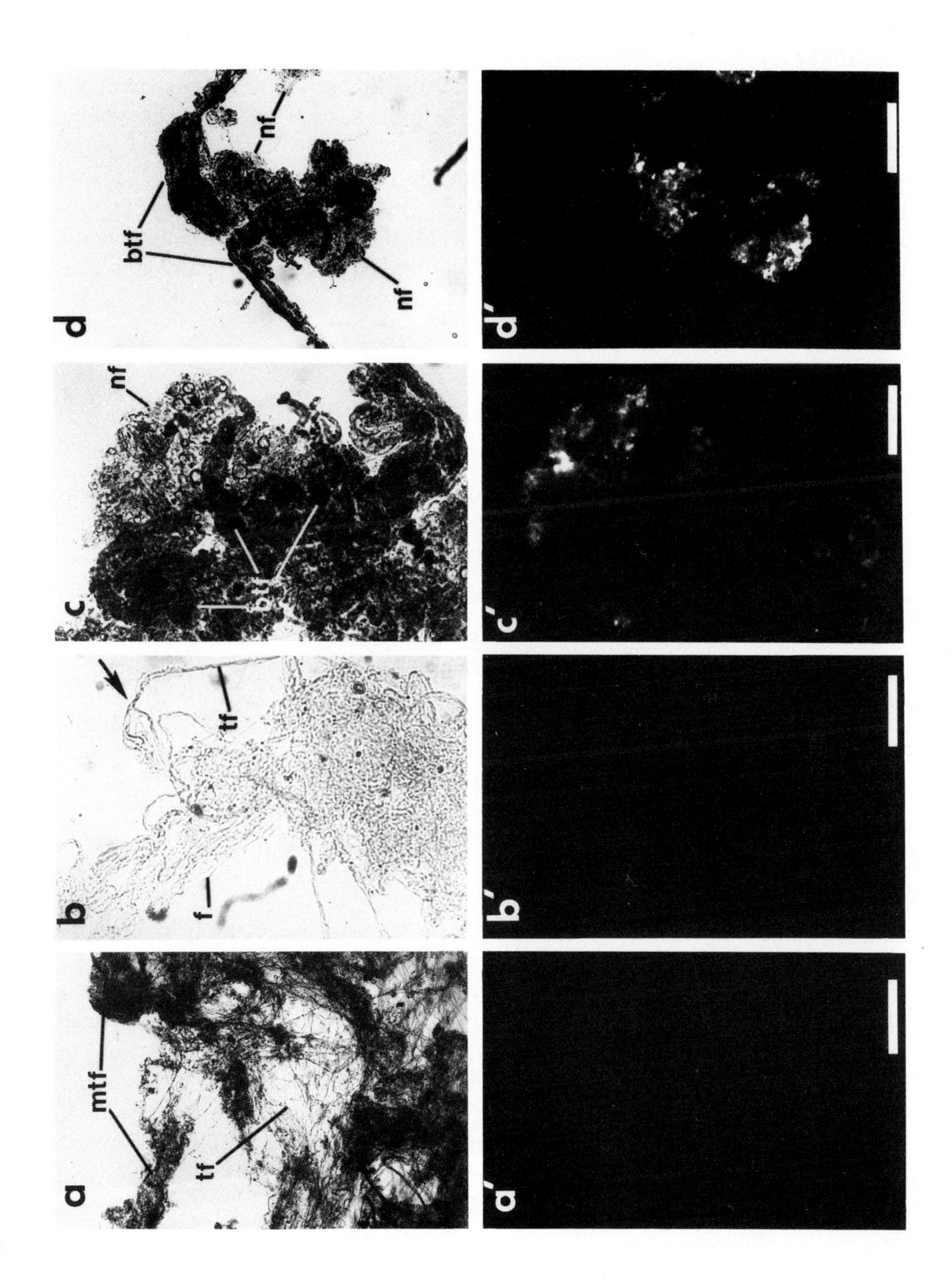
a
mtf
tf
b
f
tf
c
btf
nf
d
btf
nf
a'
b'
c'
d'

tearing than larval housing and feeding tubes (Dreesen et al. 1988). Since prepupae synthesize both larval and prepupal proteins, the most plausible explanation for these changes is that the addition of sp195 and sp140 to larval proteins and, perhaps, the replacement of sp40 and sp38, results in a modified silk with properties more suitable for pupation tubes. Recall that in preparation for ecdysis, prepupae anchor the larval cuticle to the walls of their tubes so that the pupa can emerge from within.

In summary, the expression of genes for secretory proteins is limited to salivary glands and regulated during development and in response to a variety of environmental stimuli. Hence the regulatory regions for this gene family probably consist of a hierarchy of conserved (tissue-specific) and diverged (developmental and galactose-inducible) cis-acting sequences (Lendahl and Wieslander 1987).

7 Assembly of Secretory Proteins into Silk

Silk threads have been visualized by electron (Grossbach 1977) and light (Dreesen et al. 1988) microscopy; however, the process of assembling secretory proteins into silk has been studied only recently.

7.1 Disassembly and Reassembly in vitro of Secretory Protein Complexes

Do intermediate structures form during the assembly of secretory proteins? Several lines of evidence suggest that they do. If the lumenal contents of the salivary gland are extracted by stringent denaturation and reduction, individual secretory proteins are recovered. However, less stringent extraction methods apparently yield protein complexes (Grossbach 1977; Rydlander 1981). Unlike silk, which is insoluble, these complexes must be dissociable.

Theorizing that the intrinsic properties which govern the assembly of secretory proteins in vivo might be maintained in vitro, Wellman and Case (1989) initiated a study of secretory protein assembly in vitro. Secretory proteins isolated from the glandular lumen appeared as complexes; a web-like network of strands which coalesce into smooth fibrils and multistranded beaded fibers (Fig. 11). Solvent perturbation, monitored by solution turbity and electron microscopy, indicates that these complexes are stabilized by electrostatic interactions and disulfide bonds. After disassembly, complexes can be reassembled in vitro. Moreover, a fraction of purified spIs can reassemble into complexes which, by electron microscopy and circular dichroism spectroscopy, appear similar to native complexes, as well as to complexes reassembled from an unfractionated mixture of secretory proteins (Fig. 12). Thus spIs alone can form the fibrous backbone of silk and self-assemble (Wellman and Case 1989).

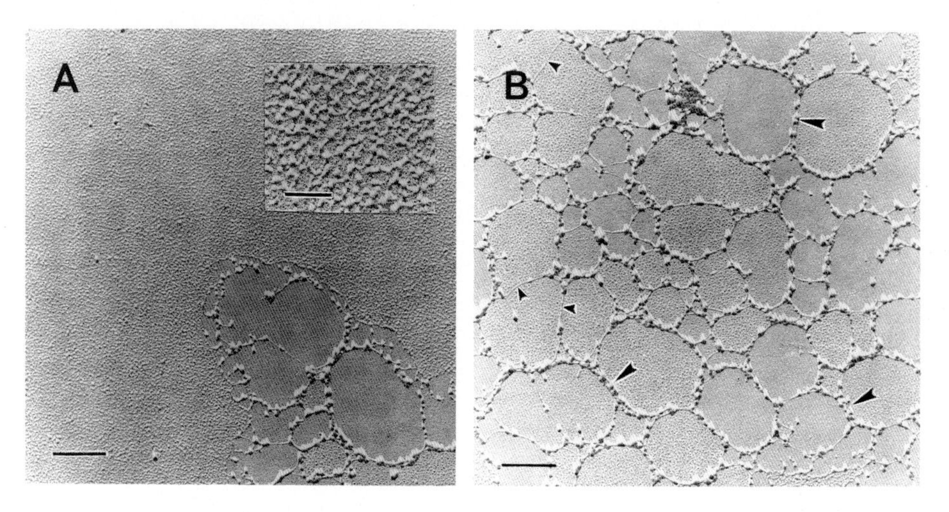

Fig. 11A, B. Electron micrographs of secretory protein complexes. Secretion was removed from salivary glands under nondenaturing conditions and prepared for electron microscopy. **A** A typical dense network of strands containing patches of fibrils and beaded fibers. The *inset* shows the network at higher magnification. **B** A more extensive area of 25–200 nm diameter fibers with 20–55 nm diameter beads (*large arrowheads*) and 5–20 nm diameter fibrils (*small arrowheads*) Bars **A** and **B** 500 nm; inset to **A** 200 nm. (Wellman and Case 1989)

7.2 Steps in the Assembly of Silk

It is now plausible to formulate a tentative outline (Fig. 13) for initial steps in the assembly of secretory proteins into silk. This outline is based primarily upon biochemical, biophysical and electron microscopic data available for spIs.

Silk assembly almost certainly begins with the intrinsic property of spIs to self-assemble into smooth fibrils and beaded fibers. The conditions that promote disassembly and reassembly of spIs in vitro are consistent with the notion put forth by several laboratories: the interactive sites are structural motifs noted in SR and C domains (Hamodrakas and Kafatos 1984, Wieslander et al. 1984). If any two spIs were aligned so that one or more of their SR domains were juxtaposed (Fig. 14), association could initiate with electrostatic interactions between ⊕ Pro ⊖ motifs. The importance of these motifs is highlighted by the fact that neutrality of these clusters of charged residues has been maintained in two ways: the negatively charged residues in spIc are Glu or Asp, whereas in spIa, spIb, and spId they are phosphorylated Ser or Thr (Edström et al. 1982; Galler and Edström 1984; Galler et al. 1984; Kao and Case 1985). These reversible intermolecular associations would be strengthened and solubility decreased by subsequent formation of disulfide bonds between Cys in C domains.

How many spI molecules associate with each other? In theory, in-phase but non-end-to-end alignment of any two molecules in parallel would provide a simple linear structure that could grow from both ends to an infinite length (Fig. 14A).

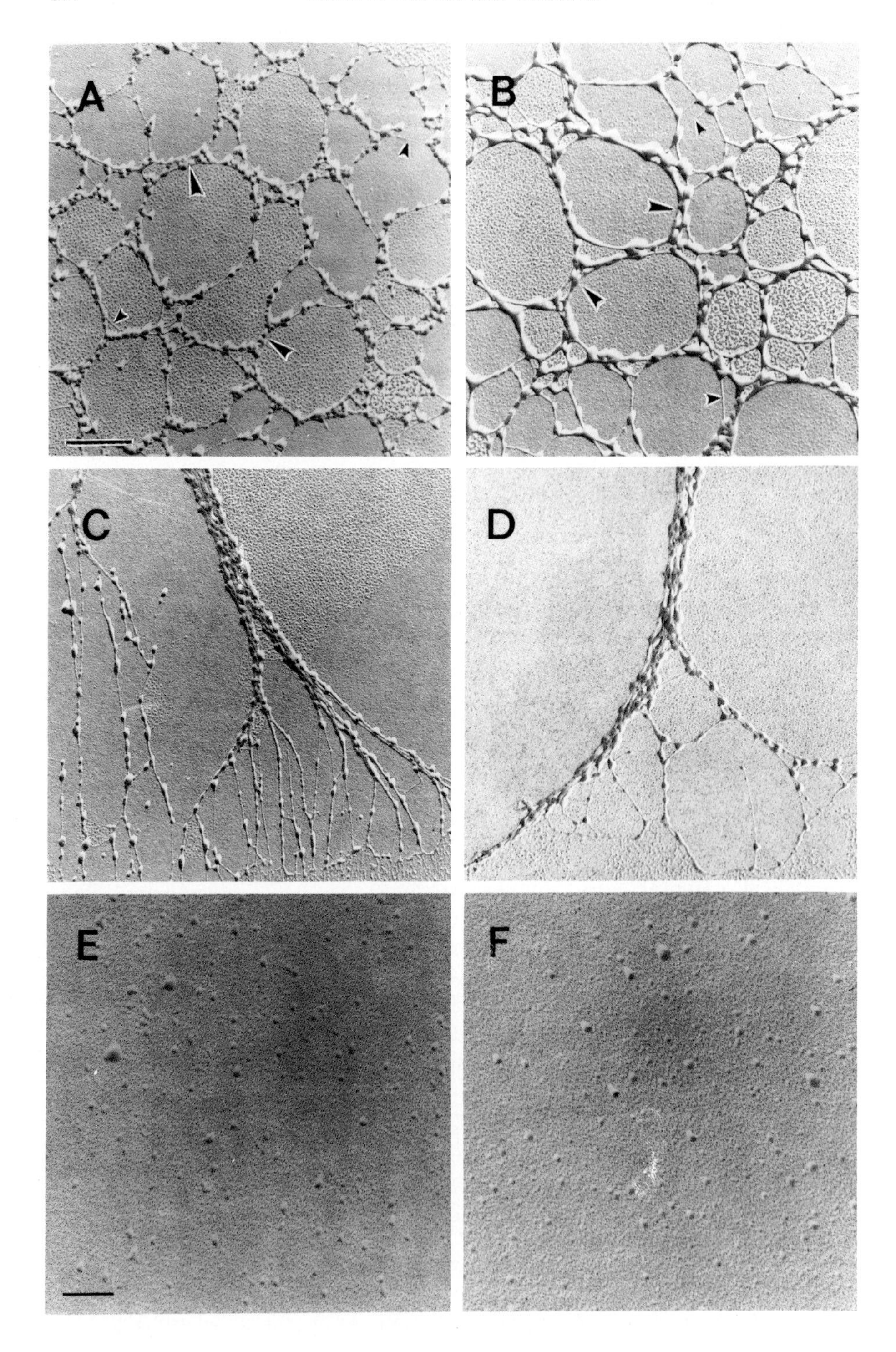
A
B
C
D
E
F

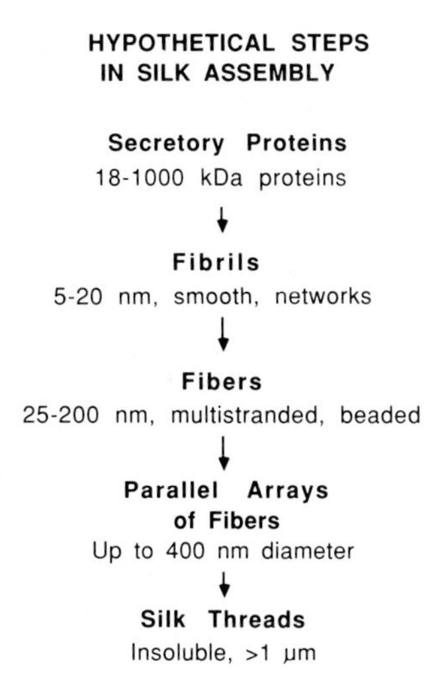

Fig. 13. Hypothetical steps in the assembly of secretory proteins into silk

This structure would be favored if all interactive sites were on one surface of each molecule; however, the resulting structures would have diameters too small to account for the fibrils and fibers observed by electron microscopy (Wellman and Case 1989). Alternatively, if spIs contain interactive sites projecting from at least two surfaces, then the number of adjacent molecules that could interact would increase, potentially yielding an ordered parallel or disordered random array of interactions (Fig. 14B and C). The type of arrays that occur may be restricted by spatial constraints encountered in the tertiary and quaternary structures of the protein. In this regard, consider the secondary structure of the C domain (Case et al. 1991; Wellman et al. 1992). If the entire C domain is represented as an α-helix with 3.6 residues per turn, all four Cys residues will project outwardly in different directions (Fig. 15). Consequently, each Cys within a single C domain could form a disulfide bond with a Cys in another C domain. Thus each Cys residue represents a potential site for propagation of a multitude of covalent intraprotein and protein-protein interactions among spIs (Fig. 16).

Interacting spI molecules probably do not lie parallel to one another. The fact that both C and SR domains are punctuated by β-turns makes it probable that repeated domains along the length of any one molecule project in a variety of directions. Also, the amino acid sequences of C domains lack the heptad repeat with hydrophobic residues at positions one and four that is characteristic of

Fig. 12A–F. Electron micrographs of complexes formed by fractionated and unfractionated secretory proteins. Secretory proteins were disassembled and reassembled in vitro with or without fractionation by centrifugation on a denaturing glycerol gradient. **A** and **C** Unfractionated secretory proteins; **B** and **D** gradient fractions containing purified spIs; **E** gradient fractions containing intermediate size secretory proteins; **F** gradient fractions containing small size secretory proteins. *Bars* 500 nm **A–D**; 200 nm **E** and **F**. (Wellman and Case 1989)

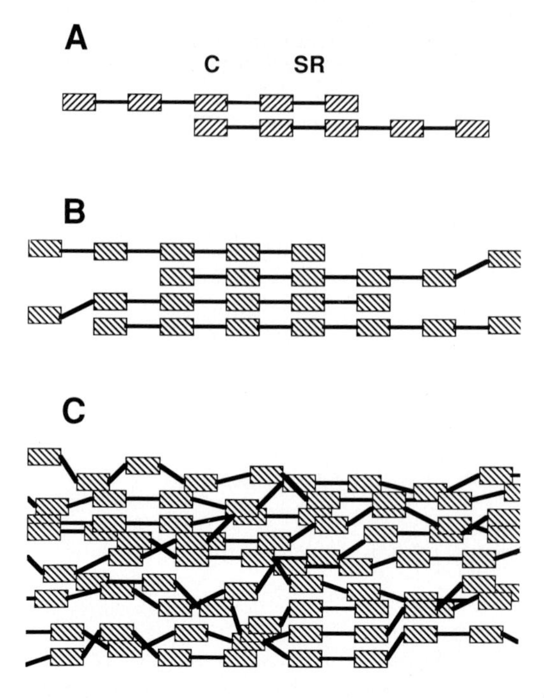

Fig. 14A–C. Theoretical organization of spI-spI contacts. Assuming that spIs align with one or more subrepeat (*SR*, *solid lines*) and constant (*C*, *cross-hatched boxes*) domains in-phase, intermolecular contacts could be: **A** limited to one surface between parallel molecules; **B** between two or more surfaces on parallel molecules; or **C** random among a network of molecules

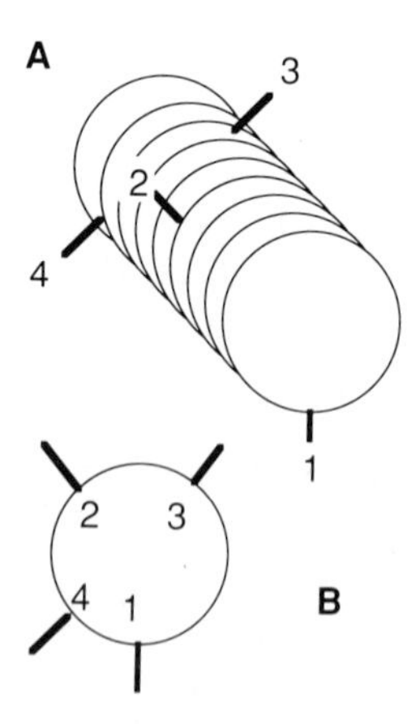

Fig. 15A, B. The location of Cys residues in the constant domain of an spI. A diagrammatic representation of a lengthwise (**A**) and end-on (**B**) view of the spatial distribution and projection of the four conserved Cys residues (*1–4*) found in a C domain from spIa. Assuming the circumference of the cylinder represents the core of an α-helix with 3.6 residues per turn, the Cys residues are located according to their position in the primary sequence

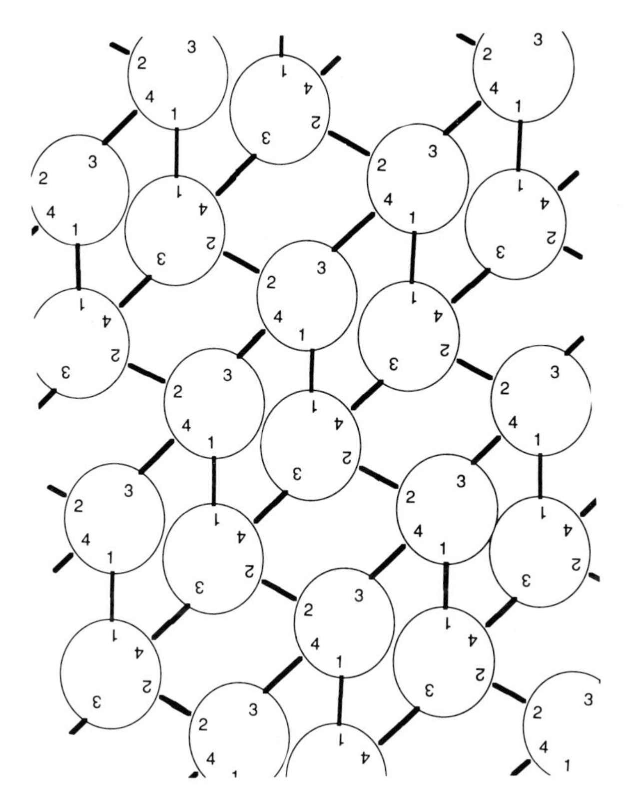

Fig. 16. Hypothetical formation of disulfide bonds between constant domains of spIs. Each *circle* represents the end-on view of an spI constant region depicted as a cylinder with its four (*1–4*) Cys residues projecting at angles corresponding to their position in an α-helix (see Fig. 15). In this illustration, each Cys forms a disulfide bond with a Cys in another constant domain. These disulfide bonds may be intramolecular or intermolecular depending upon the source of each constant region. Either way, a network of parallel bundles could occur

α-helical coiled coils (Cohen and Parry 1986). A more plausible suggestion is that interacting C domains form bundles of parallel α-helices (Figs. 15 and 16). Furthermore, the infrared spectra of synthetic peptide domains (Wellman et al. 1992) and secretion (Wellman SE, Hamodrakas SJ and Case ST, unpubl. data) suggest that bundles of helices occur. Therefore, rather than coiled coils (of α-helices or collagen-like helices such as the triple-stranded structure found in collagen fibrils), secretory proteins probably form a seemingly random, three-dimensional, web-like network due primarily to the large number of contacts made between numerous spI molecules (Figs. 14C and 16).

How then does this three-dimensional network of secretory proteins, which lacks uniform dimensions, become transformed into a reasonably uniform thread? The answer may simply be mechanical. Native and reassembled complexes have been observed in which fibrils and fibers appear oriented as parallel arrays rather than in a random networks (Fig. 12). Such arrays often associate into bundles

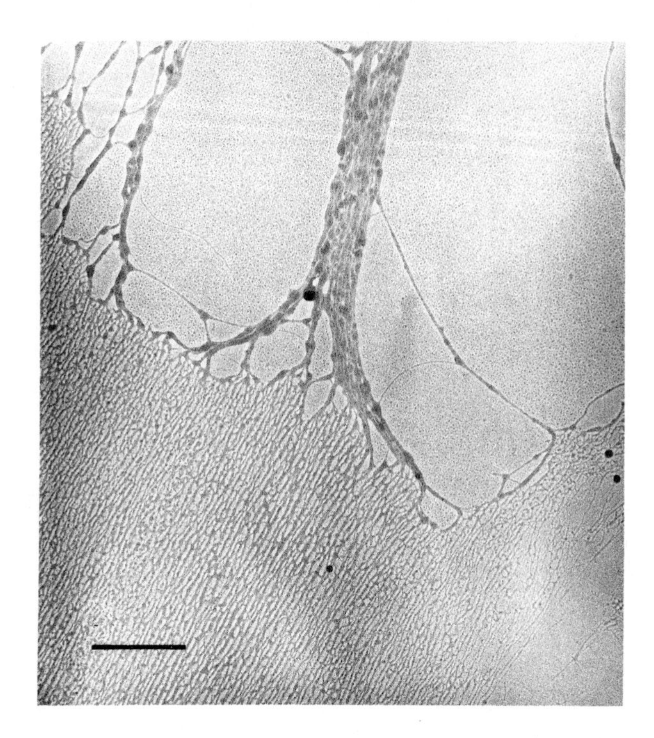

Fig. 17. A secretory protein complex containing extensive parallel arrays of beaded fibers and fibrils. This sample was prepared as described. (Wellman and Case 1989). *Bar* 500 nm

sometimes approaching 400 nm in diameter (Fig. 17). The forceful suction developed by the pumping mechansim of the salivary apparatus (Sect. 3.3) apparently draws complexes toward the lumen of the salivary duct. As complexes approach the duct, they could conceivably become compacted and oriented in parallel with the duct's long axis (Fig. 18). In fact, secretion within the lumen exhibits a higher degree of structural organization proximal to the orifice of the secretory duct (Fig. 19). If this model is correct, then the diameter of parallel arrays of fibers would simply be governed by the diameter of the salivary duct. In turn, the diameter of silk would be regulated by the diameter of the common canal. Then silk from smaller animals should have a narrower diameter than silk from larger animals.

At least two important aspects of silk assembly remain unexplained. First, parallel arrays of fibers apparently undergo additional covalent interactions that render them insoluble. Neither the nature of these interactions nor the time and place of their introduction are known. Secondly, the spatial locations and roles of the intermediate and small secretory proteins are unknown. An obvious possibility is that these are structural proteins that form cross-links between spIs. The immunofluorescent localization of sp195 in silk is consistent with this idea (Fig. 20). It is reasonable to imagine the molecular architecture and mechanical properties of a generic silk thread, comprised of spIs and sp185, being altered due to changes in the availability of cross-linking proteins such as sp140 and sp195. This may explain

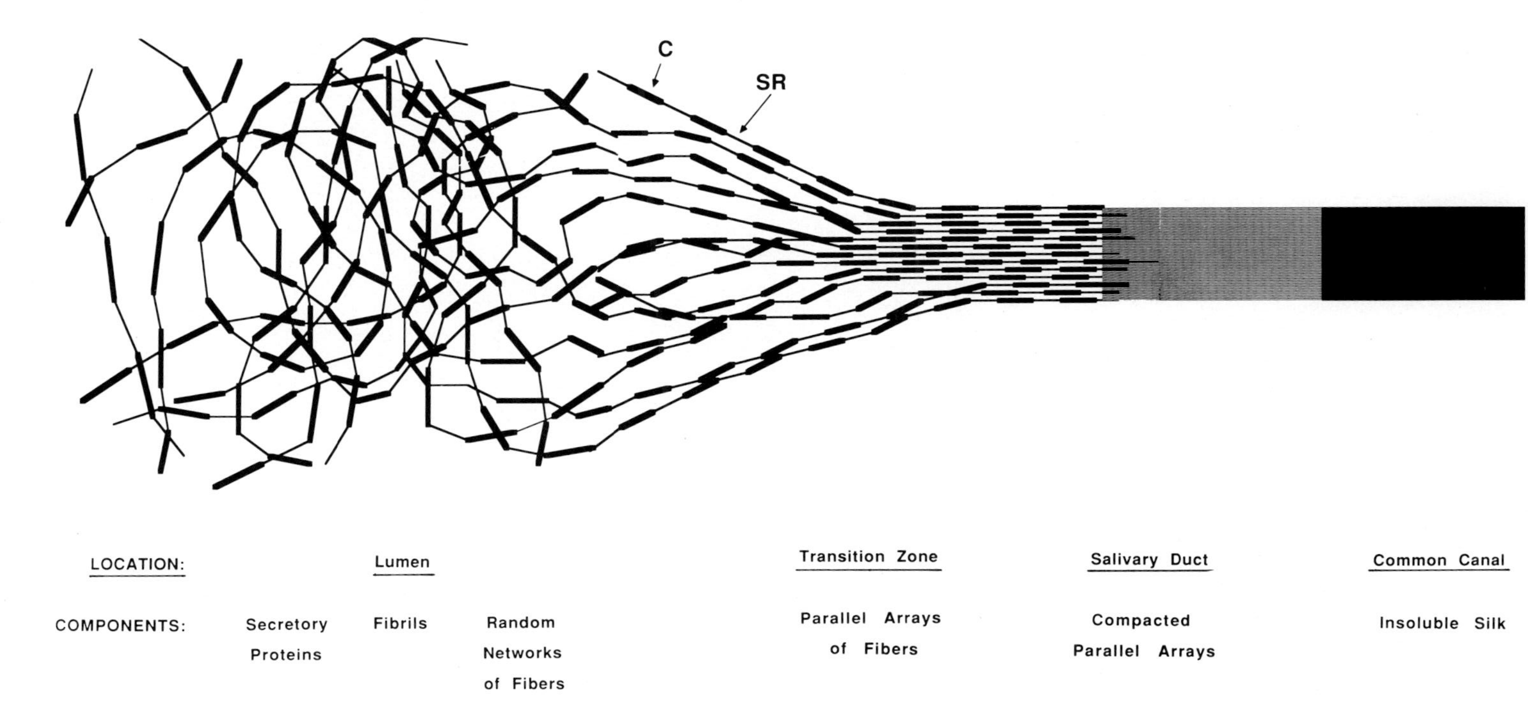

Fig. 18. A hypothetical illustration of the final steps in silk assembly. Random networks of fibers and fibrils within the lumen are presumably organized into parallel arrays of fibers by mechanical forces as secretion is sucked from the lumen into the salivary duct. Somewhere in the duct, common canal, or mouth, the complexes are compacted and rendered insoluble. The basis of this insolubility is currently unknown

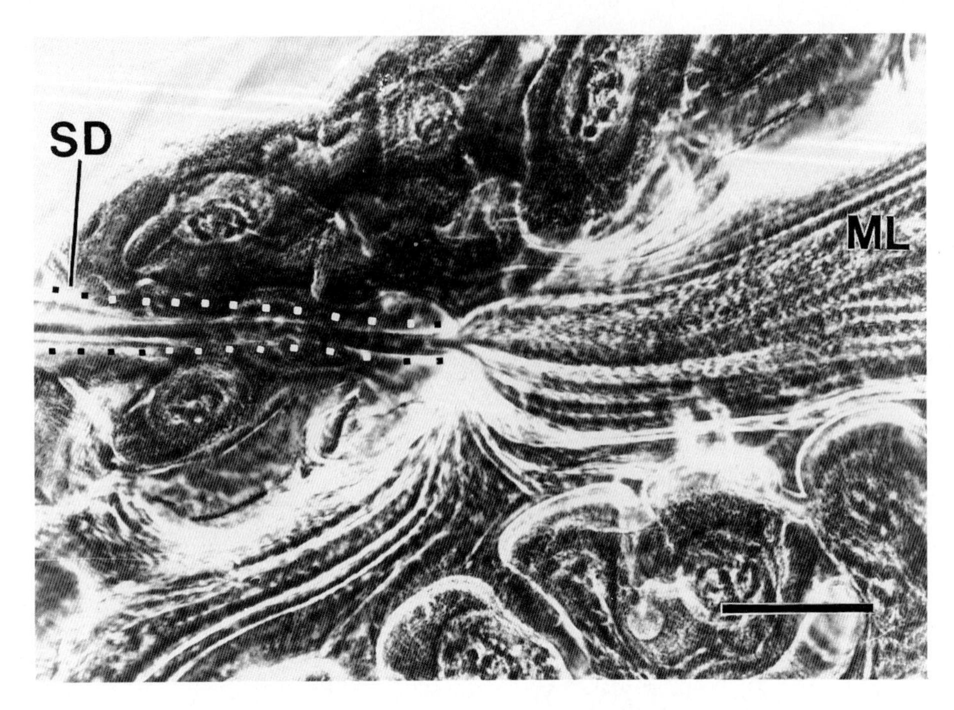

Fig. 19. The organized appearance of secretion in a transition zone proximal to the salivary duct. As secretion is drawn from the main lobe (*ML*) toward the salivary duct (*SD*) it takes on a more organized appearance of parallel arrays of fibers. *Dots* trace the outline of the secretory duct which is visible in another focal plane. This specimen was kindly provided by I.I. Kiknadze and T.E. Sebeleva. *Bar* 100 μm

why genes encoding secretory proteins can be categorized into larval and prepupal expression classes.

8 Summary

Salivary glands of *Chironomus* synthesize a family of at least ten secretory proteins that can be grouped into three size classes: the large (about 1000 kDa), intermediate (100- to 200 kDa), and small (less than 100 kDa). After synthesis, secretory proteins undergo a dramatic transformation to form a novel biopolymer. Secretory proteins accumulate in the central lumen of the gland, forming dissociable complexes that appear as a network of smooth fibrils and multistranded beaded fibers. When secretory protein complexes are extruded through the secretory duct, the fibers become oriented in parallel arrays; when these parallel arrays of fibers emerge from the mouth of larvae they are an insoluble, silk-like thread. Regulation of secretory protein-coding gene expression determines which secretory proteins are synthesized, thus, the composition of silk threads. At least two types of threads are produced: larval silk is used to construct tubes for protective housing

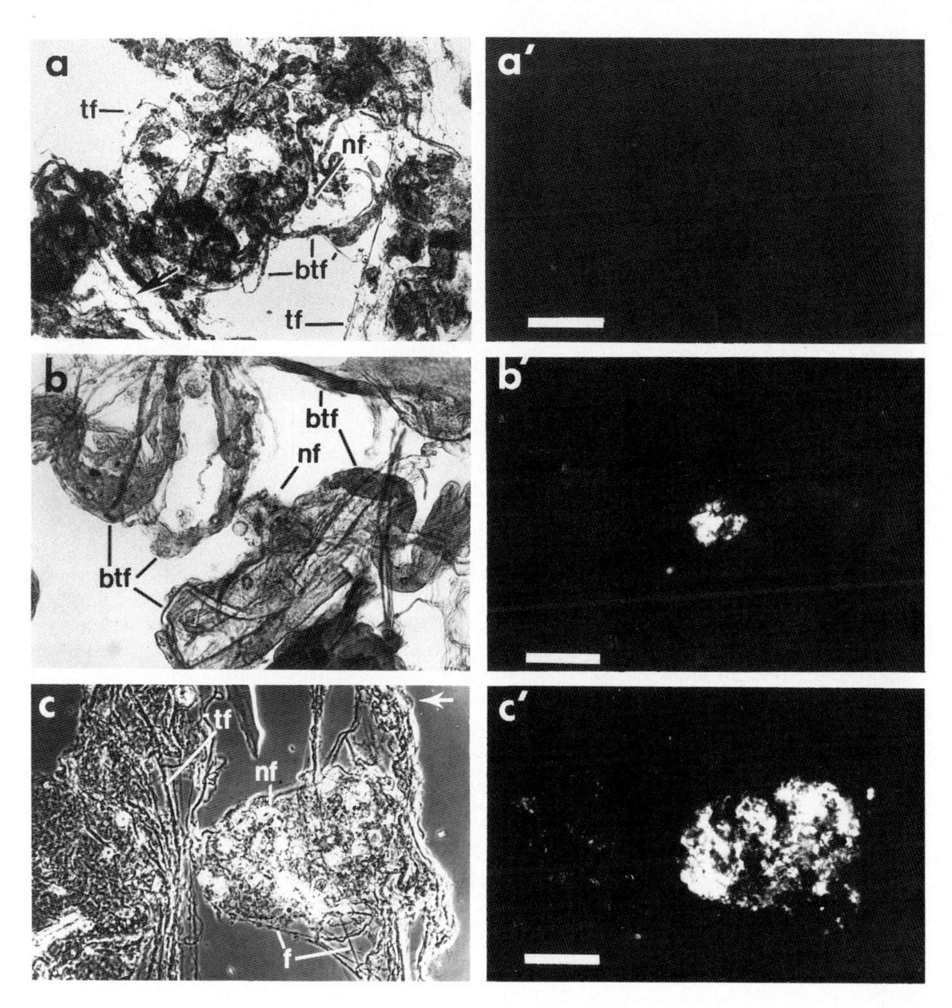

Fig. 20a–c. Immunofluorescent localization of sp195 in prepupal silk. Freshly spun silk was removed, teased apart, and incubated with (**b** and **c**) or without (**a**) a primary affinity purified rabbit antipeptide antibody specific for sp195 followed by fluorescein-conjugated goat anti-rabbit antibody. Bright-field (**a** and **b**) and phase contrast (**c**) images are presented for the same fields of view for fluorescent images (**a**′–**c**′). Bundles of thick fibers (*btf*) ranged up to 40 μm in diameter. These sometimes became stretched (*arrows*) to reveal that they are comprised of 2-μm diameter thick fibers (*tf*) which consist of several 450–900 nm fibers (*f*). The anti-sp195 antibody is primarily localized to discrete bodies within gelatinous globules of nonfibrous (*nf*) secretion and sometimes found alongside or between bundles of thick fibers. (Dreesen and Case 1987). *Bars* **a**′ 250 μm; **b**′ 100 μm; **c**′ 40 μm

and assist with feeding; prepupal silk is used to construct tubes for larval/pupal ecdysis (pupation). Variations in composition presumably contribute to different mechanical properties of larval and prepupal silk threads.

Since the macroscopic physical properties of polymerized silk most likely reflect the microscopic structure and interaction of secretory proteins, it becomes important to learn the principles which govern secretory protein assembly at the molecular level. Which secretory proteins interact and what are the sites used for intraprotein and protein-protein interactions during the assembly of this biopolymer? All eight secretory proteins characterized thus far contain tandemly repeated peptide sequences (ranging from 14–90 amino acids in length) and/or a periodic distribution of Cys residues. These motifs appear to be unique; no other biopolymer has either the repeated peptide sequences or composite structure of chironomid silk threads. The evolutionary conservation of motifs within repeats and among different secretory proteins suggests that the sequences and three-dimensional structures of the motifs may be important for assembly of secretory proteins into complexes, oriented fibers, and silk threads.

Further study of secretory protein assembly will bring us closer to understanding how this silk assembles in vivo. By learning principles that nature employs to construct such a novel composite biopolymer, it may become feasible to design and produce new classes of fibers or biomolecular materials with distinctive properties that are currently unavailable.

Acknowledgments. We both would like to thank our coworkers who, during the past decade, made significant experimental and theoretical contributions to our system. STC wishes to thank Rosemary Hoffman for critically reading this manuscript in detail; Susan Wellman for numerous stimulating discussions about secretory protein assembly; and the US Office of Naval Research (Contract No. N00014-87-K-0387), and the US Army Research Office (Grant DAAL03-91-G-0239) for financial support. LW wishes to thank the Swedish Natural Science Research Council, Torsten and Ragnar Soderbergs Stiftelser, Magnus Bergvalls Stiftelse, and Karolinska Institutet.

References

Anderson NH, Sedell JR (1979) Detritus processing by macroinvertebrates in stream ecosystems. Annu Rev Entomol 24:351–377

Balbiani EG (1881) Sur la structure du noyau des cellules salivaires chez les larves de *Chironomus*. Zool Anz 4:637–641

Bäumlein H, Pustell J, Wobus U, Case ST, Kafatos FC (1986) The 3′ ends of two genes in the Balbiani ring c locus of *Chironomus thummi*. J Mol Evol 24:72–82

Beermann W (1952) Chromomerenkonstanz und spezifische Modifikationen der Chromosomenstruktur in der Entwicklung und Organdifferenzierung von *Chironomus tentans*. Chromosoma 5:139–198

Beermann W (1955) Cytologische Analyse eines *Camptochironomus*-Artbastards. I. Kreuzungsergebnisse und die Evolution des Karyotypus. Chromosoma 7:198–259

Beermann W (1961) Ein Balbiani-Ring als Locus einer Speicheldrüsenmutation. Chromosoma 12:1–25

Beermann W (1973) Directed changes in the pattern of Balbiani ring puffing in *Chironomus*: effects of a sugar treatment. Chromosoma 41:297–326

Botella L, Grond C, Saiga H, Edström J-E (1988) Nuclear localization of a DNA-binding C-terminal domain from Balbiani ring-coded secretory protein. EMBO J 7:3881–3888

Case ST (1986) Correlated changes in steady-state levels of Balbiani ring mRNAs and secretory polypeptides in salivary glands of *Chironomus tentans*. Chromosoma 94:483–491

Case ST, Byers MR (1983) Repeated nucleotide sequence arrays in Balbiani ring 1 of *Chironomus tentans* contains internally nonrepeating and subrepeating elements. J Biol Chem 258:7793–7799

Case ST, Daneholt B (1977) Cellular and molecular aspects of genetic expression in *Chironomus tentans* salivary glands. In: Paul J (ed) Biochemistry of cell differentiation. International review of biochemistry, vol 15. University Park Press, Baltimore, pp 45–77

Case ST, Daneholt B (1978) The size of the transcription unit in Balbiani ring 2 of *Chironomus tentans* as derived from analysis of the primary transcript and 75S RNA. J Mol Biol 124:223–241

Case ST, Summers RL, Jones AG (1983) A variant tandemly repeated nucleotide sequence in Balbiani ring 2 of *Chironomus tentans*. Cell 33:555–562

Case ST, Wellman SE, Hamodrakas SJ (1991) Assembly characteristics and structural motifs in an aquatic insect's biopolymer. In: Alper M, Calvert P, Frankel R, Rieke P, Tirrell D (eds) Material synthesis based on biological processes. Proc of the materials research society symp, vol 218. Materials Research Society, Pittsburgh, pp 233–237

Churney L (1940) A contribution to the anatomy and physiology of the salivary gland system in the larva of *Chironomus* (Diptera). J Morphol 66:391–407

Cohen C, Parry DAD (1986) α-Helical coiled coils — a widespread motif in proteins. Trends Biochem Sci 11:245–248

Cortes E, Botella LM, Barettino D, Diez JL (1989) Identification of the spI products of Balbiani ring genes in *Chironomus thummi*. Chromosoma 98:428–432

Daneholt B (1972) Giant RNA transcript in a Balbiani ring. Nature New Biol 240:229–232

Daneholt B, Edström J-E (1967) The content of deoxyribonucleic acid in individual polytene chromosomes of *Chironomus tentans*. Cytogenetics 6:350–356

Daneholt B, Andersson K, Fagerlind M (1977) Large-sized polysomes in *Chironomus tentans* salivary glands and their relation to Balbiani ring 75S RNA. J Cell Biol 73:149–160

Darby RE (1962) Midges associated with California rice fields, with special reference to their ecology (Diptera: Chironomidae). Hilgardia 32:1–206

Diez JL, Cortes E, Merino Y, Santa-Cruz MC (1990) Galactose-induced puffing changes in *Chironomus thummi* Balbiani rings and their dependence on protein synthesis. Chromosoma 99:61–70

Dignam SS, Case ST (1990) Balbiani ring 3 in *Chironomus tentans* encodes a 185-kDa secretory protein which is synthesized throughout the fourth larval instar. Gene 88:133–140

Dignam SS, Yang L, Lezzi M, Case ST (1989) Identification of a developmentally regulated gene for a 140-kDa secretory protein in salivary glands of *Chironomus tentans* larvae. J Biol Chem 264:9444–9452

Dreesen TD, Case ST (1987) A peptide-reactive antibody to a Balbiani ring gene product: immunological evidence that a 6.5-kb RNA in *Chironomus tentans* salivary glands is mRNA for a 180-kDa nonfibrous component of larval secretion. Gene 55:55–65

Dreesen TD, Bower JR, Case ST (1985) A second gene in a Balbiani ring. *Chironomus* salivary glands contain a 6.5-kb poly(A) + RNA that is transcribed from a hierarchy of tandem repeated sequences in Balbiani ring 1. J Biol Chem 260:11824–11830

Dreesen TD, Lezzi M, Case ST (1988) Developmentally regulated expression of a Balbiani ring 1 gene for a 180-kD secretory polypeptide in *Chironomus tentans* salivary glands before larval/pupal ecdysis. J Cell Biol 106:21–27

Edström J-E, Rydlander L, Francke C (1980) Concomitant induction of a Balbiani ring and a giant secretory protein in *Chironomus* salivary glands. Chromosoma 81:115–124

Edström J-E, Sierakowska H, Burvall K (1982) Dependence of Balbiani ring induction in *Chironomus* salivary glands on inorganic phosphate. Dev Biol 91:131–137

Galler R, Edström J-E (1984) Phosphate incorporation into secretory protein of *Chironomus* salivary glands occurs during translation. EMBO J 3:2851–2855

Galler R, Rydlander L, Riedel N, Kluding H, Edström J-E (1984) Balbiani ring induction in phosphate metabolism. Proc Natl Acad Sci USA 81:1448–1452

Galler R, Saiga H, Widmer RM, Lezzi M, Edström J-E (1985) Two genes in Balbiani ring 2 with metabolically different 75S transcripts. EMBO J 4:2977–2982

Galli J, Wieslander L (1992) Two genes encoding secretory proteins in *Chironomus tentans* have arisen by gene duplication and exhibit different developmental expression patterns. (submitted)

Galli J, Lendahl U, Paulsson G, Ericsson C, Bergman T, Carlquist M, Wieslander L (1990) A new member of a secretory protein gene family in the dipteran *Chironomus tentans* has a variant repeat structure. J Mol Evol 31:40–50

Grond C, Saiga H, Edström J-E (1987) The sp-I genes in the Balbiani rings of *Chironomus* salivary glands. In: Hennig W (ed) Structure and function of eukaryotic chromosomes. Results and problems in cell differentiation, vol 14. Springer, Berlin Heidelberg New York, pp 69–80

Grossbach U (1969) Chromosomen-Aktivität und biochemische Zelldifferenzierung in del Speicheldrüsen von *Camptochironomus*. Chromosoma 28:136–244

Grossbach U (1977) The salivary gland of *Chironomus* (Diptera): A model system for the study of cell differentiation. In: Beermann W (ed) Biochemical differentiation in insect glands. Results and problems in cell differentiation, vol 8. Springer, Berlin Heidelberg New York, pp 147–196

Hamodrakas SJ, Kafatos FC (1984) Structural implications of primary sequences from a family of Balbiani ring-encoded proteins in *Chironomus*. J Mol Evol 20:296–303

Hardy PA, Pelling C (1980) Cell-free synthesis and immunological characterization of salivary proteins from *Chironomus tentans*. Chromosoma 81:403–417

Hertner T, Meyer B, Eppenberger HM, Mahr R (1980) The secretion proteins in *Chironomus tentans* salivary glands: electrophoretic characterization and molecular weight estimation. Wilhelm Roux's Arch Dev Biol 189:69–72

Hertner T, Eppenberger HM, Lezzi M (1983) The giant secretory proteins of *Chironomus tentans* salivary glands: the organization of their primary structure, their amino acid and carbohydrate composition. Chromosoma 88:194–200

Hickman CP (1967) Biology of the invertebrates, 2nd edn. Mosby, St. Louis, pp 442–478

Hilsenhoff WL (1966) The biology of *Chironomus plumosus* (Diptera: Chironomidae) in Lake Winnebago, Wisconsin. Ann Entomol Soc Am 59:465–473

Hodkinson ID, Williams KA (1980) Tube formation and distribution of *Chironomus plumosus* L. (Diptera: Chironomidae) in a eutrophic woodland pond. In: Murray DA (ed) Chironomidae: ecology, systemmatics and physiology. 7th Int Symp on Chironomidae. Pergamon, Oxford, England, pp 331–337

Höög C, Wieslander L (1984) Different evolutionary behavior of structurally related, repetitive sequences occurring in the same Balbiani ring gene in *Chironomus tentans*. Proc Natl Acad Sci USA 81:5165–5169

Höög C, Engberg C, Wieslander L (1986) A BR1 gene in *Chironomus tentans* has a composite structure: a large repetitive core block is separated from a short unrelated 3′-terminal domain by a small intron. Nucleic Acids Res 14:703–719

Höög C, Daneholt B, Wieslander L (1988) Terminal repeats in long repeat arrays are likely to reflect the early evolution of Balbiani ring genes. J Mol Biol 200:655–664

Ineichen H, Riesen-Willi U, Fischer J (1979) Experimental contributions to the ecology of *Chironomus* (Diptera). II. The influence of photoperiod on the development of *Chironomus plumosus* in the 4th larval instar. Oecologia 39:161–183

Ineichen H, Meyer B, Lezzi M (1983) Determination of the developmental stage of living fourth instar larvae of *Chironomus tentans*. Dev Biol 98:278–286

Kao WY, Case ST (1985) A novel giant secretion polypeptide in *Chironomus* salivary glands: implications for another Balbiani ring gene. J Cell Biol 101:1044–1051

Kao WY, Case ST (1986) Individual variations in the content of giant secretory polypeptides in salivary glands of *Chironomus*. Chromosoma 94:475–482

Kiknadze II, Lopatin OE, Kolesnikov NN, Gunderina LI (1990) The midge, *Chironomus thummi*. In: Dettlaff TA, Vassetsky SG (eds) Animal species for developmental studies. Consultants Bureau, New York, pp 133–178

Kloetzel JA, Laufer H (1969) A fine-structural analysis of larval salivary gland function in *Chironomus thummi* (Diptera). J Ultrastruct Res 29:15–36

Kloetzel JA, Laufer H (1970) Developmental changes in fine structure associated with secretion in larval salivary glands of *Chironomus*. Exp Cell Res 60:327–337

Kolesnikov NN, Karakin EI, Sebeleva TE, Meyer L, Serfling E (1981) Cell-specific synthesis and glycosylation of secretory proteins in larval salivary glands of *Chironomus thummi*. Chromosoma 83:661–677

Lendahl U, Wieslander L (1984) Balbiani ring 6 in *Chironomus tentans*: a diverged member of the Balbiani ring gene family. Cell 36:1027–1034

Lendahl U, Wieslander L (1987) Balbiani ring (BR) genes exhibit different patterns of expression during development. Dev Biol 121:130–138

Lendahl U, Saiga H, Höög C, Edström J-E, Wieslander L (1987) Rapid and concerted evolution of repeat units in a Balbiani ring gene. Genetics 117:43–49

McCafferty WP (1983) Aquatic entomology. Jones and Bartlett, Boston, pp 1–448

McLachlan AJ (1977) Some effects of tube shape on feeding of *Chironomus plumosus* L. (Diptera: Chironomidae). J Anim Ecol 46:139–146

McLachlan AJ, Cantrell MA (1976) Sediment development and its influence on the distribution and tube structure of *Chironomus plumosus* L. (Chironomidae, Diptera) in a new impoundment. Freshwater Biol 6:437–443

Meyer B, Mahr R, Eppenberger HM, Lezzi M (1983) The activity of Balbiani rings 1 and 2 in salivary glands of *Chironomus tentans* larvae under different modes of development and after pilocarpine treatment. Dev Biol 98:265–277

Miall LC, Hammond AR (1900) The structure and life-history of the harlequin fly (*Chironomus*). Clarendon, Oxford, pp 1–184

Nelson LG, Daneholt B (1981) Modulation of 75 S RNA synthesis in the Balbiani rings of *Chironomus tentans* with galactose treatment. Chromosoma 83:645–659

Neumann D (1976) Adaptations of chironomids to intertidal environments. Annu Rev Entomol 21:387–414

Oliver DR (1971) Life history of the Chironomidae. Annu Rev Entomol 16:211–230

Paulsson G, Lendahl U, Galli J, Ericsson C, Wieslander L (1990) The Balbiani ring 3 gene in *Chironomus tentans* has a diverged repetitive structure split by many introns. J Mol Biol 211:331–349

Paulsson G, Höög C, Bernholm K, Wieslander L (1992a) The Balbiani ring 1 gene in *Chironomus tentans*—sequence organization and dynamics of a coding minisatellite. J Mol Biol 225: 349–361

Paulsson G, Bernholm K, Wieslander L (1992b) Conserved and variable repeat structures in the Balbiani ring gene family in *Chironomus tentans*. J Mol Evol 35:205–216

Pelling C (1964) Ribonukleinsäure-Synthese der Riesenchromosomen. Autoradiographische Untersuchungen an *Chironomus tentans*. Chromosoma 15:71–122

Pustell J, Kafatos FC, Wobus U, Bäumlein H (1984) Balbiani ring DNA: sequence comparisons and evolutionary history of a family of hierarchically repetitive protein-coding genes. J Mol Evol 20:281–295

Rydlander L (1981) Expression of Balbiani ring genes in salivary gland cells of *Chironomus*. Thesis, Karolinska Institutet, Stockholm

Rydlander L (1984) Isolation and characterization of the two giant secretory proteins in salivary glands of *Chironomus tentans*. Biochem J 220:423–431

Rydlander L, Edström J-E (1980) Large sized nascent protein as dominating component during protein synthesis in *Chironomus* salivary glands. Chromosoma 81:85–99

Rydlander L, Pigon A, Edström J-E (1980) Sequences translated by Balbiani ring 75S RNA in vitro are present in giant secretory protein from *Chironomus tentans*. Chromosoma 81:101–113

Sadler WO (1935) Biology of the midge *Chironomus tentans fabricius*, and methods for its propagation. Cornell Univ Agric Exp Stn Mem 173:1–25

Saiga H, Grond C, Schmidt ER, Edström J-E (1987) Evolutionary conservation of the 3′ ends of members of a family of giant secretory protein genes in *Chironomus pallidivittatus*. J Mol Evol 25:20–28

Sass H (1981) Effects of DMSO on the structure and function of polytene chromosomes of *Chironomus*. Chromosoma 83:619–43

Scott KMF (1967) The larval and pupal stages of the midge *Tanytarsus* (*Rheotanytarsus*) *fucus* Freeman (Diptera: Chironomidae). J Entomol Soc S Afr 30:174–184

Serfling E, Meyer L, Rudolph A, Steiner K (1983) Secretory proteins and Balbiani ring gene activities in salivary glands of *Chironomus thummi* larvae. Chromosoma 88:16–23

Silva FJ, Botella LM, Edström J-E (1990) Functional analysis of the 3′-terminal part of the Balbiani ring 2.2 gene by interspecies sequence comparison. J Mol Evol 31:221–227

Sumegi J, Wieslander L, Daneholt B (1982) A hierarchic arrangement of the repetitive sequences in the Balbiani ring 2 gene of *Chironomus tentans*. Cell 30:579–587

Thyberg J, Sierakowska H, Edström J-E, Burvall K, Pigon A (1982) Mitochondrial distribution and ATP levels in *Chironomus* salivary gland cells as related to growth, metabolic activity, and atmospheric oxygen tension. Dev Biol 90:31–42

Walentowicz AT, McLachlan AJ (1980) Chironomids and particles: a field experiment with peat in a upland stream. In: Murray DA (ed) Chironomidae: ecology, systemmatics and physiology. 7th Int Symp on Chironomidae. Pergamon, Oxford, England, pp 179–185

Wallace JB (1980) Filter-feeding ecology of aquatic insects. Annu Rev Entomol 25:103–132

Walshe BM (1951) The feeding habits of certain chironomid larvae (subfamily Tendipedinae). Proc Zool Soc Lond 121:63–79

Weber F, Mahr R, Meyer B, Eppenberger HM, Lezzi M (1983) Cell-free translation of Balbiani ring RNA (75S) of *Chironomus tentans* salivary glands into high molecular weight products. Wilhelm Roux's Arch Dev Biol 192:200–203

Wellman SE, Case ST (1989) Disassembly and reassembly in vitro of complexes of secretory proteins from *Chironomus tentans* salivary glands. J Biol Chem 264:10878–10883

Wellman SE, Hamodrakas SJ, Kamitsos EI, Case ST (1992) Secondary structure of synthetic peptides derived from the repeating unit of a giant secretory protein from *Chironomus tentans*. Biochim Biophys Acta 1121:279–285

Wieslander L, Daneholt B (1977) Demonstration of Balbiani ring 75S RNA sequences in polysomes. J Cell Biol 73:260–264

Wieslander L, Lendahl U (1983) The Balbiani ring 2 gene in *Chironomus tentans* is built from two types of tandemly arranged major repeat units with a common evolutionary origin. EMBO J 2:1169–1175

Wieslander L, Paulsson G (1992) Sequence organization of the Balbiani ring 2.1 gene in *Chironomus tentans*. Proc Natl Acad Sci USA 89:4578–4582
Wieslander L, Sumegi J, Daneholt B (1982) Evidence for a common ancestor sequence for the Balbiani ring 1 and Balbiani ring 2 genes in *Chironomus tentans*. Proc Natl Acad Sci USA 97:6956–6960
Wieslander L, Höög C, Höög JO, Jornvall H, Lendahl U, Daneholt B (1984) Conserved and nonconserved structures in the secretory proteins encoded in the Balbiani ring genes of *Chironomus tentans*. J Mol Evol 20:304–312
Wobus U, Bäumlein H, Panitz H, Serfling E, Kafatos FC (1980) Periodicities and tandem repeats in a Balbiani ring gene. Cell 22:127–135
Wood DM, Borkent A (1989) Phylogeny and classification of the Nematocera. In: McAlpine JF (ed) Manual of neoartic Diptera 3. Agriculture Press, pp 1133–1370
Wulker W, Gotz P (1968) Die Verwendung der Imaginalscheiben zur Bestimmung des Entwicklingszustandes von *Chironomus*-Larven (Diptera). Z Morphol Oekol Tiere 62:363–388

Chapter 8

Spider Silk: a Mystery Starting to Unravel

Mike Hinman, Zhengyu Dong, Ming Xu, and Randolph V. Lewis[1]

1 Historical Aspects

Spiders are unique creatures, due to the presence of glands in their abdomen which produce silk. They are also unique in the use of this silk throughout their life span and the nearly total dependence on silk for their evolutionary success. Humans have viewed spiders with both dread and delight. The delight has been based on the beauty and precise construction of the classic orb web. There are few who would disagree about the beauty of a web glistening in the morning dew.

Although spiders have undoubtedly been studied since earliest man, the first papers using a scientific approach to spiders webs and silk appeared in the 1800s. One of the earliest was by John Blackwell (1830), describing the construction of nets (webs) by spiders. The following decades resulted in studies of the biology of the spiders and their anatomy, but little information was published about the silk itself. Benton (1907) published one of the earliest studies describing properties of the silk itself. In that same year, Fischer (1907) demonstrated the protein nature of the silk by showing the presence of predominantly amino acids. There were periods of fairly intense study prior to World War II and in the late 1950s. However, progress, especially when compared to silkworm silk, was relatively meager. Beginning in the 1970s, the laboratories of Work, Gosline, and Tillinghast revived interest in spider silk with several papers describing physical, mechanical, and chemical properties of spider silks. The structure of the spider silk protein(s) remained unknown despite the efforts of these groups and others.

2 Biological Aspects of Spider Silk Production

Spider webs are constructed from several different silks. Each of these silks is produced in a different gland. The different silks, the glands which produce them, and the use of each are listed in Table 1. The glands occur as bilaterally symmetric paired sets. Although each of the glands has its own distinctive shape and size the functional organization of all of them is similar. The majority of the gland serves as a reservoir of soluble silk protein which is synthesized in specialized cells at the

[1] Molecular Biology Department, Univ. of Wyoming, Box 3944 Laramie, WY 82071-3944, USA

Results and Problems in Cell Differentiation 19
Biopolymers
Case, S. T. (Ed.)

Table 1. Various spider silks, their uses, and glands of synthesis

Silk gland	Use	Spinneret
Major ampullate	Dragline, frame threads	Anterior
Minor Ampullate	Reinforces dragline	Median
Piriform	Attachment disk	Anterior
Aciniform	Swathing silk	Median, posterior
Cylindrical (tubuliform)	Cocoon silk	Median, posterior
Aggregate	Sticky silk glue	Posterior
Flagilliform	Thread for sticky silk	Posterior

distal end of the gland. The soluble silk is forced (pulled) down a narrow duct during which the physical and chemical changes occur which produce the solid silk fiber. There is a muscular valve at the exit to the spinneret which can control the flow rate of the fiber and may control the fiber diameter to a small degree. The silk exits through the spinnerets, of which there are three pairs, anterior, median and posterior. The exit spinneret for each of the different silks is also listed in Table 1.

Due to their size and ease of study, the major ampullate glands have received the most attention. Thus most of what is known about the synthesis of silk proteins is based on the study of that gland. However, morphological and histochemical studies of the other glands support the ideas developed for the major ampullate gland. The synthesis of the silk protein(s) takes place in specialized columnar epithelial cells (Bell and Peakall 1969) which appear to lack a Golgi apparatus. There appear to be at least two different types of cells producing protein (Kovoor 1972), which correlates with our data on the composition of the silk from these glands. The newly synthesized protein appears as droplets within the cell which are secreted into the lumen of the gland.

The state of the protein in the lumen of the gland is unknown but it must be in a state which prevents fiber formation, as the fiber is not formed until passage down the duct. This is probably accomplished by a combination of protein structure and concentration which prevents aggregation in large protein arrays. It has been shown that the silk in the gland is not birefringent, whereas the silk becomes birefringent as it passes down the duct (Work 1977b). Thus the ordered array of protein seen in the final fiber is accomplished in the duct. This ordering appears to be due to the mechanical and frictional forces aligning the protein molecules and probably altering the secondary structure to the final fiber form. Iizuka (1983) has proposed a similar mechanism for silkworm silk formation. Experimental evidence for this has been the ability to draw silk fibers directly from the lumen of the major, minor and cylindrical glands (Hinman M, pers. commun.), implying that the physical forces of drawing the solution are sufficient for fiber formation.

3 Current Information About Spider Silk

3.1 Mechanical Properties

One of the features which attracted attention to spider silk was its unique properties. The spider must be able to use the minimum amount of silk in its web to

Table 2. Comparative data on different fibers

Material	Strength	Energy to break
	($N m^{-2}$)	($J kg^{-1}$)
Dragline silk	1×10^9	1×10^5
KEVLAR	4×10^9	3×10^4
Rubber	1×10^6	8×10^4
Tendon	1×10^9	5×10^3

Data derived from Gosline et al. (1986)

catch prey in order to survive successfully. The web has to stop a rapidly flying insect nearly instantly in a manner that allows it to become entangled and trapped. To do this, the web must absorb the energy of the insect without breaking and yet not act as a trampoline to send the insect back off the web. Gosline et al. (1986) have reviewed several aspects of this and concluded that spider silk and the web are nearly optimally designed for each other.

As with any polymer, especially those made of protein, there are numerous factors which can affect the tensile strength and elasticity. These can include temperature, hydration state, and rate of extension. Even with all those caveats, it is clear that dragline silk is a unique biomaterial. As seen in Table 2, dragline silk will absorb more energy prior to breaking than nearly any commonly used material. Thus, although it is not as strong as several of the current synthetic fibers, it can outperform them in many applications.

3.2 Chemical Data

The composition of spider silks has been known to be predominantly protein since the early studies of Fischer (1907). In fact, except for the sticky spiral thread, no significant amounts of anything but protein have been detected, including sugars, minerals, and lipids. The sticky spiral thread contains a number of water-soluble compounds such as potassium phosphate, potassium nitrate, gamma amino butyramide, and several other amine-containing compounds (Anderson and Tillinghast 1980). Although not proven, it is likely these compounds serve to absorb water to insure that the sticky spiral retains its adhesive catching properties.

The various silks have significantly different amino acid compositions as do the same silks from different spiders. Table 3 lists the compositions of several different spider silks (data from Anderson 1970; Work and Young 1987; Lewis RV, unpubl.). In the major ampullate silks, the combination of Glu, Pro, Gly, and Ala comprises 80% of the silk from each species. However, the proportion of Pro is significantly different in each. As will be discussed below, these differences can be accounted for by different ratios of two proteins. The minor ampullate silks are more similar among species and differ from major ampullate silk in having significantly lower Pro values. Cylindrical (tubuliform) gland silk used for constructing egg cocoons is radically different from any of the other silks. The amount of Gly is reduced by nearly three-quarters and Ser, in particular, has increased to compensate as have

Table 3. Amino acid composition of different silks

Amino acid	Gland (species)											
	Major Amp(Ad)	Major Amp(Ag)	Major Amp(N)	Minor Amp(Ad)	Minor Amp(Ag)	Minor Amp(N)	Cylind (Ag)	Cylind (N)	Aciniform (Ad)	Pyriform (Ad)	Aggregate (Ad)	Coronate (Ad)
Asp	1.0	2.4	1.2	1.9	1.2	2.8	5.9	3.6	8.3	10.5	9.2	2.7
Thr	0.9	0.8	0.1	1.4	0.3	0.3	3.6	5.2	9.8	4.4	7.6	2.5
Ser	7.4	10.3	7.4	5.1	5.2	10.9	28.1	23.4	15.7	14.8	6.8	3.1
Glu	11.5	9.2	10.9	1.6	8.0	4.5	10.0	9.3	7.5	10.4	9.8	2.9
Pro	15.8	10.1	3.5	0.1	0.5	0.6	0.4	0.4	3.3	7.8	10.8	20.5
Gly	37.2	42.0	45.8	42.8	51.6	42.6	9.3	7.3	13.6	7.8	14.5	44.2
Ala	17.6	18.0	22.2	36.6	24.1	20.0	23.7	30.0	11.0	9.9	6.2	8.3
Val	1.2	2.2	0.8	1.7	1.2	2.5	4.9	1.8	7.1	5.4	5.8	6.7
Cys	–	–	–	–	–	–	–	–	–	–	–	–
Met	–	–	–	–	–	–	–	–	–	–	–	–
Ile	0.6	0.2	0.1	0.7	0.4	0.4	1.8	1.9	4.3	3.7	4.7	1.0
Leu	1.3	1.5	3.5	1.0	1.5	2.2	6.6	6.9	9.1	5.4	5.5	1.4
Tyr	3.9	3.9	4.3	4.7	7.9	8.4	1.5	2.1	1.7	2.2	2.2	2.6
Phe	0.5	1.0	0.3	0.4	1.0	0.9	3.2	5.2	2.4	2.3	3.8	1.1
Lys	0.5	0.5	0.3	0.4	0.5	0.5	2.0	1.0	2.2	9.0	7.4	1.4
His	–	–	–	–	–	–	–	–	0.5	2.8	2.4	0.7
Arg	0.6	1.9	1.7	1.7	2.8	2.9	1.7	3.2	4.0	3.6	3.4	1.1

Data derived from Anderson (1970), Work and Young (1987), and Lewis (unpubl.). Ad is *A. diadematus*, Ag is *A. gemmoides*, and N is *N. clavipes*

other amino acids to a much lesser extent. The coronate gland silk is also very different in having a very high proportion of Pro in relation to the other amino acids. The other three silks are more similar to each other in having no greatly predominant amino acids and having a more even distribution of the amounts of the various amino acids.

3.3 Biophysical Studies

Virtually all of the data on spider silk have been obtained from major ampullate (dragline) silk. There are two major reasons for this. First, it can be easily obtained since the spiders trail it along behind them as they move. Thus the silk can be gathered or the spider can be forced to extrude it by causing the spider to fall and wrapping the silk onto something as it does. The second reason is the combination of mechanical properties of elasticity and high tensile strength, which will be discussed in detail below. Since it is not possible to have confidence in the extrapolation of the physical data from dragline silk to the other silks because the amino acid compositions are so different. The detailed biophysical data from the other silks will have to await further experimentation.

There were several early studies of silk fibers using X-ray diffraction which provided some information, much of which was interpreted based on the structure of silkworm silk (reviewed in Fraser and MacRae 1973). These studies led to the classification of dragline silk as β-sheet group 3, 4, or 5, depending on the species. These groups are distinguished by the intersheet distance between the β-sheets. The higher the number, the larger that spacing. It was also clear that much of the structure was not β-sheet and appeared to be random coil. However, it is clear from the amino acid compositions of the different silks that large bulky groups are present and must be accommodated either in the β-sheet or in the random coil regions. The amino acid sequences of the proteins from dragline silk put limits on what the structures can be and the X-ray data may need to be reinterpreted on that basis.

Using Fourier transform infrared (FTIR) spectroscopy Dong et al. (1991) have probed the structure of the dragline silk fiber in the relaxed and extended states. The data confirms the presence of significant β-sheet-like structure which appears the same in both relaxed and extended forms (Fig. 1A, B). Dragline silk, dissolved in 4.7 M $LiClO_4$, dialyzed against water, and dried to a film, also showed predominantly β-sheet-like conformation, indicating this is a preferred secondary structure of the proteins in the solid state (Dong Z and Lewis RV, unpubl. data). However, in the extended state, the silk forms an α-helical structure which returns to the original form when the tension is released (Fig. 1A, B). This is seen by the formation of the peaks at 1557 and 1651 cm^{-1}. The parallel polarized spectra shows that the orientation is parallel to the fiber axis. These same spectral features were observed for both *Nephila clavipes* and *Araneus gemmoides* (Figs. 1 and 2). The α-helical regions appear to be coming from the random or nonoriented regions. However, minor ampullate silk, which exhibits very low elasticity, showed no such α-helix

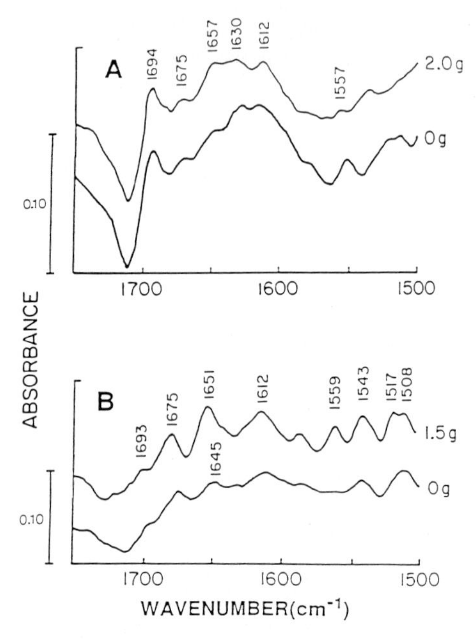

Fig. 1A, B. FTIR data from a 3-cm piece of *N. clavipes* dragline silk with no extension and with the weights indicated. **A** with perpendicular polarization; **B** with parallel polarization

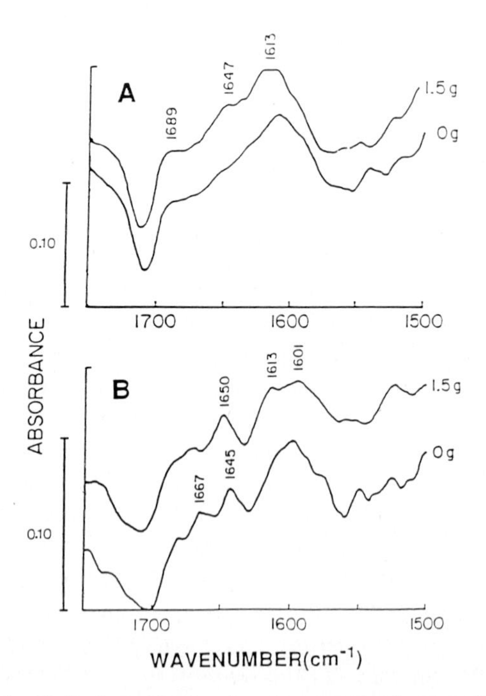

Fig. 2A, B. As in Fig. 1, using *A. gemmoides* dragline silk

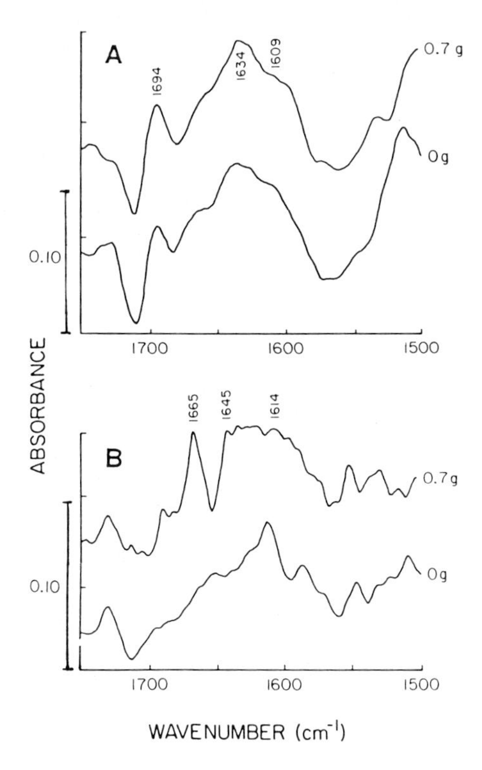

Fig. 3A, B. As in Fig. 1, using *N. clavipes* minor ampullate gland silk

formation and instead showed a breakdown of the β-sheet region into β-turns, as seen by the large increase in the 1665 cm^{-1} peak (Fig. 3). These data suggest that α-helix formation is playing an important role in the elastic function of these proteins.

In order to further study the preferred secondary structure of dragline silk proteins the sample was dissolved in 4.7 M $LiClO_4$ and dialysed against water with samples removed at various time points. The samples were analyzed by circular dichroism spectroscopy (CD). The data (Fig. 4) clearly show the proteins are in a random structure (minimum at 195 nm) prior to dialysis but, as dialysis proceeds, the protein assumes a β-sheet-like structure as seen by the strong minimum at 215 nm. The proteins also seem to be going through a state transition, as indicated by the single intersection point of all spectra. The experiment was repeated with dialysis into 0.1 and 1.0 M NaCl to determine the effect of ions on the structure. The CD spectra were taken on the samples after the completion of dialysis at various temperatures. The silk proteins seem to be undergoing an inverse temperature transition in the 0.1 M NaCl, as at low temperatures the protein is in a random configuration which changes to nearly complete β-sheet-like structure at 80° (Fig. 5A). The sample in 1.0 M NaCl undergoes the same transition but to a much lesser extent, only showing about 40% β-sheet-like structure (Fig. 5B).

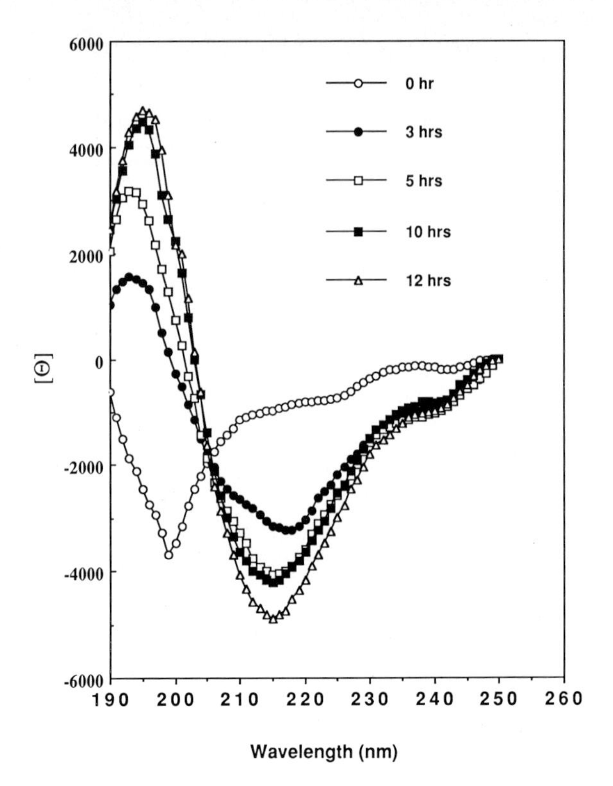

Fig. 4. CD spectra of *N. clavipes* dragline silk dissolved in 4.7 M $LiClO_4$ and dialyzed against distilled water. Samples at the indicated times following the start of dialysis were analyzed

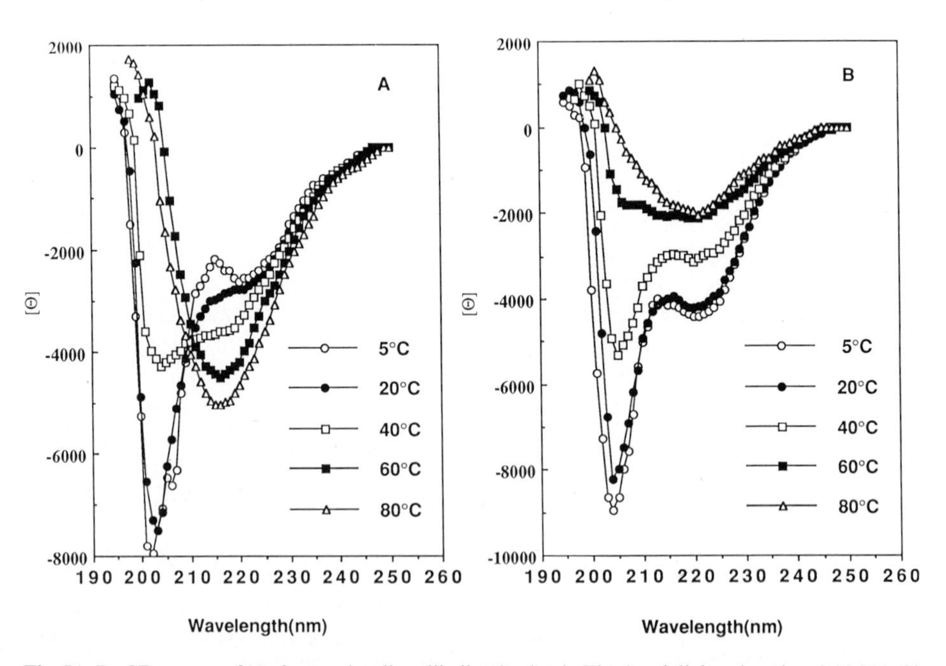

Fig. 5A, B. CD spectra of *N. clavipes* dragline silk dissolved as in Fig. 4 and dialyzed against 0.1 M NaCl (**A**) and 1.0 M NaCl (**B**). Following dialysis the samples were analyzed at the indicated temperatures

Interestingly, a model peptide based on the other major elastic protein, elastin, shows a similar inverse temperature transition to form a β-spiral (Urry 1988).

When the dialyzed silk proteins are placed in 50% trifluoroethanol (TFE), which promotes the formation of α-helices, the sample shows significant α-helix formation, as seen by the minima at 204 and 220 nm (Fig. 6). These data agree with the FTIR data described above, which indicated that the silk fiber was in a predominantly β-sheet-like conformation but the nonsheet regions could be induced to form α-helices when the fiber was extended.

Another interesting characteristic of dragline silk is its ability to supercontract. When dragline silk is placed in water unrestrained it contracts to 50–60% of its original length. This contraction results in a 1000-fold decrease in the elastic modulus and a greatly increased extensibility (Gosline et al. 1984). Although several polymers exhibit this characteristic in organic solvents, dragline silk will supercontract in water but not in organic solvents (Work 1977b, 1981, 1985; Work and Morosoff 1982; Fornes et al. 1983). The data from X-ray diffraction suggests the β-sheet regions rotate but otherwise are unchanged. Thus the water must be altering the relationship of the β-sheet regions to the nonoriented regions. This

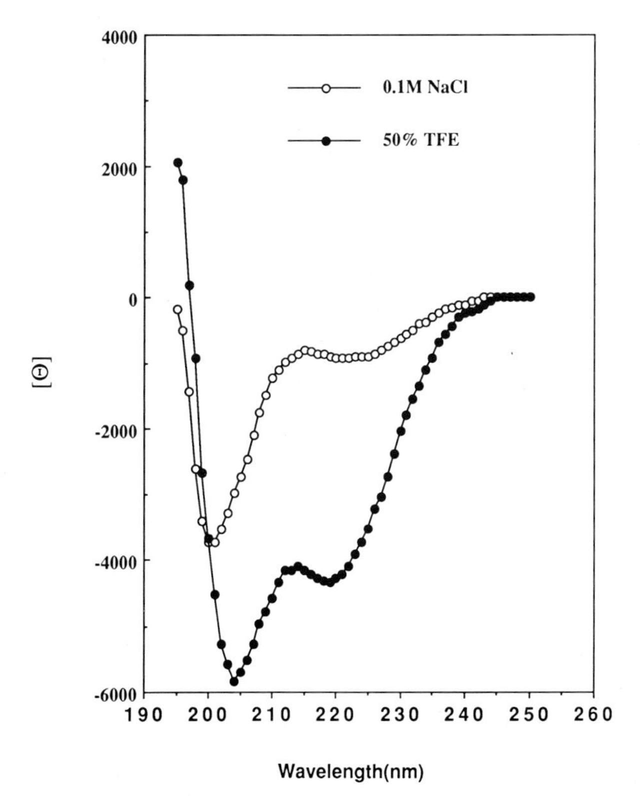

Fig. 6. CD spectra of *N. clavipes* dragline silk dialyzed against 0.1 M NaCl with and without TFE added

supercontraction is reversible and repeatable and can be used to achieve mechanical work by the fibers.

With the proposed structure of dragline silk being crystalline regions interspersed with nonoriented regions, the question arises as to the mechanism of elasticity. In the supercontracted form this appears to be predominantly an entropy-driven process (Gosline et al. 1984). They attribute about 85% of the retractive force to polymer chain conformational entropy. A later calculation estimates the size of the random chains to be about 15 amino acids (Gosline et al. 1986).

4 Recent Research

4.1 Methods

4.1.1 Pure Silk Fibers

Much of the past research on spider silk relied on obtaining samples from webs or from the trailing fibers of the spider. It has become clear that many of these samples were composed of more than one silk type. Thus, Work and Emerson (1982) designed an apparatus to forcibly silk spiders to obtain a single fiber type. Although the apparatus performed well, it was relatively complicated to construct and operate. We have designed a simple version of their instrument which consists of a variable speed drill which is connected to a sewing machine footpedal controller which regulates the drill speed. Forceps are used to take a single silk fiber, which is wrapped around a spool, from a CO_2 anesthetized spider whose legs have been taped down. The procedure is done with the aid of a microscope to insure that only a single fiber of the type desired is taken. Then, while observing under the microscope, the silk is forcibly wound onto the spool. This method is applicable to both major and minor ampullate silk. In fact, both can be obtained simultaneously using two spools separated by 2 cm or more.

Occasionally, it is possible to obtain swathing silk in the same manner but the success rate for this is relatively low in our experience. Swathing silk can be obtained from spiders fed frequently as they occasionally will wrap the prey for later use. This silk can be carefully removed from the prey for examination.

Cocoon silk has been obtained directly from fresh cocoons. However, in observing the cocoon construction by *N. clavipes*, it was seen that several fibers, probably eight, were laid down simultaneously. Thus, it is difficult to obtain single fibers, but it is possible. The amino acid compositions reported above were derived from such fibers. In addition, it was observed that after the eggs are laid, the viscous secretion that accompanies the eggs is frequently smeared over the fibers, which leads to erroneous amino acid compositions. The amino acid composition of this secretion is largely hydrophobic amino acids which are not present to such a large degree in either the cylindrical gland or the carefully obtained single fibers.

4.1.2 Protein Sequence

Initial efforts to obtain protein sequence were directed at the silk fibers themselves. When the fibers were placed on the sequencer filter, they were retained and sequence information could be obtained. The data clearly showed that no single amino terminus was present. In fact, the sequence data resembled the amino acid composition at nearly every step except for the occasional increase of Pro or Tyr. No useful sequence was obtained in this fashion.

The next approach was to solubilize the silk protein(s) and purify them by conventional means. As it was already known that these proteins were soluble only in highly disruptive agents, we used LiSCN and $LiClO_4$. The latter is particularly useful due to its lack of UV absorbance in the regions used for protein detection. However, solubilization in these reagents and several others still did not allow for purification, due to the lack of useful methods in the presence of such salts. We were unable to effectively utilize even size exclusion, due to the very broad elution profiles of these proteins. In the end we could find no useful method which would purify the proteins for sequence analysis through solubilization.

The use of enzymatic cleavage to generate fragments of the protein for sequencing was hampered by the lack of a suitable solubilizing agent in which the protein was soluble and the enzymes were active. Numerous combinations were tried, especially after solubilization in strong reagents and attempts to dialyze into less harsh reagents. These all proved unsuccessful. Proteins denatured in chaotropic agents were digested in the precipitated state with a wide variety of enzymes, but this approach also proved unsuccessful. It is easy to see why spider silks in nature are very resistant to normal degradative processes.

Attempts to use chemical cleavage were generally unsuccessful due to the lack of Met and Trp in these protein. However, later we were able to use N-bromosuccinimide (NBS) with some success in formic acid to cleave at Tyr for silks which were soluble in the concentrated acid.

Since the conventional approaches were unsuccessful, we returned to a technique employed in the early days of protein sequencing, using partial acid hydrolysis. Even this generally straightforward procedure was not without significant problems. The problems can be attributed to the repetitive nature of the proteins and the lack of significant diversity in the amino acids present. The result is peptide bonds present which have very little difference in bond strength. Thus, once hydrolysis starts, all bonds are cleaved at nearly identical rates. Trifluoroacetic acid proved to be an excellent hydrolysis reagent because the silk became soluble prior to hydrolysis. However, under conditions in which hydrolysis occurred the amino-terminus of all isolated peptides was blocked to Edman degradation and no sequence information could be obtained. This led to a procedure (Xu and Lewis 1990) in which the silk was hydrolyzed in 6 N HCl for 3–4 min at 155°C. This resulted in nearly all of the solid silk disappearing, but peptides were present. One minute longer resulted in complete hydrolysis and one minute less no cleavage at all. Other temperatures and acids were used, but none gave results that were better than these conditions. The peptides obtained were sequenced (Fig. 7) and were

SILK PEPTIDES

NEPHILA CLAVIPES

GYGPG

GQGAG

GAGQG

GYGGLG

ARANEUS GEMMOIDES

G-P-Y-G-P-G-E-E-G-P

Fig. 7. The sequence of peptides from dragline silk of the two spider species. The peptides were generated by hydrolysis of the silk in 6 M HCl at 155°C for 3–5 min

generally quite short. However, there was enough sequence to create a DNA probe for cloning studies.

4.1.3 Cloning

Oligonucleotide probes were synthesized based on the peptide sequences in Fig. 7 which were completely degenerate. In addition, the same probes were synthesized in four pools with one quarter of the total degeneracy in each pool. The cDNA libraries were constructed using RNA from the major ampullate glands of *N. clavipes*. In order to insure a maximum level of silk protein mRNA, the spiders were forcibly silked to remove as much silk as possible. After 4 h, at which time they should be maximally producing silk in the gland, they were sacrificed and the silk glands removed. Standard cDNA library construction methods were used as described in Xu and Lewis (1990)

With the anticipation that the silk mRNA would comprise a significant proportion of the total mRNA of the gland, we decided to use isolated colony screening instead of the standard plate screening used for less abundant messages. Therefore, 960 colonies were picked and grown in 96 well plates which were then dotted on filter paper. Standard screening methods were used and the positive colonies were clearly evident. In fact, overnight and 2-h autoradiography gave such large spots that it was not possible to tell which colony was positive. It was necessary to use a 15-min exposure to clearly decide which colony was the correct one. Initial screening gave 36 positives which were then analyzed by Southern blotting. We found 21 positive plasmids with the largest indicating a 4-kb insert. Each of these was grown separately, the plasmids isolated, and restriction digested to release the insert prior to another Southern blot. At this stage there were 12 positive inserts, the largest being 2.5 kb. The largest two of these were chosen for sequencing (2.1 and 1.8 kb).

It was disturbing that the large number of initial positive colonies was reduced to such a small number, but as was later found, this is a characteristic of all the silk cDNAs we have examined. They are unstable in all plasmids and cell lines we have tested. Some plasmids and cells have greater stability than others, but all have a significant deletion rate which can lead to problems if it is not taken into account. It is also clear that the larger the insert the higher the deletion rate, although there appear to be islands of stable sizes, which accounts for our finding the two larger colonies. We have transiently observed larger inserts in several libraries but have been unable to maintain them long enough to sequence them. In fact, this deletion rate can be seen in the colonies where initially white colonies start to turn blue with time. We have even used this as a marker during initial screening to identify likely silk cDNA-containing colonies.

4.1.4 DNA Sequencing

As might be expected, the problem with insert deletion led to numerous problems with sequencing as well. Unfortunately, this only became clear in retrospect after completion of the sequence of the first protein. Due to the likely repetitive nature of the DNA, we felt random fragmentation and sequencing might not be the most efficient approach. We therefore chose to create nested deletions to obtain the total sequence. This choice has proven to be the correct one, as there were regions of over 200 bases which were identical between repeats. These large regions of identity also led to the need to have a large number of overlapping sequences to insure the correct placement of each fragment.

The problem with insert deletions manifested itself in the presence of only a few stable sizes of inserts no matter what time point of exonuclease digestion was used. As we found on the second protein, the solution was to examine a very large number (24–48) of colonies from each time point and choose a wide variety of insert sizes to insure adequate coverage of the region. We also observed compression regions which led to problems with accurate sequencing. Only with careful analysis of both strands could these be resolved (Hinman M and Lewis RV, unpubl. data).

4.2 Dragline Silk Protein 1

The DNA sequence of Protein 1 is presented here (Fig. 8, data from Xu and Lewis 1990). The clone contained the entire 3′-region through the poly(A) tail which has been truncated in the figure. The polyadenylation signal has been underlined in the figure. Of interest is the high GC content in the coding region and the clear change to a more normal base content in the untranslated region (starting at nucleotide 2154).

The amino acid sequence is shown in Fig. 9. It should be noted that the complete sequence has not been obtained. At this stage, it is not clear what the molecular weight of the entire protein is. A MW of over 300 000 has been proposed

```
CAA GGG GCA GGT GCA GCA GCA GCA GCA GCT GGA GGT GCC GGA CAA GGA GGA TAT GGA GGT CTT GGT GGA CAA GGA GCT GGT CAA GGT GGA   90
TAT GGA GGT CTT GGT GGA CAA GGT GCC GGA CAA GGA GCT GGT GCA GCC GCC GCA GCA GCA GCT GGT GGT GCC GGA CAA GGA GGA TAT GGA  180
GGT CTT GGA AGC CAA GGT GCT GGA CGA GGT GGA CAA GGA GCT GGA GCA GCC GCT GCA GCT GCG GGT GGT GCC GGA CAA GGA GGT TAT GGA  270
GGT CTT GGA AGT CAA GGT GCA GGA CGA GGT GGA TTA GGT GGA CAA GGG GCA GGT GCA GCA GCC GCT GCA GCA GCT GGA GGT GCC GGA CAA  360
GGA GGA TAT GGA GGC CTT GGA AAC CAA GGT GCT GGA CGA GGT GGA CAA GGT GCA GCA GCA GCA GCA GCT GGA GGT GCT GGA CAA GGA GGA  450
TAT GGA GGT CTT GGA AGC CAA GGT GCA GGA CGA GGT GGA TTA GGT GGA CAA GGT GCA GGT GCA GCA GCA GCA GCA GCC GGA GGT GCT GGA  540
CAA GGC GGA TAC GGT GGT CTT GGT GGA CAA GGT GCC GGA CAA GGA GGC TAT GGA GGA CTT GGA AGC CAA GGT GCT GGA CGA GGA GGA TTA  630
GGT GGA CAA GGT GCA GGT GCA GCA GCA GCA GCA GCA GCT GGA GGT GCC GGA CAA GGA GGA CTA GGT GGA CAA GGT GCT GGA CAA GGA GCT  720
GGA GCA TCC GCT GCA GCA GCT GGT GGT GCC GGA CAA GGA GGA TAT GGA GGT CTT GGA AGC CAA GGT GCT GGA CGA GGT GGA GAA GGT GCA  810
GGC GCA GCC GCA GCA GCA GCC GGA GGT GCT GGA CAA GGA GGA TAC GGT GGT CTT GGT GGA CAA GGT GCC GGA CAA GGA GGC TAT GGA GGA  900
CTT GGA AGC CAA GGT GCT GGA CGA GGA GGA TTA GGT GGA CAA GGT GCA GGT GCA GCA GCA GCT GGA GGT GCC GGG CAA GGA GGA CTA GGT  990
GGA CAA GGT GCT GGA CAA GGA GCT GGA GCA GCC GCT GCA GCA GCT GGT GGT GCC GGA CAA GGA GGA TAT GGA GGT CTT GGA AGC CAA GGT 1080
GCA GGA CGA GGT GGA TTA GGT GGA CAA GGG GCA GGT GCA GTA GCC GCT GCA GCA GCT GGA GGT GCC GGA CAA GGA GGA TAT GGA GGT CTT 1170
GGA AGC CAA GGT GCT GGA CGA GGT GGA CAA GGA GCT GGA GCA GCC GCT GCA GCA GCT GGT GGT GCC GGA CAA AGA GGT TAT GGA GGT CTT 1260
GGA AAT CAA GGT GCA GGA CGA GGT GGA TTA GGT GGA CAA GGG GCA GGT GCA GCA GCC GCT GCA GCA GCT GGA GGT GCC GGA CAA GGA GGA 1350
TAT GGA GGC CTT GGA AAC CAA GGT GCT GGA CGA GGT GGA CAA GGT GCA GCA GCA GCA GCT GGA GGT GCC GGA CAA GGA GGA TAT GGA GGT 1440
CTT GGA AGC CAA GGT GCT GGA CGA GGT GGA CAA GGT GCA GGC GCA GCC GCA GCA GCA GCC GTA GGT GCT GGA CAA GAA GGA ATA CGT GGA 1530
CAA GGT GCC GGA CAA GGA GGC TAT GGA GGA CTT GGA AGC CAA GGT TCT GGT CGA GGA GGA TTA GGT GGA CAA GGT GCA GGT GCA GCA GCA 1620
GCA GCA GCT GGA GGT GCT GGA CAA GGA GGA TTA GGT GGA CAA GGT GCT GGA CAA GGA GCT GGA GCA GCC GCT GCA GCA GCT GGT GGT GTT 1710
AGA CAA GGA GGA TAT GGA GGT CTT GGA AGC CAA GGT GCT GGA CGA GGT GGA CAA GGT GCA GGC GCA GCC GCA GCA GCA GCC GGA GGT GCT 1800
GGA CAA GGA GGA TAT GGT GGT CTT GGT GGA CAA GGT GTT GGC CGA GGT GGA TTA GGT GGA CAG GGT GCA GGC GCA GCG GCA GCT GGT GGT 1890
GCT GGA CAA GGA GGA TAT GGT GGT GTT GGT TCT GGG GCG TCT GCT GCC TCT GCA GCT GCA TCC CGG TTG TCT TCT CCT CAA GCT AGT TCA 1980
AGA CTT TCA TCA GCT GTT TCC AAC TTG GTT GCA ACT GGT CCT ACT AAT TCT GCG GCC TTG TCA AGT ACA ATC AGT AAC GTG GTT TCA CAA 2070
ATT GGC GCC AGC ATC CTG GTC TTT CTG GAT GTG ATG TCC TCA TTC AAG CTC TTC TCG AGG TTG TTT CTG CTC TTA TCC AGA TCT TAG GTT 2160
CTT CCA GCA TCG GCC AAG TTA ACT ATG GTT CCG CTG GAC AAG CCA CTC AGA TCG TTG GTC AAT CAG TTT ATC AAG CCC TAG GTT AAA TGT 2250
AAA ATC AAG AGT TGC TAA AAC TTA ATG AAC TCG GGC TGT TTA TTT GTG TTA GGT TTT AAA ATA TTT TCA ATA AAT ATT ATG CAT ATA AAA 2340
```

Fig. 8. The DNA sequence of *N. clavipes* dragline silk Protein 1. The polyadenylation signal sequence is *underlined* and the stop codon is at base 2157

Gln Gly Ala Gly Ala Ala Ala Ala Ala Ala Gly Gly Ala Gly Gln Gly Gly Tyr Gly Gly Leu Gly Gly Gln Gly Ala Gly Gln Gly Gly 30

Tyr Gly Gly Ley Gly Gly Gln Gly Ala Gly Gln Gly Ala Gly Ala Ala Ala Ala Ala Ala Ala Gly Gly Ala Gly Gln Gly Gly Tyr Gly 60

Gly Leu Gly Ser Gln Gly Ala Gly Arg Gly Gly Gln Gly Ala Gly Ala Ala Ala Ala Ala Ala Gly Gly Ala Gly Gln Gly Gly Tyr Gly 90

Gly Leu Gly Ser Gln Gly Ala Gly Arg Gly Gly Leu Gly Gly Gln Gly Ala Gly Ala Ala Ala Ala Ala Ala Ala Gly Gly Ala Gly Gln 120

Gly Gly Tyr Gly Gly Leu Gly Asn Gln Gly Ala Gly Arg Gly Gly Gln Gly Ala Ala Ala Ala Ala Ala Gly Gly Ala Gly Gln Gly Gly 150

Tyr Gly Gly Leu Gly Ser Gln Gly Ala Gly Arg Gly Gly Leu Gly Gly Gln Gly Ala Gly Ala Ala Ala Ala Ala Ala Gly Gly Ala Gly 180

Gln Gly Gly Tyr Gly Gly Leu Gly Gly Gln Gly Ala Gly Gln Gly Gly Tyr Gly Gly Leu Gly Ser Gln Gly Ala Gly Arg Gly Gly Leu 210

Gly Gly Gln Gly Ala Gly Ala Ala Ala Ala Ala Ala Ala Gly Gly Ala Gly Gln Gly Gly Leu Gly Gly Gln Gly Ala Gly Gln Gly Ala 240

Gly Ala Ser Ala Ala Ala Ala Gly Gly Ala Gly Gln Gly Gly Tyr Gly Gly Leu Gly Ser Gln Gly Ala Gly Arg Gly Gly Glu Gly Ala 270

Gly Ala Ala Ala Ala Ala Ala Gly Gly Ala Gly Gln Gly Gly Tyr Gly Gly Leu Gly Gly Gln Gln Ala Gly Gln Gly Gly Tyr Gly Gly 300

Leu Gly Ser Gln Gly Ala Gly Arg Gly Gly Leu Gly Gly Gln Gly Ala Gly Ala Ala Ala Ala Gly Gly Ala Gly Gln Gly Gly Leu Gly 330

Gly Gln Gly Ala Gly Gln Gly Ala Gly Ala Ala Ala Ala Ala Ala Gly Gly Ala Gly Gln Gly Gly Try Gly Gly Leu Gly Ser Gln Gly 360

Ala Gly Arg Gly Gly Leu Gly Gly Gln Gly Ala Gly Ala Ala Ala Ala Ala Ala Ala Gly Gly Ala Gly Gln Gly Gly Tyr Gly Gly Leu 390

Gly Ser Gln Gly Ala Gly Arg Gly Gly Gln Gly Ala Gly Ala Ala Ala Ala Ala Ala Gly Gly Ala Gly Gln Arg Gly Tyr Gly Gly Leu 420

Gly Asn Gln Gly Ala Gly Arg Gly Gly Leu Gly Gly Gln Gly Ala Gly Ala Ala Ala Ala Ala Ala Ala Gly Gly Ala Gly Gln Gly Gly 450

Tyr Gly Gly Leu Gly Asn Gln Gly Ala Gly Arg Gly Gly Gln Gly Ala Ala Ala Ala Ala Gly Gly Ala Gly Gln Gly Gly Tyr Gly Gly 480

Leu Gly Ser Gln Gly Ala Gly Arg Gly Gly Gln Gly Ala Gly Ala Ala Ala Ala Ala Ala Val Gly Ala Gly Gln Glu Gly Ile Arg Gly 510

Gln Gly Ala Gly Gln Gly Gly Tyr Gly Gly Leu Gly Ser Gln Gly Ser Gly Arg Gly Gly Leu Gly Gly Gln Gly Ala Gly Ala Ala Ala 540

Ala Ala Ala Gly Gly Ala Gly Gln Gly Gly Leu Gly Gly Gln Gly Ala Gly Gln Gly Ala Gly Ala Ala Ala Ala Ala Ala Gly Gly Val 570

Arg Gln Gly Gly Tyr Gly Gly Leu Gly Ser Gln Gly Ala Gly Arg Gly Gly Gln GLy Ala Gly Ala Ala Ala Ala Ala Ala Gly Gly Ala 600

Gly Gln Gly Gly Tyr Gly Gly Leu Gly Gly Gln Gly Val Gly Arg Gly Gly Leu Gly Gly Gln Gly Ala Gly Ala Ala Ala Ala Gly Gly 630

Ala Gly Gln Gly Gly Tyr Gly Gly Val Gly Ser Gly Ala Ser Ala Ala Ser Ala Ala Ala Ser Arg Leu Ser Ser Pro Gln Ala Ser Ser 660

Arg Val Ser Ser Ala Val Ser Asn Leu Val Ala Ser Gly Pro Thr Asn Ser Ala Ala Leu Ser Ser Thr Ile Ser Asn Val Val Ser Gln 690

Ile Gly Ala Ser Ile Leu Val Phe Leu Asp Val Met Ser Ser Phe Lys Leu Phe Ser Arg Leu Phe Leu Leu Leu Ser Arg Ser End

Fig. 9. The amino acid sequence of dragline silk Protein 1

by Candelas and Cintron (1981), but this may be an overestimate as, in our experience, these proteins run abnormally slow in polyacrylamide gels containing SDS. Northern blot analysis gives sizes ranging from 3.5 to 12 kb for the mRNA from the gland. So it is still not clear what the entire protein size is. The complete carboxyl-terminal region has been sequenced and shows a nonrepetitive region starting around residue 650 although elements of the repeat are still present through the end of the coding region.

The same sequence is shown in Fig. 10, arranged to show the repeating units more clearly. The first obvious feature is the very low number of substitutions in the repeats. Another interesting feature is the large number of deletions from the consensus sequence. These deletions are almost all in multiples of three for currently unknown reasons. We have broken the sequence into three segments. The first nine amino acids are conserved in sequence but the number of deletions is very high in this segment. The second segment is the $GAG(A)_n$ segment. This region is highly conserved with few substitutions and some variation in the number of Ala residues present. The third segment is the last 15 amino acids, which is very highly conserved in sequence with virtually no substitutions and very few deletions. The only position showing any variation is the antepenultimate residue which can be Gly, Ser, or Asn but no others. The sequence can be thought of as a $(GGX)_n(A)_n$ repeat or alternatively a $(GXG)_n(A)_n$ with X being Gln, Tyr, or Leu. A search of the protein sequence database found no matching sequences to any six amino acids in the sequence except the $(Ala)_n$ region. Thus, it appears these are, to date, unique combinations of amino acids.

SPIDER SILK PROTEIN 1 REPEATS

```
---------......................G..
---------------------..........G..
------.........................S..
...---.........................S..
AGRGGLGGQGAGAAAAAAAGGAGQGGYGGLGNQG
...---...--.......-............S..
.........-........-............G..
---------------------..........S..
........................---....G..
------A......S....-............S..
...---..E.........-............G..
---------------------..........S..
................---.....---....G..
------A...........-............S..
.............V.................S..
...---............-.....R......N..
...............................N..
...---...--......--............S..
...---............-V....E.IR---G..
---------------------..........S..
S.................-.....---....G..
------A...........-..VR........S..
...---............-............G..
V...............---..........V.S-.
----------.S..S....
```

Fig. 10. The alignment of repeats in dragline silk Protein 1

4.3 Dragline Silk Protein 2

From Fig. 10 it can be seen that no Pro is present in this sequence, yet the silk is 3.5% Pro and we isolated a major Pro-containing peptide from the silk (Fig. 7). We synthesized a new probe based on the Pro-containing peptide and use it to rescreen our initial library. Over 20 positives were detected and, following the same procedure as for Protein 1, we sequenced the largest, which was about 2.1 kb. The DNA sequence encoding that protein is shown in Fig. 11. Again, note the high GC content in the coding region and the shift to a more normal distribution in the noncoding region. This clone has the poly(A) tail and the polyadenylation signal sequence as well.

The amino acid sequence is shown in Fig. 12. The sequence has a number of similarities to Protein 1 including the breakdown of the repetitive sequence near the carboxy-terminal end. As with Protein 1 we clearly do not have the entire coding region for this protein. Northern blots have indicated a similar distribution of sizes for the mRNA from the gland, so it is not clear what the true size of the protein is either.

The sequence has been arranged to highlight the repetitive segments in Fig. 13. The repeat sequence of Protein 2 can be broken into three segments as well. In this case the first segment, the first 20 amino acids, is very highly conserved with only a couple of substitutions and one insertion. The second segment, the polyalanine region, is longer than in Protein 1 and has substitutions of Ser only. There is some difference in the number of Ala residues but less than was seen for Protein 1. The final 15 amino acids are characterized by very few substitutions but a large number of deletions. Interestingly, the deletions are virtually all in multiples of five instead of three. This reflects the repeat which could be written as (GPGQQ)(GPGGY)GP(SGPGS)$(A)_n$(GPGGY)(GPGQQ)(GPGGY). The reason for the pentamer deletions may be more clear as discussed below in the predicted structure of this protein. The sequence database again shows no identical protein sequences for these peptide segments, although some similar groups of amino acids were detected as discussed in more detail below.

4.4 Codon Usage

There is an incredible codon usage bias in both of these proteins as can be seen in Figs. 14 and 15. The skew is away from using C or G in the wobble base. This is particularly seen in Gln and Gly where there is over 90% use of A and T in the third position. This is probably not overly surprising since the majority of codons already have C and G in the first two bases. In order to prevent long stretches of C's and G's resulting in stem-loop formations which would be unstable, the wobble base is restricted to A or T. This is in contrast to the silkworm silk DNA, which shows no strong codon bias (Mita 1988).

```
CCT GGA GGA TAT GGA CCA GGA CAA CAA GGC CCA GGA GGA TAT GGC CCT GGA CAA CAA GGA CCA TCT GGA CCT GGC     75
AGT GCC GCT GCA GCA GCA GCA GCC GCC GCA GCA GGA CCT GGA GGA TAT GGC CCT GGA CAA CAA GGA CCC GGA GGA    150
TAT GGA CCA GGA CAA CAA GGA CCC GGA AGA TAT GGA CCA GGA CAA CAA GGA CCA TCT GGA CCT GGC AGT GCC GCT    225
GCA GCC GCA GCA GGA TCT GGA CAA CAA GGC CCA GGA GGA TAT GGA CCA CGT CAA CAA GGT CCA GGA GGT TAT GGA    300
CAA GGA CAA CAA GGA CCA TCT GGA CCA GGC AGT GCA GCC GCA GCC TCA GCC GCA GCC TCA GCA GAA TCT GGA CAA    375
CAA GGC CCA GGA GGT TAT GGA CCA GGT CAA CAA GGC CCA GGA GGT TAT GGA CCA GGT CAA CAA GGT CCT GGA GGA    450
TAT GGA CCA GGA CAA CAA GGA CCA TCT GGA CCA GGT AGT GCC GCT GCA GCA GCC GCC GCC GCA TCA GGA CCT GGA    525
CAA CAA GGA CCA GGA GGA TAT GGA CCA GGT CAA CAA GGT CCT GGA GGA TAT GGA CCA GGA CAA CAA GGA CCA TCT    600
GGA CCA GGT AGT GCC GCT GCA GCC GCC GCC GCC GCA TCA GGA CCT GGA CAA CAA GGA CCA GGA GGA TAT GGA CCA    675
GGT CAA CAA GGT CCA GGA GGT TAT GGA CCA GGA CAA CAA GGA CTA TCT GGA CCA GGC AGT GCA GCT GCA GCA GCC    750
GCA GCA GGA CCT GGA CAA CAA GGA CCC GGA GGA TAT GGA CCA GGA CAA CAA GGA CCA TCT GGA CCC GGT AGT GCC    825
GCT GCA GCA GCA GCC GCC GCA GCA GGA CCT GGA GGA TAT GGC CCT GGA CAA CAA GGA CCC GGA GGA TAT GGA CCA    900
GGA CAA CAA GGA CCA TCT GGA GCA GGC AGT GCA GCA GCA GCA GCC GCA GCA GGA CCT GGA CAA CAA GGA TTA GGA    975
GGT TAT GGA CCA GGA CAA CAA GGT CCA GGA GGA TAT GGA CCA GGA CAA CAA GGT CCA GGA GGA TAT GGA CCA GGT   1050
AGT GCA TCT GCA GCA GCA GCC GCA GCA GGA CCT GGA CAA CAA GGA CCA GGA GGA TAT GGA CCT GGA CAA CAA GGA   1125
CCA TCT GGA CCA GGC AGT GCA TCT GCA GCA GCA GCC GCA GCC GCA GCA GGA CCA GGA GGA TAT GGA CCA GGA CAA   1200
CAA GGT CCA GGA GGA TAT GCA CCA GGA CAA CAA GGA CCA TCT GGA CCA GGC AGT GCA TCT GCA GCA GCA GCC GCA   1275
GCC GCA GCA GGA CCA GGA GGA TAT GGA CCA GGA CAA CAA GGT CCA GGA GGA TAT GCA CCA GGA CAA CAA GGA CCA   1350
TCT GGA CCA GGC AGT GCA GCA GCA GCA GCA GCT GCT GCA GCA GGA CCT GGT GGA TAT GGA CCA GCG CAA CAG GGA   1425
CCA TCT GGT CCT GGA ATC GCA GCT TCA GCT GCT TCA GCA GGA CCT GGA GGT TAT GGA CCA GCA CAA CAA GGA CCA   1500
GCT GGA TAT GGG CCT GGA AGC GCA GTA GCA GCC TCT GCC GGT GCA GGA TCT GCA GGT TAT GGG CCA GGT TCT CAA   1575
GCT TCC GCT GCA GCT TCT CGT CTG GCT TCT CCA GAT TCA GGC GCT AGA GTT GCA TCA GCT GTT TCT AAC TTG GTA   1650
TCC AGT GGC CCA ACT AGC TCT GCT GCC TTA TCA AGT GTT ATC AGT AAC GCT GTG TCT CAA ATT GGC GCA AGT AAT   1725
CCT GGT CTC TCT GGT TGC GAT GTC CTC ATT CAA GCT CTC TTG GAA ATC GTT TCT GCT TGT GTA ACC ATC CTT TCT   1800
TCA TCC AGC ATT GGT CAA GTT AAT TAT GGA GCG GCT TCT CAG TTC GCC CAA GTT GTC GGC CAA TCT GTT TTG AGT   1875
GCA TTT TAA TTG AAA AAT TTA TTA AAA TAT GCA TGG ATT TTC TAG CCT GGG CAA CTA ATT GCT CGT ACT ATG TAA   1950
TTT TTT TTT AAA TAA ATT CTT TGC AAC TTC
```

Fig. 11. The DNA sequence of *N. clavipes* dragline silk Protein 2

```
Pro Gly Gly Tyr Gly Pro Gly Gln Gln Gly Pro Gly Gly Tyr Gly Pro Gly Gln Gln Gly Pro Ser Gly Pro Gly  25
Ser Ala Ala Ala Ala Ala Ala Ala Ala Ala Ala Gly Pro Gly Gly Tyr Gly Pro Gly Gln Gln Gly Pro Gly Gly  50
Tyr Gly Pro Gly Gln Gln Gly Pro Gly Arg Tyr Gly Pro Gly Gln Gln Gly Pro Ser Gly Pro Gly Ser Ala Ala  75
Ala Ala Ala Ala Gly Ser Gly Gln Gln Gly Pro Gly Gly Tyr Gly Pro Arg Gln Gln Gly Pro Gly Gly Tyr Gly  100
Gln Gly Gln Gln Gly Pro Ser Gly Pro Gly Ser Ala Ala Ala Ala Ser Ala Ala Ala Ser Ala Glu Ser Gly Gln  125
Gln Gly Pro Gly Gly Tyr Gly Pro Gly Gln Gln Gly Pro Gly Gly Tyr Gly Pro Gly Gln Gln Gly Pro Gly Gly  150
Tyr Gly Pro Gly Gln Gln Gly Pro Ser Gly Pro Gly Ser Ala Ala Ala Ala Ala Ala Ala Ala Ser Gly Pro Gly  175
Gln Gln Gly Pro Gly Gly Tyr Gly Pro Gly Gln Gln Gly Pro Gly Gly Tyr Gly Pro Gly Gln Gln Gly Pro Ser  200
Gly Pro Gly Ser Ala Ala Ala Ala Ala Ala Ala Ala Ser Gly Pro Gly Gln Gln Gly Pro Gly Gly Tyr Gly Pro  225
Gly Gln Gln Gly Pro Gly Gly Tyr Gly Pro Gly Gln Gln Gly Leu Ser Gly Pro Gly Ser Ala Ala Ala Ala Ala  250
Ala Ala Gly Pro Gly Gln Gln Gly Pro Gly Gly Tyr Gly Pro Gly Gln Gln Gly Pro Ser Gly Pro Gly Ser Ala  275
Ala Ala Ala Ala Ala Ala Ala Ala Gly Pro Gly Gly Tyr Gly Pro Gly Gln Gln Gly Pro Gly Gly Tyr Gly Pro  300
Gly Gln Gln Gly Pro Ser Gly Ala Gly Ser Ala Ala Ala Ala Ala Ala Ala Gly Pro Gly Gln Gln Gly Leu Gly  325
Gly Tyr Gly Pro Gly Gln Gln Gly Pro Gly Gly Tyr Gly Pro Gly Gln Gln Gly Pro Gly Gly Tyr Gly Pro Gly  350
Ser Ala Ser Ala Ala Ala Ala Ala Ala Gly Pro Gly Gln Gln Gly Pro Gly Gly Tyr Gly Pro Gly Gln Gln Gly  375
Pro Ser Gly Pro Gly Ser Ala Ser Ala Ala Ala Ala Ala Ala Ala Ala Gly Pro Gly Gly Tyr Gly Pro Gly Gln  400
Gln Gly Pro Gly Gly Tyr Ala Pro Gly Gln Gln Gly Pro Ser Gly Pro Gly Ser Ala Ser Ala Ala Ala Ala Ala  425
Ala Ala Ala Gly Pro Gly Gly Tyr Gly Pro Gly Gln Gln Gly Pro Gly Gly Tyr Ala Pro Gly Gln Gln Gly Pro  450
Ser Gly Pro Gly Ser Ala Ala Ala Ala Ala Ala Ala Ala Ala Gly Pro Gly Gly Tyr Gly Pro Ala Gln Gln Gly  475
Pro Ser Gly Pro Gly Ile Ala Ala Ser Ala Ala Ser Ala Gly Pro Gly Gly Tyr Gly Pro Ala Gln Gln Gly Pro  500
Ala Gly Tyr Gly Pro Gly Ser Ala Val Ala Ala Ser Ala Gly Ala Gly Ser Ala Gly Tyr Gly Pro Gly Ser Gln  525
Ala Ser Ala Ala Ala Ser Arg Leu Ala Ser Pro Asp Ser Gly Ala Arg Val Ala Ser Ala Val Ser Asn Leu Val  550
Ser Ser Gly Pro Thr Ser Ser Ala Ala Leu Ser Ser Val Ile Ser Asn Ala Val Ser Gln Ile Gly Ala Ser Asn  575
Pro Gly Leu Ser Gly Cys Asp Val Ler Ile Gln Ala Leu Leu Glu Ile Val Ser Ala Cys Val Thr Ile Leu Ser  600
Ser Ser Ser Ile Gly Gln Val Asn Tyr Gly Ala Ala Ser Gln Phe Ala Gln Val Val Gly Gln Ser Val Leu Ser  625
Ala Phe
```

Fig. 12. The protein sequence of dragline silk Protein 2

SILK PROTEIN 2 REPEATS

```
                                                 ....
GPGQQGPGGYGPGQQGP--SGPGSAAAAAAAAAA----GPGGYGPGQQGPGGY
........R.....................----....-----.S........
..R.........................S...S.ESGQQ..............
...............................S--....-----..........
...............................S--....------.........
................L..............---....---------------
.................................-....----------.....
.....................A.........---....-----......L...
.................GGY.....S......--....---------------
.........................S............----------.....
..........A..............S............----------.....
..........A......................-....----------.....
```

Fig. 13. The alignment of repeats in dragline silk Protein 2

Codon	AA	Ct.	Percent	Codon	AA	Ct.	Percent
TTT	F	2	0.3	ATT	I	1	0.1
TTC	F	2	0.3	ATC	I	2	0.3
TTA	L	10	1.4	ATA	I	1	0.1
TTG	L	4	0.6	ATG	M	1	0.1
TCT	S	8	1.1	ACT	T	2	0.3
TCC	S	5	0.7	ACC	T	0	0.0
TCA	S	6	0.8	ACA	T	1	0.1
TCG	S	1	0.1	ACG	T	0	0.0
TAT	Y	18	2.5	AAT	N	2	0.3
TAC	Y	2	0.3	AAC	N	4	0.6
TAA	*	0	0.0	AAA	K	0	0.0
TAG	*	1	0.1	AAG	K	1	0.1
TGT	C	0	0.0	AGT	S	4	0.6
TGC	C	0	0.0	AGC	S	11	1.5
TGA	*	0	0.0	AGA	R	4	0.6
TGG	W	0	0.0	AGG	R	1	0.1
CTT	L	20	2.8	GTT	V	6	0.8
CTC	L	2	0.3	GTC	V	1	0.1
CTA	L	2	0.3	GTA	V	2	0.3
CTG	L	3	0.4	GTG	V	2	0.3
CCT	P	2	0.3	GCT	A	54	7.5
CCC	P	0	0.0	GCC	A	35	4.9
CCA	P	0	0.0	GCA	A	95	13.2
CCG	P	0	0.0	GCG	A	4	0.6
CAT	H	0	0.0	GAT	D	1	0.1
CAC	H	0	0.0	GAC	D	0	0.0
CAA	Q	67	9.3	GAA	E	2	0.3
CAG	Q	1	0.1	GAG	E	0	0.0
CGT	R	1	0.1	GGT	G	121	17.0
CGC	R	0	0.0	GGC	G	12	1.7
CGA	R	15	2.1	GGA	G	169	23.5
CGG	R	1	0.1	GGG	G	6	0.8

Fig. 14. Codon usage for the coding sequence of dragline silk Protein 1

4.5 Structural Predictions Based on Primary Sequences

The hydropathy profiles for both proteins (Fig. 16) also clearly demonstrate the repetitive nature of both proteins. They also indicate the much more extensive hydrophilic nature of Protein 2 although Protein 1 has such regions which are not

Codon	AA	Ct	Percent	Codon	AA	Ct.	Percent
TTT	F	1	0.2	ATT	I	3	0.5
TTC	F	1	0.2	ATC	I	4	0.6
TTA	L	2	0.3	ATA	I	0	0.0
TTG	L	3	0.5	ATG	M	0	0.0
TCT	S	30	4.8	ACT	T	1	0.2
TCC	S	3	0.5	ACC	T	1	0.2
TCA	S	10	1.6	ACA	T	0	0.0
TCG	S	0	0.0	ACG	T	0	0.0
TAT	Y	30	4.8	AAT	N	2	0.3
TAC	Y	0	0.0	AAC	N	2	0.3
TAA	*	1	0.2	AAA	K	0	0.0
TAG	*	0	0.0	AAG	K	0	0.0
TGT	C	1	0.2	AGT	S	17	2.7
TGC	C	1	0.2	AGC	S	3	0.5
TGA	*	0	0.0	AGA	R	2	0.3
TGG	W	0	0.0	AGG	R	0	0.0
CTT	L	1	0.2	GTT	V	7	1.1
CTC	L	3	0.5	GTC	V	2	0.3
CTA	L	2	0.3	GTA	V	3	0.5
CTG	L	1	0.2	GTG	V	1	0.2
CCT	P	22	3.3	GCT	A	23	3.7
CCC	P	5	0.8	GCC	A	32	5.1
CCA	P	59	9.4	GCA	A	75	11.9
CCG	P	0	0.0	GCG	A	2	0.3
CAT	H	0	0.0	GAT	D	2	0.3
CAC	H	0	0.0	GAC	D	0	0.0
CAA	Q	72	11.5	GAA	E	2	0.3
CAG	Q	2	0.3	GAG	E	0	0.0
CGT	R	2	0.3	GGT	G	30	4.8
CGC	R	0	0.0	GGC	G	19	3.0
CGA	R	0	0.0	GGA	G	143	22.8
CGG	R	0	0.0	GGG	G	2	0.3

Fig. 15. Codon usage for the coding sequence of dragline silk Protein 2

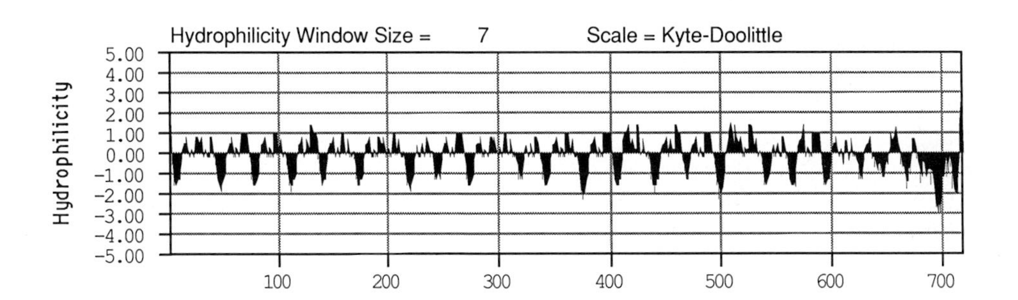

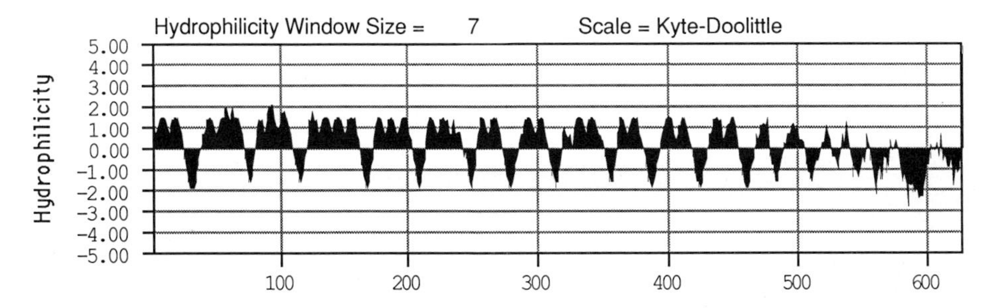

Fig. 16. Hydropathy profiles for dragline silk Proteins 1 (*upper panel*) and 2 (*lower panel*)

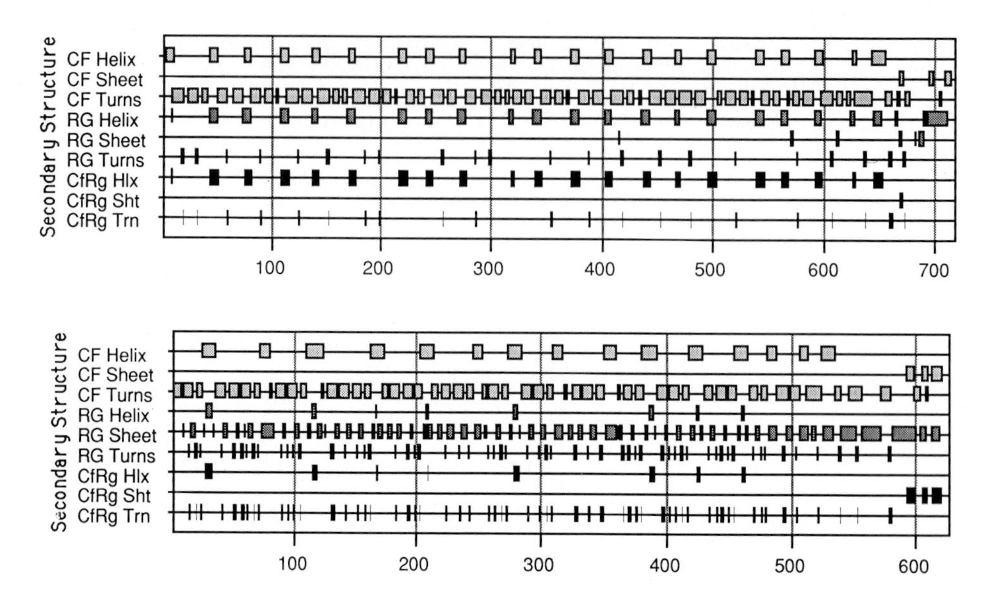

Fig. 17. Secondary structure prediction profiles of dragline silk Proteins 1 (*upper panel*) and 2 (*lower panel*)

as hydrophobic. The polyalanine regions stand out in both proteins as hydrophobic regions. The pattern of alternating hydrophobic and hydrophilic regions breaks down toward the carboxyl-terminus of both proteins confirming the breakdown of both the repetitive sequences and the codon bias.

Although structure prediction on sequences which are not present in the database clearly has to be studied judiciously the data are interesting (Fig. 17). Neither protein shows any predicted tendency to form beta sheets which are believed to be the major secondary structure in the fibers as seen by both X-ray diffraction and FTIR. However, both proteins are predicted to have a tendency to form turns in the Gly-rich regions and α-helices in the polyalanine regions. The helical tendency of alanine is well known and thus it is not surprising to find this type of structure predicted. The fact that, upon stretching, the fibers demonstrate the presence of α-helical structure, which disappears upon release of tension, is consistent with this prediction. The presence of β-turns is both more difficult to predict and to detect using biophysical methods. Thus, their presence cannot be established as yet but, as described below, they could be consistent with the biophysical data.

4.6 Peptide Studies

In order to study the possible structures of the two proteins, peptides corresponding to the consensus repeats were synthesized (Fig. 18). These peptides were chosen based on the most common amino acid at each position of the various

SILK PROTEIN 1 PEPTIDE (30P)

AGRGGQGAGAAAAAAGGAGQGGYGGLGGQG

SILK PROTEIN 2 PEPTIDE (47P)

**GPGQQGPGGYGPGQQGPSGPGSAAAAAAAAAAGP
GGYGPGQQGPGGY**

Fig. 18. Synthetic peptides corresponding to the consensus repeat sequences of Proteins 1 (*30P*) and 2 (*47P*)

repeats. They were synthesized using the Fmoc chemistry on an Applied Biosystems solid phase peptide synthesizer. Purification was by reverse phase high performance liquid chromatography and the structure of the peptide was confirmed by amino acid analysis and sequencing.

Both of these peptides were examined using FTIR and CD. The CD studies showed a strong concentration dependence on the structure of the silk protein 1 peptide. Below 2 mg/ml, the CD indicated no structure other than random coil. Above this concentration, there was a marked temperature dependence on structure as well (Fig. 19). At 5 °C, the CD indicated a possible multi-strand helix, which is reasonable in view of the large amount of Gly in the peptide. As the temperature increases, the spectra shifts to a β-sheet-like structure at 80 °C which strongly resembles the spectra of the dissolved and dialyzed silk protein at high temperature (Fig. 4). The shoulder in the 225–230 nm range is due to the Tyr residues and is responsible for the broad peaks observed for this peptide. When the peptide is placed in TFE there is a marked shift to a polyalanine α-helix spectra which is the only structure seen. This is surprising since the Ala residues account for only 20% of the peptide and yet they seem to have induced the whole peptide to assume the α-helical structure. FTIR spectroscopy was done on the dried peptide which showed predominantly β-sheet-like structure (data not shown). One interesting feature of the concentration study was that at 10 mg/ml, the peptide began to precipitate after 16 h. This precipitate also showed a β-sheet-like structure in the FTIR spectrum and had solubility characteristics similar to the native silk protein suggesting that this 30 amino acid peptide could mimic reasonably well the protein in the fiber. The final conclusion about the studies of this peptide to data is that it must have the flexibility, probably due to the Gly residues, to assume a wide variety of structures with little difference in energy.

The suggested structure of silk protein 2 was a series of β-turns. Since these are difficult to detect conclusively in either CD or FTIR spectroscopy, it is not surprising that the CD spectra of this protein provided no solid evidence for any structure. The spectra do not match at all types I, II, or III β-turns (Brahms and Brahms 1980). However, the spectra (Fig. 20) is similar to a cyclo-polyalanine peptide shown in the above reference. This peptide is thought to be a type IV turn,

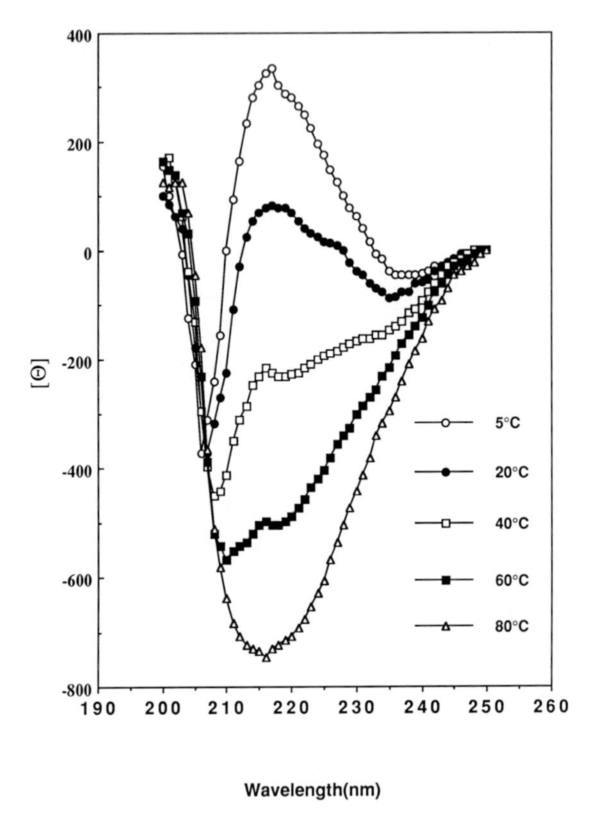

Fig. 19. CD spectra of 30P at the temperatures indicated

which is rare in most globular proteins but may be more likely in the fiber environment.

4.7 Proposed Structure

It is always with trepidation that one tries to predict the structure of any protein without X-ray or two-dimensional nuclear magnetic resonance data but with proteins which show no sequence homology to other proteins, it is even more difficult. To further complicate the situation with dragline silk there are two proteins, not one, and they exist in an environment of low water, a situation which has not been explored by protein structural research to any large degree. Finally, the data to date on the proteins are somewhat contradictory as to the secondary structure of the basic protein repeat elements. Despite these factors, we will propose a structure based on available data and on analogies to proteins showing some similarity to the silk sequences. There are testable elements to this structure which we hope to examine to determine if it is correct.

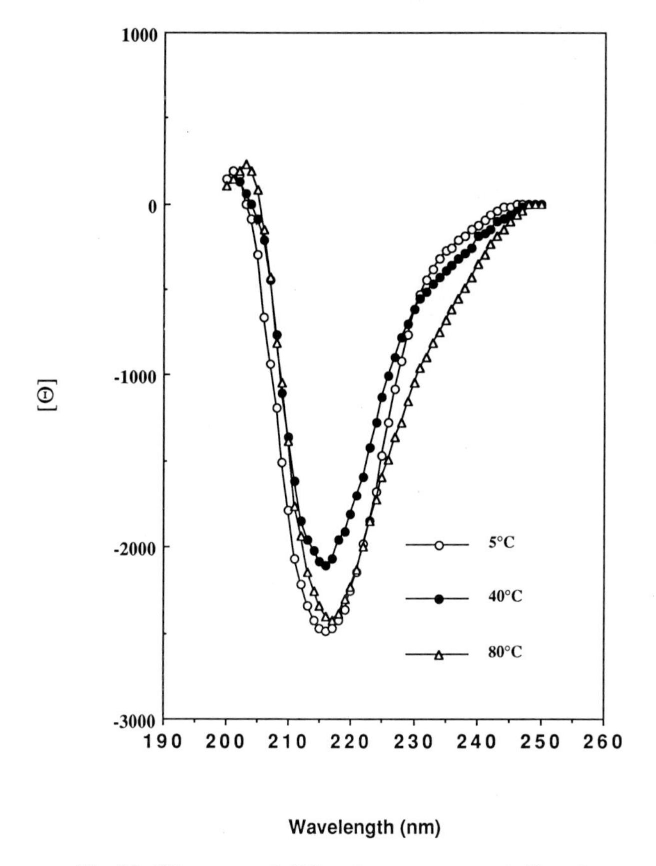

Fig. 20. CD spectra of 47P at the temperatures indicated

Protein 1 seems to have little tendency to form a single thermodynamically stable structure. Rather, it can assume a variety of secondary structures based on other extrinsic factors. Thus we turned to Protein 2 to establish a preliminary structure. In view of the large number and spacing of Pro residues there is no chance the protein can assume either a α-helical or a typical β-sheet structure. Since the repeat distance of the Pro is 5 residues it cannot form a cross-β-structure either since that would require an even number of residues.

When other Pro-rich proteins are examined, there are two that have some similarity in sequence and spacing of Pro residues to silk Protein 2. The first is gluten, an insoluble protein from wheat which is thought to be responsible for the elasticity of dough (Field et al. 1987). The other is synaptophysin, an integral membrane protein of synaptic vesicles which can bind calcium (Buckley et al. 1987). Gluten has some repeat sequences which are Pro-Gly-Gln-Gly-Gln-Gln, and synaptophysin is Tyr-Gly-Pro-Gln-Gly. Both of these proteins are thought to form β-turn helices or a β-spiral. In addition, when the prediction of β-turn is examined in more detail (Wilmot and Thornton 1988), the prediction table gives the sequence

Gly-Pro-Gly-Gln, one of the highest possible scores for a β-turn with Gly-Pro-Gly-Gly just slightly below this score. These are both for Type II turns.

The polyalanine region can clearly form an α-helix, as we showed with the CD studies and others have noted as well (Marqusee et al. 1989). However, these studies are all done in the presence of waters of hydration. The structure of these types of peptides in the absence of substantial water, as in the fiber, are unknown.

Finally, as noted above, the X-ray data and our FTIR studies suggest the presence of β-sheet structure as the predominant secondary structure. Thus, to assimilate these various data into a consistent model, we have utilized a variation of the linked β-turns which would allow for a more planar β-sheet-like structure to be present. In this model the Pro-containing regions form a series of linked β-turns which are not twisted to the extent that they form the spiral in the relaxed state. In fact, they resemble cross β-sheet except that the cross length is much shorter. Due to the more planar nature of the turns, they can form β-sheet-like intermolecular hydrogen bonds to another protein molecule. This would be consistent with our peptide studies where only at higher concentrations is structure observed.

If this is the structure for Protein 2, how does Protein 1 fit into the fiber structure? As noted above, Protein 1 can adapt to a wide variety of structures, depending on the extrinsic conditions. Thus, we can see Protein 1 adapting to the β-turn scaffolding provided by Protein 2 due to its turn forming potential and its flexibility. This interaction would be stabilized by the intermolecular hydrogen bonds between protein molecules. This model would predict that a 1:1 ratio of proteins is unneeded since once the basic structure is set the other proteins would adapt to it. This is consistent with the fact that various spider species have different Pro content yet similar physical properties for dragline silk. Preliminary data from *Araneus gemmoides* peptides indicate that the protein sequences in that dragline silk are similar to those in *N. clavipes* except that the proportion of Pro peptides is much higher. By fitting the amino acid compositions of these two proteins in varying ratios to the composition data from different species we see a very reasonable fit (data not shown).

Both proteins have the polyalanine stretches which would probably be aligned between all proteins molecules since they would not fit the turn structure. In the relaxed state these stretches seem to be in a more random coil conformation which is altered to a α-helical structure when tension is applied.

The mechanism of elasticity based on this structure would involve both regions. The formation of α-helical stretches of Ala from a random coil structure is unable to account for the 30–35% extension. Thus we would propose that the linked β-turns begin to form extended β-spiral structures when tension is applied. This is easily sufficient to account for the observed extension of these fibers. The fact that neither X-ray nor FTIR data show any changes in the β-sheet structure is consistent since the molecules stretch in concert they would maintain most of the intermolecular hydrogen bonds.

The mechanism for retraction is more difficult to account for, but, if the data of Gosline et al. (1984) for super-contracted silk applies to the more normal dry fiber, then entropy must play a major role. One way to account for this is to utilize the small amount of water which is present in the fiber. If that water were sequestered from the hydrophobic Leu and Tyr groups in the relaxed state but forced to interact with those side chains following extension the entropy driven retractive force would be accounted for.

There is no question this structure and mechanism differ from the traditional thinking about spider silk. Yet the actual protein sequences suggest the traditionally held views cannot be correct. Only further structural and biophysical studies will resolve the question of the structure and mechanism of these proteins.

References

Anderson CM, Tillinghast EK (1980) GABA and taurine derivatives on the adhesive spiral of the orb web of *Argiope* spiders and their possible behavioral significance. Physiol Entomol 5:101–106

Anderson SO (1970) Amino acid composition of spider silks. Comp Biochem Physiol 35:705–711

Bell AL, Peakall DB (1969) Changes in the fine structure during silk protein productions in the ampullate gland of the spider *Araneus sericatus*. J Cell Biol 42:284–295

Benton JR (1907) The strength and elasticity of spider thread. Am J Sci xxiv:75–78

Blackwell J (1830) On the manner in which geometric spiders construct their Nets. Zool J V:181–188

Brahms S, Brahms J (1980) Determination of protein secondary structure in solution by vacuum ultraviolet circular dichroism. J Mol Biol 138:149–178

Buckley KM, Floor E, Kelly RB (1987) Cloning and sequence analysis of cDNA encoding p38, a major synaptic vesicle protein. J Cell Biol 105:2447–2456

Candelas GC, Cintron J (1981) A spider fibroin and its synthesis. J Exp Zool 216:1–6

Dong Z, Lewis RV, Middaugh CR (1991) Molecular mechanism of spider silk elasticity. Arch Biochem Biophys 284:53–57

Field M, Tatham AS, Shewry PR (1987) The structure of a high-Mr subunit of durum-wheat (*Triticum durum*) gluten. Biochem J 247:215–221

Fischer E (1907) About spider silk. Hoppe-Seyler's Z Physiol Chem 53:440–450

Fornes RE, Work RW, Morosoff N (1983) Molecular orientation of spider silks in the natural and supercontracted states. J Polym Sci 21:1163–172

Fraser RDB, MacRae TP (1973) Silks. In: Conformation in fibrous proteins. Academic Press, New York, pp 292–343

Gosline JM, Denny MW, DeMont ME (1984) Spider silk as rubber. Nature 309:551–552

Gosline JM, DeMont ME, Denny MW (1986) The structure and properties of spider silk. Endeavour 10:37–43

Iiazuka E (1983) III. The physico-chemical properties of silk fibers and the fiber spinning process. Experientia 39:449–454

Kovoor J (1972) Etude histochimique et cytologique des glandes sericigènes de quelques Argiopidae. Ann Sci Nat Zool Biol Anim 14:1–40

Marqusee S, Robbins VH, Baldwin RW (1989) Unusually stable helix formation in short alanine-based peptides. Proc Natl Acad Sci USA 86:5286–5290

Mita K, Ichimura S, Zarne M, Jones TC (1988) Specific codon usage pattern and its implications on the secondary structure of silk fibroin mRNA. J Mol Biol 203:917–925

Urry DW (1988) Entropic elastic processes in protein mechanisms. I. Elastic structure due to an inverse temperature transition and elasticity due to internal chain dynamics. J Prot Chem 7:1–34

Wilmot CM, Thornton JM (1988) Analysis and prediction of the different type of β-turn in proteins. J Mol Biol 203:221–232

Work RW (1977a) Mechanisms of major ampullate silk fiber formation by orb-web-spinning spiders. Trans Am Microsc Soc 100:1–20

Work RW (1977b) Dimensions, birefringences and force-elongation behavior of major and minor ampullate silk fibers from orb-web-spinning spiders — the effects of wetting on these properties. Text Res J 47:650–662

Work RW (1981) A comparative study of the supercontraction of major ampullate silk fibers of orb-web-spinning spiders (Araneae). J Arachnol 9:299–308

Work RW (1985) Viscoelastic behaviour and wet supercontraction of major ampullate silk fibers of certain orb-web-building spiders (Araneae). J Exp Biol 118:379–404

Work RW, Emerson PD (1982) An apparatus and technique for the forcible silking of spiders. J Arachnol 10:1–10

Work RW, Morosoff N (1982) A physico-chemical study of the supercontraction of spider major ampullate silk fibers. Text Res J 52:349–356

Work RW, Young CT (1987) The amino acid compositions of major and minor ampullate silks of certain orb-web-building spiders (Araneae, Araneidae). J Arachnol 15:65–80

Xu M, Lewis RV (1990) Structure of a protein superfiber: spider dragline silk. Proc Natl Acad Sci USA 87:7120–7124

Chapter 9

The Nature and Role of Liquid Crystalline Order in Silk Secretions

Christopher Viney[1]

1 Background: Silk

1.1 What Is Silk?

Natural silks are extracellular fibers that are spun by a variety of organisms, for highly specialized applications. Their formation specifically by *spinning* is the single feature that distinguishes silks from other extracellular fibrous material such as hair or byssus. Spinning entails secretion of material by epithelial cells into the lumen of a gland, where it is stored in readiness for subsequent molding into fiber through an active or passive spinneret. While the majority of silks consist of protein, some examples based on chitin have been identified among cockroaches and praying mantids (Rudall 1962).

It is recognized that at least three broad classes of protein are capable of forming silks in nature: *fibroin*, and, occasionally, *sericin* and *collagen*. The distinction between these types of protein is made on the basis of differences in the structure of the solid fiber as determined by X-ray diffraction (Rudall 1962). In particular, the classification takes both the preferred conformation of the chains in the solid fiber and how the chains are packed into account.

1.2 Silk-Producing Organisms

While silkworms and spiders are the most widely known producers of silk, they are certainly not unique in this ability. Several insect and arachnid orders produce silk for specific purposes that include construction of habitats, protection of eggs or pupating larvae, capturing and restraining prey, and aids used in locomotion. Silkworm silk is well known because of its 4000-year history of use in garments (Hyde 1984). Spider silk is familiar because spiders are cosmopolitan. The spiders that have been most carefully studied with respect to the production and properties of their silk, are the orb weavers *Nephila clavipes* and *Araneus diadematus*. The former is favored for study because of its conveniently large size: mature spiders

[1] Center for Bioengineering WD-12, University of Washington, Seattle, WA 98195, USA

Results and Problems in Cell Differentiation 19
Biopolymers
Case, S. T. (Ed.)

produce relatively large quantities of silk, and they are relatively easy to dissect. Results obtained from *N. clavipes* silk are one focus of attention in the original studies described later in this chapter. The garden spider *A. diadematus*, is smaller, but it is extremely common, and so its silk is readily available. It certainly should *not* be assumed that either of these spiders necessarily has the strongest silk (or silk optimized with respect to any other property); most natural silks have yet to undergo appropriate characterization.

Orb-weaving spiders such as *N. clavipes* and *A. diadematus* are the most highly sophisticated users of silk. Individual animals can produce as many as five different types of silk, each with properties tailored to a particular application. The different silks are: dragline or frame silk (the radial component of webs), capture threads (the sticky spiral of webs), cocoon silk, attachment discs (that anchor the dragline or frame silk), and prey-wrapping silk. Results pertaining to the first three types of silk in this list will be described in Section 5.

1.3 Rationale for Studying Silk

From the point of view of molecular biology, silk production represents a highly specialized adaptation of the protein synthesis mechanism. This specialization is inferred from the apparent absence of Golgi bodies in *Araneus sericatus* spider ampullate glands (Bell and Peakall 1969) and the observation of Golgi body behavior in *N. clavipes* ampullate glands (Plazaola and Candelas 1991). The silk is transported from the endoplasmic reticulum in the form of secretory granules, which move to the cell membrane and empty into the lumen of the gland. Given that spider spinnerets have been found in Middle Devonian fossil records (Shear et al. 1989), the development of this particular specialization has been in progress for at least 380 million years.

From the point of view of the materials engineer or bioengineer, silk (and the process by which it is spun) exhibits a spectrum of impressive characteristics that cannot at present be achieved with artificial polymers. Materials properties of interest (Fig. 1) include the unusual combination of high strength (load per unit cross-section needed to cause tensile failure), stiffness (tensile load to achieve unit elastic extension), and toughness (total energy absorbed in causing failure) of dragline silk. A few synthetic polymers such as Kevlar, a polyaramid, have a slightly higher strength than *N. clavipes* dragline, but their toughness is significantly lower (Gosline et al. 1986; Kaplan et al. 1991). (Since data from multiple sources were used, the comparison made here cannot be stated more quantitatively; one would first have to be certain that each type of mechanical test was performed on samples of identical dimensions and under identical conditions.) Other desirable materials properties of silk are its durability in a variety of chemical and biological environments, and its biodegradability by proteolytic enzymes as evidenced by the fact that spiders can recycle their web silk (Peakall 1971).

The way in which silk secretions are processed into fiber is also of great interest. Natural silk fibers are spun at ambient temperatures from aqueous solutions, whereas synthetic polymers are spun at elevated temperatures and/or from more

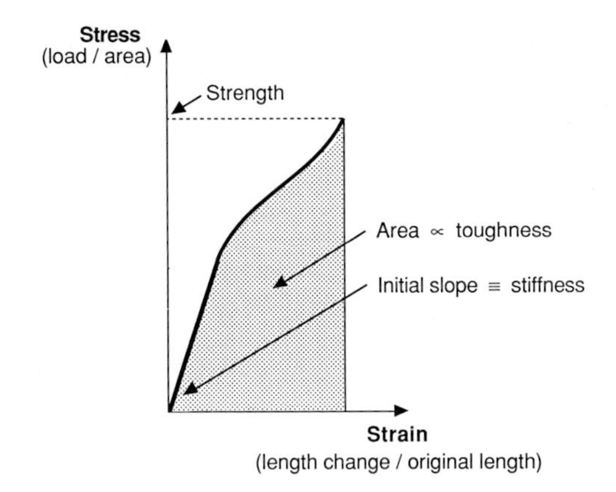

Fig. 1. Schematic definition of the physical parameters *stress*, *strain*, *strength*, *stiffness*, and *toughness* as used in this chapter

aggressive solvents. For example, Kevlar fiber is spun from solution in concentrated sulfuric acid maintained at elevated temperature and pressure. The concept of a water-soluble system that converts to insoluble fiber upon shearing through a spinneret is also novel to industrial polymer processing. Finally, cribellate spiders can produce silk thread that is very much finer than any polymer fibers spun industrially. These spiders are equipped with a spinning organ (the cribellum) that is situated forward of the spinnerets; it can contain over 40000 spigots. Silk is combed out of the cribellum by rhythmic movements of an array of hairs on the hindmost legs. The resultant mass of silk consists of individual filaments that are a mere 0.01 μm in diameter (Foelix 1982), and is used to ensnare prey.

Several aspects of the natural silk-spinning process must be understood if one hopes to produce similar materials on a manufacturing scale. Firstly, one has to determine how the choice of amino acid sequence affects the accessible molecular conformations and hence the properties exhibited by several well-characterized silks. Progress is severely limited because, to date, the only primary structures published for silks are partial amino acid sequences of *Bombyx mori* (silkworm) silk (Tsujimoto and Suzuki 1979) and *N. clavipes* dragline silk (Xu and Lewis 1990). Secondly, a DNA fragment that codes for the selected polypeptide must be introduced into an appropriate expression system. Materials science has already begun to benefit from bacterial expression systems that enable laboratory-scale syntheses of model polypeptides associated with predictable conformations (Tirrell et al. 1991a; b). Thirdly, it is necessary to understand the conditions under which silk solutions are spun in nature, so that the microstructures of natural silk can be duplicated or improved upon under artificial conditions. Relevant issues to resolve include: how does the molecular conformation change during spinning, is there a particular concentration at which soluble silk converts to insoluble fiber, and what rates of shear should be used? Because as-secreted silk is initially water-soluble, while silk fiber is not, there is a particular challenge to understand the molecular conformational changes that occur during spinning. The fact that some spiders and insect larvae can produce solid silk structures under water (Levi 1987; Case et al. 1991) very clearly demonstrates that solidification is not a process of "drying out"; rather, it is one in which the pattern of inter- and intramolecular bonding is altered.

A detailed understanding of the silk-spinning process therefore requires interdisciplinary studies that intersect the fields of biochemistry, molecular biology, biotechnology and materials processing. Some recent insights related to materials processing are provided in the present chapter.

2 Background: Liquid Crystals

2.1 Relevance to Silk Spinning

It has recently been established that natural silk secretions pass through a liquid crystalline state of molecular order en route to the formation of solid fibers (Kerkam et al. 1991a; b). This background section describes what a liquid crystal is, what types of molecule form liquid crystalline phases, what conditions stabilize these phases, and how liquid crystalline order is exploited in nature.

2.2 Definition of the Liquid Crystalline State

The term liquid crystal aptly describes a class of materials that can exhibit both liquid- and solid-like properties. For example, they may flow freely like a liquid; yet at a local scale they exhibit anisotropic physical properties such as optical birefringence, and their X-ray diffraction patterns indicate substantial long-range order (de Gennes 1979). Thus, the *molecular order* must be intermediate between that of a liquid and that of a solid. Liquid crystals are formally classified as *mesophasic* materials that exhibit at least long-range *orientational* order (Fig. 2). The underlying consistent molecular characteristic of liquid crystalline materials is that the molecules must have anisotropy of structure and/or bonding ability, to

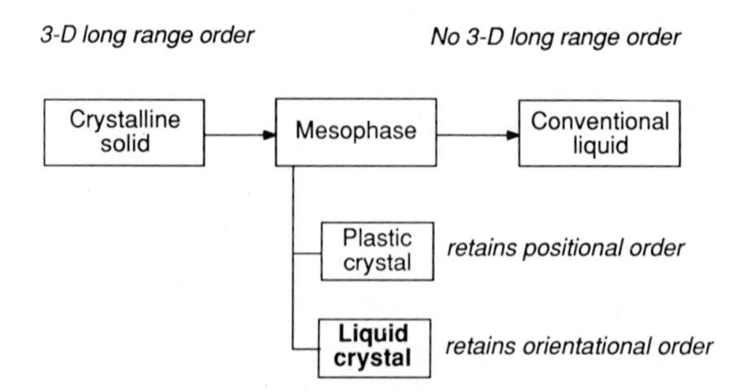

Fig. 2. The relationship of liquid crystals to other states of condensed matter

provide the driving force for long-range alignment. *Structural (geometrical) anisotropy is dominant* in leading to the formation of liquid crystalline phases (Frenkel 1988).

2.3 Molecules that Form Liquid Crystalline Phases

Many low molecular weight liquid crystalline materials consist of molecules that are intrinsically rod-like. Main-chain polymeric liquid crystals can also consist of molecules that are entirely rod-like (e.g., DNA), but they more usually consist of shorter rod-like sequences that are linked by flexible or semi-flexible moieties. Simple homopolypeptides can acquire rod-like characteristics if the molecule forms an α-helix (Fasman 1991). In the case of some proteins, α-helical segments may be separated by relatively flexible segments that have no defined secondary structure, so the molecules then resemble those of the main chain polymeric liquid crystals just described. Silk fibroins in *B. mori* secretions are reported to contain only a small amount of α-helix and no other secondary structure (Magoshi et al. 1985b). However, we cannot at this stage be certain that the α-helical rod-like sequences in silk are responsible for their liquid crystalline behavior. Arguments presented below suggest that the structural anisotropy may alternatively result from higher-level associations between molecules or molecular segments.

The likelihood of rigid, rod-like molecules forming stable liquid crystalline phases can be increased by making it difficult for perfect register to develop between adjacent rods, i.e., by inhibiting the ability of the material to crystallize. This consideration is especially relevant if the potential for extensive inter-rod hydrogen bonding exists. Thermodynamically, this is justified as follows:

The equilibrium temperature for the transition between crystalline solid and liquid crystal is given by:

$$T_{eq} = \frac{\Delta H}{\Delta S},$$

where ΔH and ΔS are, respectively, the enthalpy change and entropy change for the transition. Because the liquid crystalline phase is more ordered than a conventional melt, ΔS is relatively small, tending to raise the transition temperature. However, this effect can be offset by a suitably small value of ΔH, which can be achieved if the material is not highly crystalline. Alternatively, an aggressive, highly polar solvent must be used to break up the intermolecular hydrogen bonds; this is why hot, concentrated sulfuric acid is necessary to produce liquid crystalline solutions of Kevlar. When a protein forms liquid crystalline phases in water, the hydrogen bonding is likely to be *intra*molecular, e.g., stabilizing α-helices. The tendency of molecules in this conformation to crystallize will be further inhibited by the nonperiodic sequence of amino acid residues along the molecules. (In contrast, silk in the β-pleated sheet crystalline form is insoluble in water; the hydrogen bonding is then *inter*molecular.)

2.4 Classification Schemes for Liquid Crystalline Materials

A variety of classifications for liquid crystalline materials exist. The following list is not exhaustive, but contains various descriptions that can usefully be applied to natural silk secretions.

1. Molecules do not have to be rod-like in order to form liquid crystalline phases. Structural anisotropy is also provided by disc-like or lath-like geometries. One distinguishes therefore between *calamitic*, *sanidic*, and *discotic* liquid crystals, respectively, based on the shape of the geometrically anisotropic unit. It is important to realize that these adjectives do not necessarily apply to the molecule as a whole, but only to the anisotropic moieties (separated by flexible spacers) that become ordered in the liquid crystalline state. Indeed, even apparently isometric methanes of the general formula C-[(flexible spacer)-(calamitic mesogen)]$_4$ can form liquid crystalline phases (Eidenschink 1990). If α-helical regions in silk fibroin are indeed responsible for the liquid crystalline properties of silk secretion, silk would be described as calamitic in this scheme. However, we cannot yet rule out the possibility that the anisotropic structures are more complex molecular aggregates with an as yet undetermined shape.

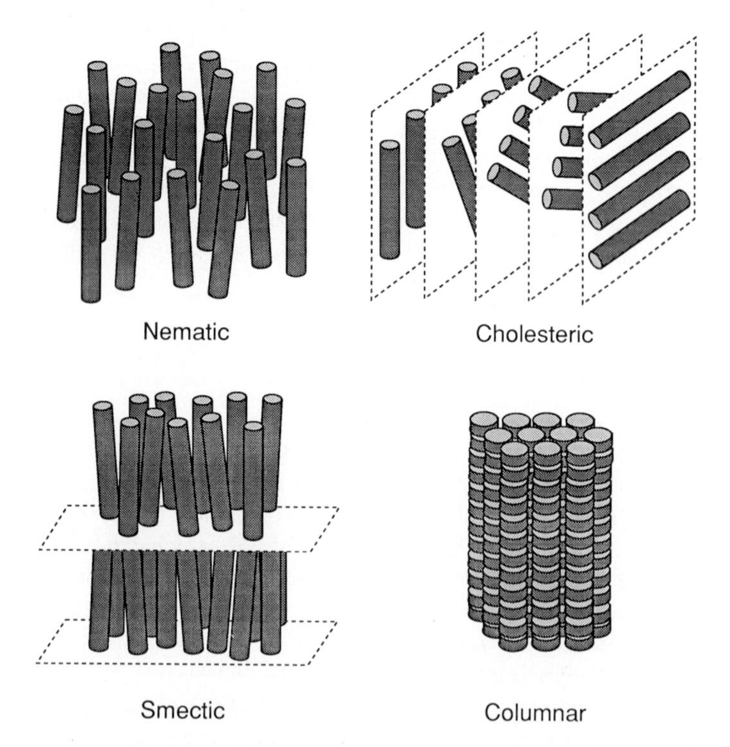

Fig. 3. Some examples of the different types of long range order exhibited by liquid crystalline materials. (Molecules in the smectic liquid crystal shown here are aligned parallel to the layer normal on average, and their positions within the layers are disordered; these specific characteristics distinguish the subclass termed smectic A)

2. Another classification scheme is based on the detailed type of long-range order that is present in the liquid crystalline phase (Fig. 3). In *nematic* phases, such as those formed by natural silk secretions, there is long-range orientational order of rods or laths. (If a lath is capable of spinning freely about an axis at typical molecular rotation frequencies, it effectively behaves like a rod.) A subclass of the nematic state is the *cholesteric* state, in which a twist axis is superimposed on the nematic, normal to the preferred direction of molecular orientation. In *smectic* phases, the molecules are constrained to lie approximately in layers. Several subclasses of smectic are identified on the basis of additional positional order that may develop within the layers, and how the molecules are tilted relative to the layer normal (Gray and Goodby 1984). In *columnar* phases, there is long-range order of discs into columns, that may or may not be arranged approximately on a periodic template. Because properties, including processability, depend on molecular order, this classification allows prediction of material behavior. For example, nematic order is associated with the lowest viscosity in liquid crystalline phases, because there is no positional order.
3. The transition from rigid solid to liquid crystal can sometimes be achieved by the addition of heat alone, in which case the material is said to be *thermotropic*. A liquid crystal is *lyotropic* if the formation of liquid crystalline phases requires the presence of a solvent. Natural silk secretions are therefore lyotropic.
4. When polypeptide molecules form a liquid crystalline phase, the structurally anisotropic unit is larger than an individual amino acid monomer. The basic mesogenic unit is therefore not an individual amino acid. For this reason, it is appropriate to speak of a *macromolecular* liquid crystal, but not a *polymeric* liquid crystal, since the latter term is only strictly correct when individual monomers are mesogenic. In cases where the molecule contains only one mesogenic unit and the molecular weight is low ($\leqslant$ 1 kDa), one refers to a *small molecule* liquid crystal.

2.5 Origin of Nematic Liquid Crystalline Behavior

A thermodynamic basis for predicting the existence of nematic liquid crystalline phases can be developed by considering a simple system initially containing solvent, into which we introduce monodisperse rod-like molecules. To simplify the model, we assume athermal conditions, where forces between molecular species can be ignored. In other words, there are no enthalpic contributions to the free energy of mixing when rods are added; the only contributions are entropic. If a few rods are now added to the system, replacing solvent, they will not impinge on one another significantly, because their chance encounters will be infrequent. Their orientations and positions will therefore be random. More and more rods are added, until it becomes impossible to add another one without causing some degree of continual net alignment of the rods. While alignment reduces the orientational entropy of the system, any one rod acquires a greater freedom for translations without running into another rod. Thus, while orientational entropy is

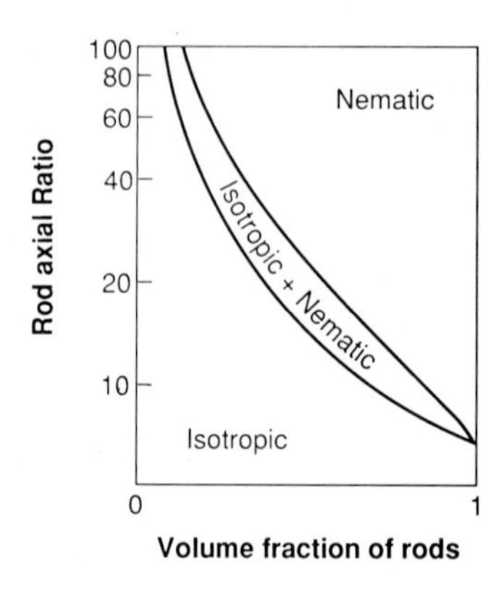

Fig. 4. The existence of a single nematic phase, a single isotropic phase, or a two-phase mixture of isotropic and nematic material, predicted as a function of rod axial ratio and concentration. [Behavior of an athermal system; prediction used a simple lattice model (Flory and Ronca 1979a)]

decreased, translational entropy is gained; on balance, the onset of long-range molecular alignment above a critical rod concentration can lead to an increase in the configurational entropy of the system.

Aside from the stated simplifications, the foregoing argument omits one important consideration. According to Gibbs' Phase Rule (Atkins 1978), a two-component system cannot be made to undergo an abrupt transition between single phases simply by changing the composition at constant temperature, at least not under equilibrium conditions. Instead, the concentration ranges over which the random (isotropic) and aligned (nematic) phases are stable must be separated by a concentration range over which the system is biphasic, where it consists of a *mixture* of isotropic and nematic regions. At the critical concentration where nematic order first appears, only some of the molecules in the sample develop a net alignment; the number of molecules with net alignment increases with further increases in concentration, until eventually a second critical concentration is reached where the entire sample is nematic (Flory and Ronca 1979a). There are various theoretical models that quantify entropy and enable the lower and upper concentration limits of biphasic behavior to be calculated as a function of rod axial ratio (Chick and Viney 1991). In this way, it is possible to construct a "phase diagram" of the type shown in Fig. 4.

2.6 Accounting for the Effect of Intermolecular Forces

Theoretical models describing phase separation in two-component rigid-rod/solvent systems can be modified to take account of temperature-dependent intermolecular forces (Flory 1956; Flory and Ronca 1979b). The forces are characterized by a dimensionless *interaction parameter*, which represents

$$\frac{1}{kT}[(\text{energy of solvent molecule immersed in an otherwise pure system of rods}) - (\text{energy of a solvent molecule in pure solvent})].$$

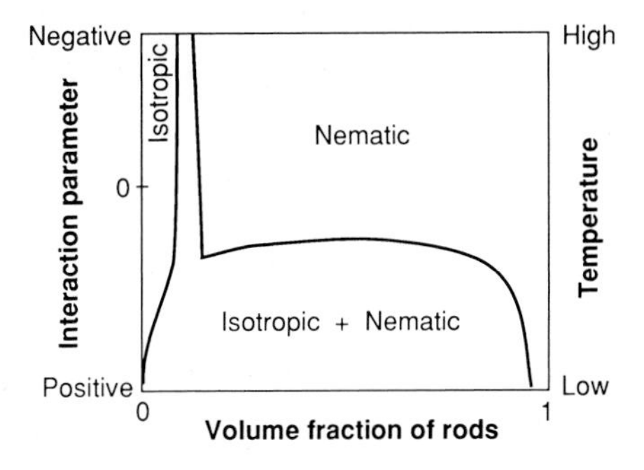

Fig. 5. Phase diagram showing the existence of a single nematic phase, a single isotropic phase, or a two-phase mixture of isotropic and nematic material, predicted as a function of temperature (or interaction parameter) and concentration for rods with an axial ratio of 100. [Behavior of a non-athermal system; prediction used a simple lattice model (Flory 1956)]

Under normal circumstances, the solubility of rods in the solvent will increase with temperature; i.e., a solvent molecule will then have lower energy when surrounded by rods, compared to its energy in bulk solvent. The interaction parameter therefore becomes large and negative at high temperatures. Similarly, the interaction parameter becomes large and positive at low temperatures. Interaction parameter and temperature therefore plot in opposite directions on the vertical axis of phase diagrams (Fig. 5). We note that at high temperatures, where intermolecular attractions become insignificant relative to the energy of thermal motion, the biphasic regime becomes narrow as in Fig. 4, i.e., the system acquires athermal behavior. The onset of the biphasic regime is displaced towards higher concentrations as the axial ratio decreases, which again is consistent with Fig. 4.

Attempts to relate the behavior of real proteins to the predictions shown in Figs. 4 and 5 run into two difficulties:

1. The phase diagram in Fig. 5 predicts that the pure polymer can form a liquid crystalline phase above a certain temperature. Real proteins will denature at temperatures below that required to form a thermotropic nematic phase. However, since we are at present interested in the nematic behavior of silk secretions containing significant amounts of solvent, the far right-hand side of the phase diagram is irrelevant.
2. The models assume that the rod axial ratio does not depend on either the system concentration or the temperature. Protein conformations may, in fact, be sensitive to both.

In spite of these reservations, two reliable predictions for the behavior of natural silk do emerge from Figs. 4 and 5:

1. The axial ratio of the rod-like segments have to exceed some minimum value if liquid crystalline phases are to form at all. While there is some disagreement between the predictions of different theories as to what this limiting ratio is, experimental studies on model compounds (Viney et al. 1990) suggest that the value is close to 4.3.

2. If biphasic behavior is found to be limited to a narrow concentration range, the system is exhibiting athermal behavior, and one can use Fig. 4 to estimate the *axial ratio* of the structurally anisotropic units at the biphasic concentrations.

2.7 Microstructures

While there is local alignment in a nematic sample (typically over distances of a few microns), there is no spontaneous global alignment on the scale of a bulk sample. Topologically, this requires the existence of singularities in the pattern of molecular orientation. By way of analogy, one can note the impossibility of combing one's hair without introducing a parting or a whorl, or that it is necessary for a cyclone or anticyclone to always exist somewhere in the pattern of air flow across the earth's surface. Singularities in the pattern of molecular orientations in a liquid crystal are called *disclinations*, and are illustrated in Fig. 6. The number of such defects in a liquid crystalline microstructure is important, because it affects the rheology of the fluid material and also the final properties after it has solidified.

Phase diagrams tell us whether a particular phase will form, and can also be interpreted to tell us how much of that phase will appear in a microstructure. They do not provide information about the distances over which order is preserved, i.e., the number density of defects in a nematic phase.

Studies using model compounds (Dannels et al. 1991) have shown that a number of factors control the scale of nematic microstructures:

- the number of chain ends,

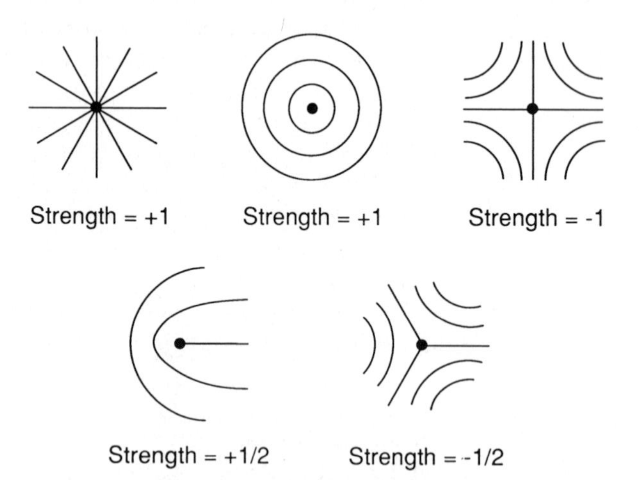

Fig. 6. Schematic topology of molecular orientation around disclinations in a nematic liquid crystal. Molecular orientations are everywhere tangential to the lines shown; positions of disclinations are marked with a *dot*. In three dimensions, each disclination would be a line singularity that is normal to the plane of the diagram at these marked positions. The significance of the strength of a disclination is explained in Section 4.5.2

- whether or not the molecules are longer than the constituent structural elements responsible for their nematic behavior, and
- the polydispersity of the system.

All these considerations affect the number of molecular-scale building blocks available for constructing micron-scale disclinations. Fine microstructures are promoted by shorter, more polydisperse molecules with a contour length greater than the persistence length. While there is no absolute quantitative link between microstructural scale and any one of these molecular parameters/characteristics, differences in scale can at least be ascribed qualitatively to differences in one or more of these attributes. We will return to this point when comparing the microstructural scale of different types of silk.

3 How Liquid Crystalline Order Is Exploited in Nature

Liquid crystalline phases are formed by all four classes of biological macromolecules — proteins, poly(nucleic acids), polysaccharides, and lipids.

3.1 Proteins

As this chapter demonstrates, the silk secretions of at least some organisms form liquid crystalline phases as they are processed to solid fiber. The organisms derive at least two benefits from this. The viscosity of a liquid crystalline polymer solution can be several orders of magnitude below that of a conventional polymer solution at the same concentration (Dobb and McIntyre 1984). In this way, the energy required to expel solution through the duct and spinneret is minimized. [Exactly *where* solid spider dragline first begins to form depends on the extent of opening of a valve situated in the duct that leads to the spigot (Wilson 1962)]. Also, the organism is able to achieve a high degree of molecular extension and alignment in the fibrous product, despite only a limited ability for post spinning draw (head movement in silkworms, combing silk with legs by spiders). Alignment of molecules in a particular direction results in high strength and stiffness when loads are applied parallel to that direction. Because liquid crystalline phases are observed at intermediate concentrations in natural secretions of silks other than those in which the highest strength and stiffness are required, we can argue that *ease of processing is the principal benefit* of the phase being present.

Collagen exhibits liquid crystalline order in compact bone, and forms cholesteric gels in vitro (Giraud-Guille 1987). Whether or not liquid crystalline ordering plays a role in the hierarchical structuring of collagen to form tissues such as bone is not known, but it is likely that a nematic phase would be facilitate the spinning of collagen into "silk" fibers by organisms like the gooseberry sawfly larva (*Nematus ribesii*) (Rudall 1962).

Some natural materials based on cuticulin (a lipoprotein), for example the cuticle of scarab beetles (Caveney 1971) and the scales of many types of butterfly (Ghiradella 1984), exhibit molecular order that is isomorphous to the order in nematic liquid crystalline materials. Formation of a nematic phase may therefore play a role in the synthesis of these materials.

3.2 Poly(Nucleic Acids)

A variety of liquid crystalline phases have been observed in solutions of DNA in vitro (Strzelecka et al. 1988; Livolant et al. 1989) and in vivo (Rill et al. 1989). Here, too, there appear to be two direct benefits of long-range molecular orientation. Firstly, the genetic material can be packed at high density, yet retain the fluidity necessary to enable replication and transcription. If DNA were stored by cells in a crystalline form, additional energy would be needed to give the molecules the freedom of mobility needed for these processes. Secondly, liquid crystalline order in vivo would help the separated strands of replicating DNA from becoming too dissociated in space, thus enabling easy reversion to the double helix conformation when replication is completed.

3.3 Polysaccharides

A very large number of natural and derivatized polysaccharides form nematic and cholesteric phases in vitro (Gray 1983; Livolant 1986), but there is no hard evidence as to whether they do so in vivo. Also, in the case of those "silks" that contain chitin (Rudall 1962), spinning would be facilitated by the formation of nematic phases.

3.4 Lipids

The local order of lipids in membranes and vesicles is similar to that in smectic liquid crystals. Theories that model the forces acting on and within smectic layers have been used successfully to predict a variety of environmentally-induced changes in membrane structure. In this way, it is possible to model cellular processes such as exocytosis, endocytosis, budding, cell division, and fusion (Chapman 1991; Lipowsky 1991; Verhas 1988).

4 Experimental Identification and Characterization of Liquid Crystalline Phases

Several techniques are available for recognizing the existence of liquid crystalline phases, characterizing their molecular order, and determining the underlying molecular conformations that cause liquid crystalline order to develop.

4.1 Spectroscopy

Typically, spectroscopic studies would be used to probe the conformation of individual molecules. Nuclear magnetic resonance (NMR) and Raman spectroscopy have shown that *B. mori* and *Philosamia cynthia ricini* (wild silkworm) silk is principally random coil in vitro, with only a small (unquantified) fraction of material adopting an α-helical conformation (Asakura 1986; Zheng et al. 1989). In the solid fiber, the conformation is predominantly β-sheet. Most significantly, the conformation in regenerated aqueous solutions of *B. mori* fibroin has been found to remain unchanged over a wide range of concentrations, provided that the solutions are not subjected to shear.

4.2 Thermal Analysis

Thermal analysis (differential thermal analysis, differential scanning calorimetry) is commonly used in the characterization of new potentially thermotropic liquid crystalline compounds, to confirm the existence of intermediate phases between the solid and isotropic liquid states. A liquid crystalline compound exhibits multiple melting endotherms on heating, representing the enthalpy needed to achieve the transition to each successively more disordered phase. Thermal analysis therefore also indicates the temperature ranges over which each phase is stable, though it does not enable structural identification of the phase. The technique has only limited use in the case of lyotropic proteins, because heating denatures the molecular conformation and affects the polymer-solvent interaction parameter. Also, it is not possible to change the sample concentration in a controlled manner.

4.3 X-Ray Diffraction

X-ray diffraction is another otherwise commonly used characterization technique that is not easily adapted to the study of liquid silk secretions. To maintain their concentration during the several hours typically required for diffraction data collection, fluid samples can be confined to thin-walled, weakly scattering Lindemann capillaries made from glass with a high Li_2O content. Unfortunately, silk secretions are too viscous to enable loading without centrifugation, which in turn generates shear stresses that cause the material to solidify. X-ray studies of the molecular order in silk are therefore confined to the solid state (Asakura et al. 1985).

4.4 Turbidity

One of the simplest indicators that a sample is liquid crystalline, is if a pure melt or solution is turbid. The turbidity arises because light is scattered at disclinations, i.e., because the molecular orientation (and hence the mean refractive index) near the core of the defects changes on a scale that is comparable to the wavelength of light. However, this simple test is not reliable in the case of more viscous, gel-like systems, where a variety of sub-microscopic density variations might lead to scattering. When silk secretions are allowed to leak from a sectioned gland onto a glass slide, we do observe turbidity, which at least merits further investigation.

4.5 Transmitted Polarized Light Microscopy

The most useful technique for easily identifying liquid crystalline phases in natural silk, is transmitted polarized light microscopy. The technique relies on the optical anisotropy of liquid crystalline material to generate contrast. The resultant patterns observed in the microscope are characteristic of the allowed topologies of molecular order, and thus of the type of liquid crystalline phase.

4.5.1 Origin of Contrast Between Crossed Polars

The distinguishing feature of a transmitted polarized light microscope is the presence of plane-polarizing elements below and above the sample; these are the *polarizer* and the *analyzer*, respectively. Light enters the specimen from below. The polars are said to be crossed when their vibration directions are orthogonal.

When light is propagated through optically anisotropic material, it travels in the form of two distinct rays, that have orthogonal vibration directions. This can be demonstrated experimentally with a single-crystal calcite ($CaCO_3$) prism or rhomb, which, if placed over a pinhole in a sheet of paper on a light-box, will cause two images of the pinhole to be observed. A piece of Polaroid sheet from a pair of sunglasses can be used to demonstrate that the light constituting one image is orthogonally polarized to that constituting the other. One also notices that the two images appear to be at different depths in the material, which means that they must be associated with different refractive indices. In the case of thin samples ($\leqslant 10\,\mu m$ thick, i.e., 1000 times thinner than the calcite rhomb typically used in the above demonstration) between glass microscope slides, the spatial separation of the two images is too small to be observed.

In thin samples of most nematic materials, the tendency is for molecules to become oriented within the plane of the glass substrates, unless special surface treatments are carried out on the glass. The material is optically anisotropic, and, in the great majority of cases, the two vibration directions for light lie approximately parallel and perpendicular to the local direction of molecular orientation.

We now consider what happens when this liquid crystalline material is viewed in white light between crossed polars (Viney 1990). If the initially incident light is polarized (vibrates) in some arbitrary direction, it will be resolved into two components along the local vibration directions of the material. Because light in the two ensuing rays experiences different refractive indices, light constituting one ray lags behind that constituting the other, i.e., a phase difference develops between the two rays. Dependent on the magnitude of the refractive index difference, and on the thickness of the sample, light of certain frequencies may emerge from the top of the sample with a whole wavelength of path difference between the two rays. For these frequencies, interference between light in the two rays then generates a resultant that is indistinguishable in its polarization state from light originally incident on the lower surface of the specimen. These frequencies are therefore not transmitted by the crossed analyzer, so the complementary color is seen. The specimen therefore appears locally bright, and it may be tinged with color.

This result of observing the specimen between crossed polars is quite different if the initially incident light is specifically polarized (vibrates) parallel to one of the vibration directions in the specimen. Then all the incident light is resolved into that one vibration direction (it can have no component at 90° to itself), and it therefore has no component to interfere with on exiting the top specimen surface. Thus all the light (not just certain spectral components) is extinguished at the analyzer, i.e., the particular region of specimen appears dark.

4.5.2 Liquid Crystalline Textures

In a sample without global molecular alignment, the in-plane orientation of the molecules will vary from point to point. Thus, the positions of extinction observed between crossed polars will change as the crossed polars are synchronously rotated. The pattern of contrast obtained from a liquid crystalline specimen between crossed polars is known as texture. By observing how the positions of extinction move when the crossed polars are rotated, one can determine how the preferred direction of molecular orientation varies across the sample (Demus and Richter 1978). Some points in the microstructure may be defined by the confluence of two or four extinction bands for all orientation of the crossed polars. They indicate the positions of disclinations. If a disclination is associated with four extinction bands, it is assigned a strength of ± 1, while a disclination with only two bands is designated as having strength $\pm 1/2$. The sign denotes whether rotation of the crossed polars causes the extinction bands to rotate around the disclination in the same (+) or opposite (–) sense as the polars. The topology of molecular orientation in the vicinity of disclinations is shown schematically in Fig. 6. Because disclinations of strength $\pm 1/2$ are topologically compatible with nematic order only (Gray and Goodby 1984), their presence in a microstructure is definitive proof that a liquid crystalline phase is nematic.

5 Evidence for Liquid Crystalline Phases in Natural Silk

5.1 Initial Clues

The stiffness (up to 30 GN.m^{-2}) and breaking strength (up to 2 GN.m^{-2}) of spider dragline silk (Gosline et al. 1986; Zemlin 1968) are superior to the corresponding properties of highly drawn nylon (2.4 GN.m^{-2} and 0.7 GN.m^{-2}, respectively (Billmeyer 1984)); see also Sect. 1.3 above. For polymer fibers, high strength and stiffness indicate a microstructure in which there is significant molecular extension and alignment. In conventional polymer fiber processing, extension and alignment introduced at the spinnerets are lost during the time taken for the fiber to solidify, so a subsequent post-spinning draw is required to recover these properties. For this reason, liquid crystalline polymers have become increasingly attractive to industry (Weiss and Ober 1990): molecules align spontaneously into domains that are several microns across; then it is only necessary to align whole domains, and not individual molecules, to generate global alignment in a fiber.

It has been reported previously that the fluid contents of silkworm glands (Magoshi et al. 1985a, c; Li and Yu 1989) and honey bee larva silk glands (Lucas and Rudall 1968) are optically birefringent, evidenced by the fact that these materials transmit light when viewed between crossed polars. Section 5 above described that birefringence (optical anisotropy) is a characteristic of liquid crystalline fluids. However, it is not *uniquely* a characteristic of liquid crystals, and can be exhibited by conventional polymer solutions or melts that are made to flow, or that are subjected to meniscus or other external forces. The formation of specific *textures* between crossed polars, rather than the observation of birefringence, is a definitive test for liquid crystallinity.

Natural silk secretions are fluid and easily spun despite their high concentration [approx. 30% by weight for silkworm secretions (Iizuka 1985)]. Relatively low viscosity at high concentrations is also a distinguishing characteristic of liquid crystalline fluids (Dobb and McIntyre 1984).

A number of synthetic polypeptides, i.e., materials that are chemically related to silk proteins, are known to form liquid crystalline phases in a variety of solvents (Elliott and Ambrose 1950; Fasman 1991).

5.2 Sample Preparation

Our studies were performed on silk secretions of *B. mori* silkworms and *N. clavipes* orb-weaving spiders (Fig. 7).

B. mori (various strains) were kindly provided by Marian Goldsmith, University of Rhode Island (Kingston, RI), or purchased from Carolina Biological Supply Company. They were raised on mulberry leaves or on artificial diet (Carolina). Silk glands were removed from larvae in the fifth instar, i.e., during the approximately 8-day period that occurs between the final molt and the onset of cocoon spinning.

N. clavipes were obtained from Angela Choate, University of Florida (Gainesville, FL). They were raised on a diet of crickets (Fluker's Cricket Farm, Baton Rouge, LA) and *Drosophila* fruit flies. Studies were confined to the contents of ampullate (dragline), flagelliform (capture thread), and tubuliform (cocoon silk) glands.

The silk glands were dissected into Ringer's solution [14.0 g NaCl, 0.2 g KCl, 0.2 g $NaHCO_3$, 0.4 g $CaCl_2$, 1000 ml distilled water (Cokendolpher and Brown 1985)]. Glands were transferred to glass microscope slides and rinsed with distilled

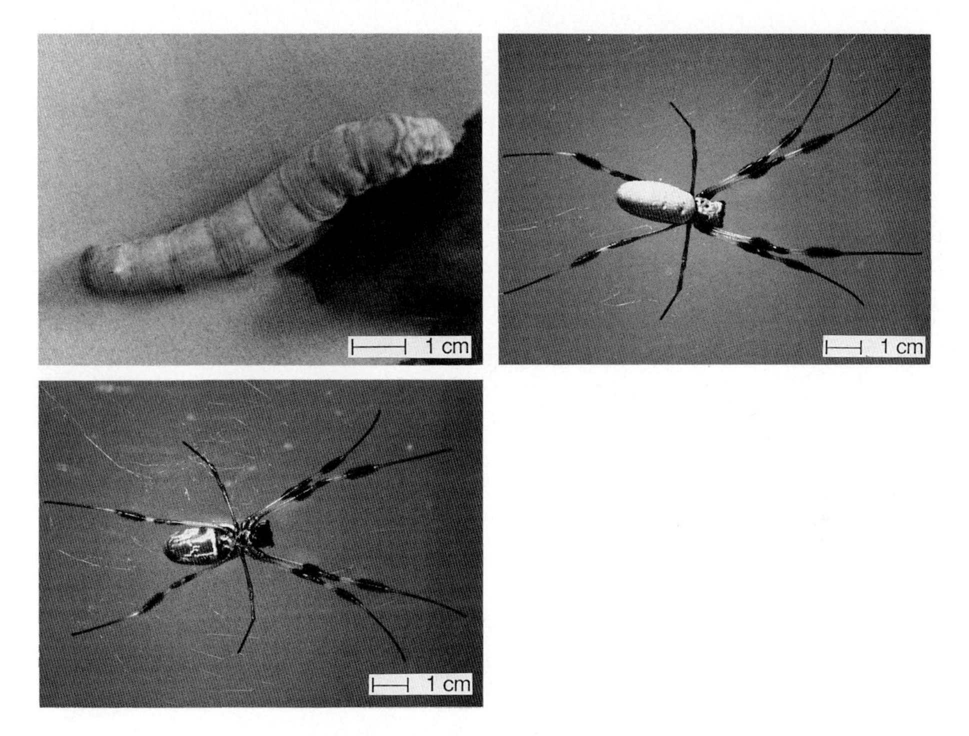

Fig. 7. Silk-producers from which glands were removed for microstructural study. *Top* Fifth-instar larva of *B. mori*, feeding on mulberry leaf; *Middle* and *bottom* mature female *N. clavipes* orb-weaving spider, in the hub of her web (dorsal and ventral views, respectively)

water. Excess surface water was then allowed to drain away. The glands were transferred to dry microscope slides, and were sectioned with a scalpel. Droplets of silk secretion from the core of middle division of *B. mori* glands, and from the middle region of *N. clavipes* glands, were allowed to leak onto the slides. (By using material from the *core* of the silkworm gland, we hope to obtain a sample that contains little or no sericin (Magoshi et al. 1985a).) The glands were then discarded, and glass cover slips were placed gently over the droplets. These specimens were observed and photographed between crossed polars in a transmitted light microscope.

5.3 Nematic Microstructures

The undisturbed silk secretions, held between a glass microscope slide and cover slip and viewed in transmission between crossed polars, are optically isotropic. After some increase in their concentration, due to evaporation, they exhibit microstructures (Fig. 8) that are uniquely characteristic of the nematic liquid crystalline state. Specifically, disclinations of strength $\pm 1/2$ are found in all

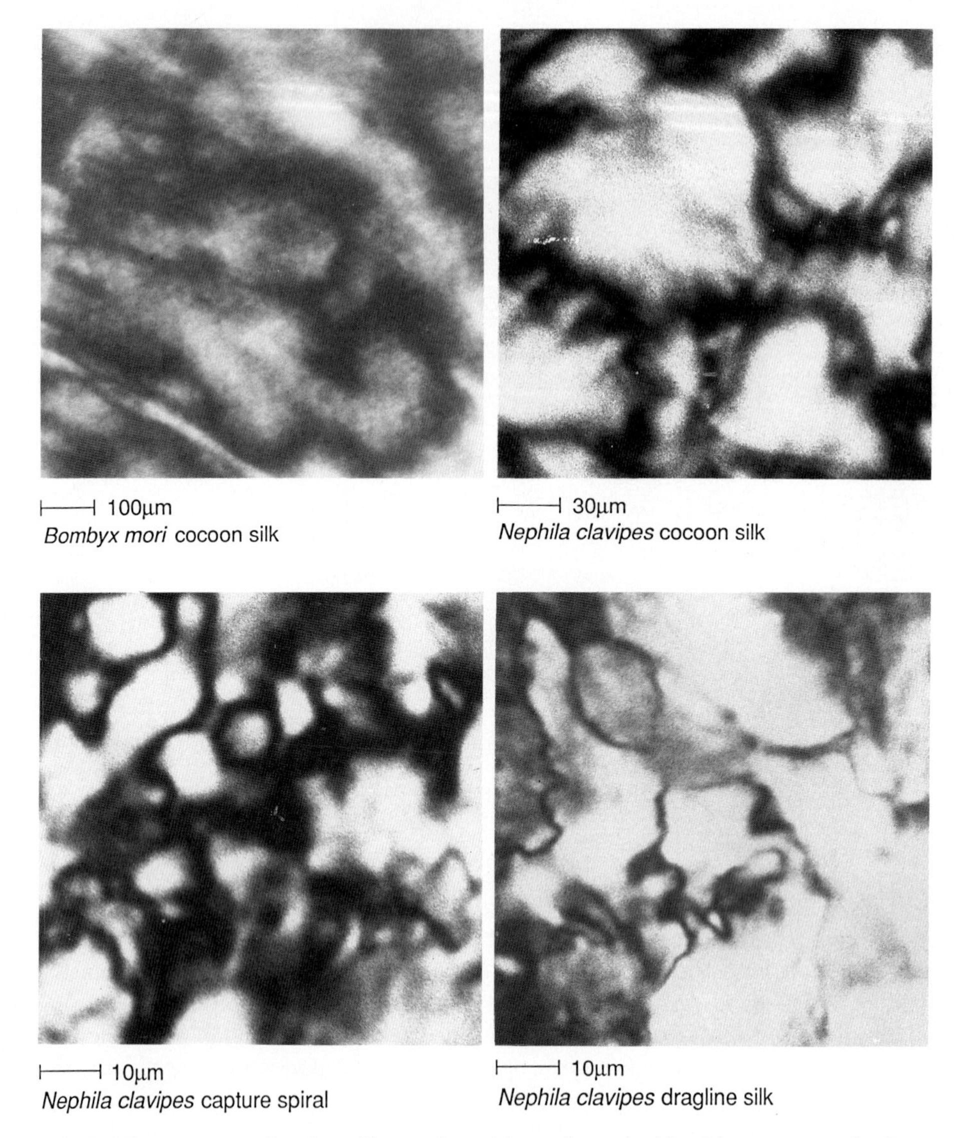

Fig. 8. Microstructures of various silk secretions; thin specimens ($\leqslant 10\,\mu m$) between crossed polars

the microstructures, which, as described in Section 4.5.2, is only consistent with nematic topologies.

In all cases, liquid crystalline microstructures begin to develop within 10 min of placing a thin specimen on the microscope stage. The fact that some time has to elapse before a nematic texture is seen could suggest that the as-secreted material is already liquid crystalline, but needs to coarsen before the microstructure is resolvable. However, the nematic microstructures first appear at the *edges* of

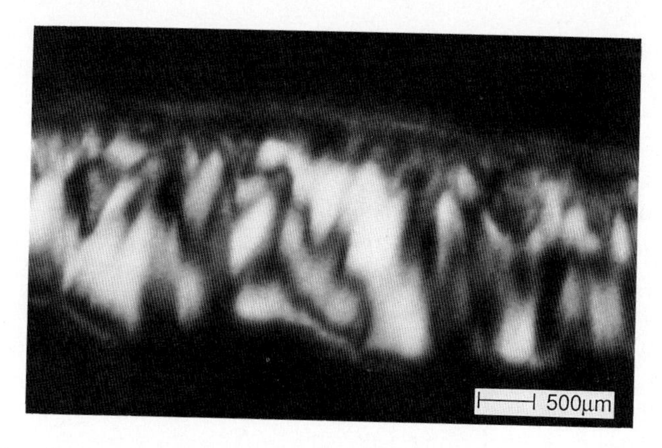

Fig. 9. Formation of nematic liquid crystalline phase at the edge of a thin specimen of *Bombyx mori* silk secretion; material confined between glass microscope slide and cover slip. The still isotropic interior of the droplet is at the *bottom of the picture*

specimens (Fig. 9), i.e., where evaporation is greatest, and they do not change in scale after they first appear.

No biphasic (mixture of isotropic and nematic) material is resolved between the isotropic interior and liquid crystalline perimeter of the sample in Fig. 9. From this, we deduce that biphasic behavior is confined to a very narrow concentration range, and therefore that the material has the thermodynamic characteristics of an athermal system (Sect. 2.6).

5.4 Microstructural Scale

The most striking difference between the microstructures of the various silk secretions concerns their scale. Scale decreases in the order

$$\textit{B. mori}\ \text{silk} >>> \textit{N. clavipes}\ \text{cocoon silk} >> \textit{N. clavipes}\ \text{capture thread} > \textit{N. clavipes}\ \text{dragline} \quad (1)$$

According to studies of model nematic compounds (Sect. 2.7), finer microstructures are promoted by increased polydispersity of molecular weight, by a molecular contour length that is greater than the persistence length, and by shorter rod-like molecules. Polydispersity is unlikely to be relevant to the microstructures of silk secretions, since the mechanism of protein synthesis ensures a monodisperse product. All the silk proteins will contain flexible sequences, and thus have a contour length that is significantly greater than their persistence length, so the silk secretions are comparable in this respect. One must therefore contemplate differences in the lengths of rod-like sequences as the cause for the observed wide variation in microstructural scale.

On this basis, the length of the rod-like sequences should decrease in the order that the microstructural scale decreases, i.e., in the order indicated by the inequality

(1) above. However, there are two self-consistent arguments that the rod-like sequence lengths *do not* decrease in this order. These arguments are now presented.

5.4.1 Computer Predictions of Secondary Structure

The literature contains only partial information on nucleotide and deduced amino acid sequences in *B. mori* and *N. clavipes* dragline, and none on the *N. clavipes* capture thread or cocoon silk. One must acknowledge the possibility that the anisotropic structural components responsible for liquid crystalline behavior do not occur in these published fragments, but it is likely that they are at least representative of the respective silk structures.

Structural analysis of published sequences was performed with the program GENEPRO (Riverside Scientific Enterprises, Seattle, WA; version 4.20). This program uses the method of Chou and Fasman (1978) to compute secondary structure, and the algorithm of Kyte and Doolittle (1982) to calculate hydropathy. A window six amino acids long was used in searching polypeptide sequences for α-helices, and a seven-unit window was used in scanning their hydropathic character. The following polypeptide sequences were studied:

1. The sequence coded by the *B. mori* fibroin gene, consisting of 176 amino acids from the presumed initiation sequence to the beginning of the repetitious Gly-Ala peptide region, omitting an intervening sequence of 970 base pairs that contains multiple stop codons in all three reading frames and therefore is unlikely to code for protein (Tsujimoto and Suzuki 1979)
(sequence A)

2. The *B. mori* 59-mer repeat (Lucas and Rudall 1968)
Gly-Ala-Gly-Ala-Gly-[Ser-Gly-$(\text{Ala-Gly})_2]_8$-Ser-Gly-Ala-Ala-Gly-Tyr
(sequence B)

3. The *B. mori* 59-mer repeat (Strydom et al. 1977)
Gly-Ala-Gly-Ala-Gly-Ser-Gly-Ala-Ala-Gly-[Ser-Gly-$(\text{Ala-Gly})_2]_8$-Tyr
(sequence C)

4. A simple model polypeptide for *B. mori* fibroin (Rawn 1989)
$(\text{Ser-Gly-Ala-Gly})_n$
(sequence D)

5. The published partial sequence for *N. clavipes* dragline silk (Xu and Lewis 1990), consisting of 719 amino acids; this sequence lacks an initiation (Met) codon, so its relationship to the protein coding section of the gene is not clear
(sequence E)

Confidence in the structural predictions of Chou–Fasman is subject to a number of caveats. Firstly, the algorithm is entirely empirical, calculating protein sequence/structure relationships on a probabilistic basis from a library of known structures. The library contains proteins with a variety of tertiary structures, which may compromise its statistical reliability for specifically fibrous proteins.

Secondly, one may expect that the conformation of the polymer will depend on its environment — for example, its concentration in water, which changes during the spinning process. Hence, the computer predictions may not be valid for the polymer structure throughout all stages of the spinning process. However, the hydropathy search indicated no hydrophilic or hydrophobic regions in sequences B–D, only six (contiguous) hydrophilic residues and seven (contiguous) hydrophobic residues out of the 176 residues in sequence A, and only six (contiguous) hydrophobic residues out of the 719 residues in sequence E. Thus the conformations may, in fact, not be especially sensitive to the polymer concentration in water; this is supported experimentally in the case of regenerated *B. mori* silk (Asakura 1986).

Thirdly, there is the more serious concern that the shear flow associated with spinning, coupled with rapid solidification times, probably prevents the polymer from ever reaching its equilibrium conformation. This concern is supported by our observations that the optical birefringence of *N. clavipes* dragline changes (relaxes) over a period of days.

While we cannot obtain an accurate conformational analysis from this type of approach, we do at least expect that the order-of-magnitude difference in microstructural scale of *N. clavipes* dragline and *B. mori* silk should result from significantly different α-helix lengths in the two types of silk, if α-helices are indeed the rod-like structural unit responsible for liquid crystalline behavior in these materials.

The Chou-Fasman algorithm predicted no α-helix-forming potential for the *B. mori* sequences B–D. The sequence A is predicted to contain one amino acid hexamer, one tetramer, and three trimers with significant likelihood of forming an α-helix; in total, 21 residues out of 176 (12%) in sequence A are predicted to have α-helix-forming potential. The *N. clavipes* dragline sequence E is predicted to contain one hexamer, five pentamers, 12 tetramers and one trimer with α-helical conformations; a total of 88 residues out of 719 (also 12%) are attributed α-helix-forming ability. Thus the *N. clavipes* dragline sequence is predicted to contain a greater proportion of longer α-helical regions, in which case the microstructure of this silk should be slightly coarser (not the observed order of magnitude finer) than *B. mori* silk. One must therefore seriously question whether α-helical regions are indeed responsible for the development of liquid crystalline order in silk.

5.4.2 Comparison with Phase Diagrams for Athermal Behavior

It was observed (Sect. 5.3) that, for all the silks studied, liquid crystalline microstructures begin to develop within 10 min of placing a thin specimen on the microscope stage. With the assumption that the initial concentrations are comparable, the critical concentrations for mesophase formation must then be comparable. Since the materials also exhibit athermal behavior (the concentration range for coexistence of the isotropic and liquid crystalline phases is very narrow), Fig. 4 suggests that the structural units responsible for liquid crystallinity have a similar axial ratio in the different silks. Also, because liquid crystalline silk secretions have a concentration in excess of 26% by volume, Fig. 4 suggests that the numerical value of this axial ratio should be significantly greater than that of any rods implied by our analysis of molecular secondary structure (Viney et al. 1992).

We have already noted that the α-helices predicted for *N. clavipes* dragline silk on are average longer than those predicted for *B. mori* silk. The predicted *N. clavipes* α-helices consist of largely alanine residues, with some serine and valine. The α-helices in *B. mori* silk all contain glutamic acid, aspartic acid, lysine, isoleucine, or methionine; these amino acids have bulky sidegroups that will significantly raise the helix diameter (by a factor of at least three, based on a simple estimate of excluded volume). On this basis a large difference in α-helix axial ratio is predicted for the two types of silk, which contradicts the experimental indications that the axial ratios should be approximately similar.

6 Summary and Challenges

Natural silk secretions form liquid crystalline phases at concentrations higher than those prevailing in the silk glands. The principal benefit to the spider or silkworm appears to be ease of processing. A simple model of the liquid crystalline phase, where α-helical regions of the silk proteins provide the structurally anisotropic units needed for nematic phase formation, is incompatible with observed differences in microstructural scale, and with indications that the anisotropic units have similar axial ratios. Thus the nature of the nematogenic units remains elusive.

Individual protein molecules do not contain any other types of sequence that are recognizably nematogenic, and the protein molecular conformation apparently does not depend on the concentration of protein in water. It follows that the nematic phase must be based on structures that arise from the self-assembly of individual molecules into more complex aggregates as their concentration increases. The complexity of the process is exacerbated by shear, which can alter the molecular conformation and does contribute to the precipitation of insoluble fiber; however, shear is not necessary for the formation of the nematic phase.

At present, progress towards understanding the production of silk fiber in vivo is most severely limited by the paucity of protein sequence data, and by the limitations of computer modelling of protein secondary structure.

Acknowledgments. Useful insights were gained through interactions with Keven Kerkam, Lisa Gilliland, Terry Lybrand, David Kaplan, Stephen Lombardi, and Wayne Muller. The author's research has been supported by the US Army Research, Development, and Engineering Center (Natick, MA), by the American Chemical Society (Petroleum Research Fund), and by the Washington Technology Centers (Seattle, WA).

References

Asakura T (1986) Structure of *Bombyx mori* silk fibroin in aqueous solution. Makromol Chem Rapid Comm 7:755–759

Asakura T, Kuzuhara A, Tabeta R, Saito H (1985) Conformational characterization of *Bombyx mori* silk fibroin in the solid state by high-frequency ^{13}C cross polarization-magic angle spinning NMR, X-ray diffraction, and infrared spectroscopy. Macromol 18:1841–1845

Atkins PW (1978) Physical chemistry, 1st edn. Oxford University Press, Oxford
Bell AL, Peakall DB (1969) Changes in the fine structure during silk protein production in the ampullate gland of the spider *Araneus sericatus*. J Cell Biol 42:284–295
Billmeyer FW (1984) Textbook of polymer science, 3rd edn. Wiley, New York
Case ST, Wellman SE, Hamodrakas S (1991) Assembly characteristics and structural motifs in an aquatic insect's biopolymer. In: Alper M, Calvert P, Frankel R, Rieke P, Tirrell D (eds) Materials synthesis based on biological processes. Materials Research Society, Pittsburgh, PA, pp 233–237
Caveney S (1971) Cuticle reflectivity and optical activity in scarab beetles: the role of uric acid. Proc R Soc Lond B178:205–225
Chapman D (1991) Lyotropic mesophases in biological systems. In: Bahadur B (ed) Liquid crystals: applications and uses. World Scientific, Singapore, pp 185–223
Chick LA, Viney C (1991) Entropy of rodlike particles: continuum versus lattice representations. Mol Cryst Liq Cryst 204:123–132
Chou PY, Fasman GD (1978) Empirical predictions of protein conformation. Annu Rev Biochem 47:251–276
Cokendolpher JC, Brown JD (1985) Air-dry method for studying chromosomes of insects and arachnids. Entomol News 96:114–118
Dannels CM, Viney C, Twieg RJ, Chang MY (1991) Factors affecting microstructural scale in liquid crystalline polymers. Mol Cryst Liq Cryst 198:341–350
de Gennes PG (1979) The physics of liquid crystals, 1st edn. Oxford University Press, Oxford
Demus D, Richter L (1978) Textures of liquid crystals, 1st edn. Verlag Chemie, Weinheim
Dobb MG, McIntyre JE (1984) Properties and applications of liquid-crystalline main-chain polymers. Adv Polym Sci 60/61:61–98
Eidenschink R (1990) Thermotropic liquid crystals formed by isometric molecules. The 13th Int Liquid Crystal Conf, 22–27 July 1990, Vancouver, BC, vol 2, p 348
Elliott A, Ambrose EJ (1950) Evidence of chain folding in polypeptides and proteins. Discussions Faraday Soc 9:246–251
Fasman GD (1991) Monolayer studies of synthetic poly(α-amino acids). In: Alper M, Calvert P, Frankel R, Rieke P, Tirrell D (eds) Materials synthesis based on biological processes. Materials Research Society, Pittsburgh, PA, pp 49–55
Flory PJ (1956) Phase equilibria in solutions of rod-like particles. Proc R Soc Lond A234:73–89
Flory PJ, Ronca G (1979a) Theory of systems of rodlike particles. 1. Athermal systems. Mol Cryst Liq Cryst 54:289–310
Flory PJ, Ronca G (1979b) Theory of systems of rodlike particles. 2. Thermotropic systems with orientation-dependent interactions. Mol Cryst Liq Cryst 54:311–330
Foelix RF (1982) Biology of spiders, Ist edn. Harvard University Press, Cambridge, MA
Frenkel D (1988) Structure of hard-core models for liquid crystals. J Phys Chem 92:3280–3284
Ghiradella H (1984) Structure of iridescent lepidopteran scales: variations on several themes. Ann Entomol Soc Am 77:637–645
Giraud-Guille (1987) Cholesteric twist of collagen *in vivo* and *in vitro*. Mol Cryst Liq Cryst 153:15–30
Gosline JM, DeMont ME, Denny MW (1986) The structure and properties of spider silk. Endeavour 10:37–43
Gray DG (1983) Liquid crystalline cellulose derivatives. J Appl Polym Sci: Appl Polym Symp 37:179–192
Gray GW, Goodby JWG (1984) Smectic liquid crystals: textures and structures, 1st edn. Leonard Hill, Glasgow
Hyde N (1984) The queen of textiles. Nat Geogr 165:3–49
Iizuka E (1985) Silk thread: mechanism of spinning and its mechanical properties. J Appl Polym Sci: Appl Polym Symp 41:173–185
Kaplan DL, Lombardi SJ, Muller WS, Fossey SA (1991) Silks: chemistry, properties and genetics. In: Byrom D (ed) Biomaterials: novel biological processes. Stockton Press, New York, pp 3–53
Kerkam K, Kaplan DL, Lombardi SJ, Viney C (1991a) Liquid crystalline characteristics of natural silk secretions. In: Alper M, Calvert PD, Frankel R, Rieke PC, Tirrell DA (eds) Materials synthesis based on biological processes. Materials Research Society, Pittsburgh, pp 239–244
Kerkam K, Viney C, Kaplan DL, Lombardi SJ (1991b) Liquid crystallinity of natural silk secretions. Nature 349:596–598
Kyte J, Doolittle RF (1982) A simple method for displaying the hydropathic character of a protein. J Mol Biol 157:105–132
Levi HW (1987) Spiders and their kin, 3rd edn. Golden Press, New York
Li G, Yu T (1989) Investigation of the liquid-crystal state in silk fibroin. Makromol Chem Rapid Comm 10:387–389

Lipowsky R (1991) The conformation of membranes. Nature 349:475–481
Livolant F (1986) Cholesteric liquid crystalline phases given by three helical biological polymers: DNA, PBLG and xanthan. A comparative analysis of their textures. J Phys 47:1605–1616
Livolant F, Levelut AM, Doucet J, Benoit JP (1989) The highly concentrated liquid-crystalline phase of DNA is columnar hexagonal. Nature 339:724–726
Lucas F, Rudall KM (1968) Extracellular fibrous proteins: the silks. In: Florkin M, Stotz EH (eds) Comprehensive biochemistry: extracellular and supporting structures. Elsevier, Amsterdam, pp 475–558
Magoshi J, Magoshi Y, Nakamura S (1985a) Crystallization, liquid crystal, and fiber formation of silk fibroin. J Appl Polym Sci: Appl Polym Symp 41:187–204
Magoshi J, Magoshi Y, Nakamura S (1985b) Physical properties and structure of silk: 10. The mechanism of fiber formation from liquid silk of silkworm *Bombyx mori*. Polym Comm 26:309–311
Magoshi J, Magoshi Y, Nakamura S (1985c) Physical properties and structure of silk: 9. Liquid crystal formation of silk fibroin. Polym Comm 26:60–61
Peakall DB (1971) Conservation of web proteins in the spider, *Araneus diadematus*. J Exp Zool 176:257–264
Plazaola A, Candelas GC (1991) Stimulation of fibroin synthesis elicits ultrastructural modifications in spider silk secretory cells. Tissue Cell 23:277–284
Rawn JD (1989) Proteins, energy, and metabolism, 1st edn. Neil Patterson Publ, Burlington, NC
Rill RL, Livolant F, Aldrich HC, Davidson MW (1989) Electron microscopy of liquid crystalline DNA: direct evidence for cholesteric-like organization of DNA in Dinoflagellate chromosomes. Chromosoma 98:280–286
Rudall KM (1962) Silk and other cocoon proteins. In: Florkin M, Mason HS (eds) Comparative biochemistry: a comprehensive treatise. Academic Press, New York, pp 397–433
Shear WA, Palmer JM, Coddington JA, Bonamo PM (1989) A Devonian spinneret: early evidence of spiders and silk use. Science 246:479–481
Strydom DJ, Haylett T, Stead RH (1977) The amino-terminal sequence of silk fibroin peptide Cp — a reinvestigation. Biochem Biophys Res Comm 79:932–938
Strzelecka TE, Davidson MW, Rill RL (1988) Multiple liquid crystal phases of DNA at high concentrations. Nature 331:457–460
Tirrell DA, Fournier MJ, Mason TL (1991a) Genetic engineering of polymeric materials. Mat Res Soc Bull 16:23–28
Tirrell DA, Fournier MJ, Mason TL (1991b) Protein engineering for materials applications. Curr Opinion Struct Biol 1:638–641
Tsujimoto Y, Suzuki Y (1979) The DNA Sequence of *Bombyx mori* fibroin gene including the 5′ flanking, mRNA coding, entire intervening and fibroin protein coding regions. Cell 18:591–600
Verhas J (1988) A liquid crystal model for early cell division. Liq Cryst 3:1183–1190
Viney C (1990) Transmitted polarised light microscopy, McCrone Research Institute, Chicago
Viney C, Twieg RJ, Dannels CM, Chang MY (1990) Nematic phases of rigid rod polytolans. Mol Cryst Liq Cryst Lett 7:147–151
Viney C, Kerkam K, Gilliland LK, Kaplan DL, Fossey S (1992) Molecular order in spider silk secretions. In: Sirota EB, Weitz D, Witten T, Israelachvili J (eds) Complex fluids. Materials Research Society, Pittsburgh, PA pp 89–94
Weiss RA, Ober CK (eds) (1990) Liquid-crystalline polymers. American Chemical Society, Washington DC
Wilson RS (1962) The control of dragline spinning in the garden spider. Q J Microsc Sci 104:557–571
Xu M, Lewis RV (1990) Structure of a protein superfiber: spider dragline silk. Proc Natl Acad Sci USA 87:7120–7124
Zemlin JC (1968) A study of the mechanical behavior of spider silks. US Army Natick Laboratories, Natick MA
Zheng S, Li G, Yao W, Yu T (1989) Raman spectroscopic investigation of the denaturation process of silk fibroin. Appl Spectrosc 43:1269–1272

Chapter 10

Micromechanics of Natural Composites

Joseph E. Saliba[1]

1 Introduction

Composite materials have been used for many years in structural applications which require advantageous properties such as light weight, high stiffness, and the ability to tailor the material to suit specialized requirements (Shaffer 1964; Adams 1974; Chamis 1974; Jones 1975). Design procedures, analytical prediction methods, failure analysis techniques, and manufacturing processes have been developed to a high degree of sophistication (Porter 1966; Shibley 1969; Hayashi 1970).

Design details such as fiber volume ratio, fiber cross-section, number of layers, and layer-stacking geometry are selected largely to facilitate available analysis methods and to adjust to practical fabrication capabilities (Foye and Baker 1971). If any optimization is considered in the design of structural components, it is done within these predictive and manufacturing constraints. Therefore, structural components composed of composite materials which are commonly utilized in the aerospace, automotive, construction, and other industries are classified as "man-made" (Ward and McCarthy 1989; Allan et al. 1990).

"Natural" composites, on the other hand, are not formed in compliance with imposed artificial conditions of practicality (Vincent and Gravell 1986; Allan et al. 1989; Vincent 1991). Rather, they are formed through a process which is based on reaction to natural, environmental effects. It has been suspected that such natural composites, having been formed by an evolutionary selection process, are optimum structures that best suit an organism within its environment (Saliba et al. 1990).

Just about every structure found in nature including bone, wood, mollusc shell, and insect cuticle is a composite (Gunderson et al. 1991). Nature is able to combine and organize relatively weak constituent materials into structures that are strong, stiff, tough, and lightweight. Constituent material selection is restricted for use in natural structures, so unique designs and material combinations are used to compensate. The opposite is true for man-made composites, the selection of constituent materials seems endless but we are still limited by design and processing constraints. Therefore, man-made composites may benefit from the analysis and application of novel design concepts used by nature to possibly solve current problems and create new designs and structures for the future.

[1] Civil Engineering Department, University of Dayton Research Institute, 300 College Park, Dayton, OH 45469-0243, USA

Results and Problems in Cell Differentiation 19
Biopolymers
Case, S. T. (Ed.)

The drift from the development of simple isotropic structural material to an orthotropic or anisotropic composite member requires more sophisticated analysis and design methods. While such analysis and design techniques for composite materials are based on older methods used for metals, new problems and criteria are unique to composites alone. Such unique problems appear in determining the complex stress state present on the microscopic scale (Piehler 1965; Stevenson 1970). The problem is evident in the stress concentration phenomena (Hedgepeth 1961; Foye 1970) induced when high strength fibers with high modulus are embedded in the relatively low strength fibers and modulus found in a matrix material (Tsai et al. 1966a). Such stress concentration problems increase the complexity of the design problem tremendously (Fichter 1969) and shatter any analysis with the most common design methods, such as allowable stress design or the ultimate design methods. Thus, two different approaches are commonly used in the mechanics of fiber-reinforced composites.

Macromechanics is concerned with the overall response of a composite laminate to mechanical and thermal loads (Halpin and Tsai 1969; Christensen 1979). Orthotropic or anisotropic behavior is assumed, and classical elasticity approaches and methods are then used to compute stresses, strains, and deflection in the laminate (Ekvall 1956; Tsai et al. 1966b).

On the other hand, micromechanics is concerned with the interaction of the constituent materials on a microscopic scale. Thus, micromechanics is an attempt to mathematically describe such characteristics as matrix yielding (Chamis 1967; Greszczuk 1971), crack initiation, and thermal and elastic behavior at the local scale. Such understanding of the interaction between components is extremely useful, as mentioned before, in determining particular failure initiation modes in a fiber-reinforced composite material.

The three dominant theoretical methods used in micromechanics to compute single-ply strengths and stiffnesses from constituent properties are the strength-of-materials, statistical, and advanced methods (Hashin and Rosen 1964; Pickett 1965). These methods are commonly used to predict the longitudinal tensile (Lifshitz and Rotem 1970) and compressive (Foye 1966) strength, the matrix-strain magnification and stress-concentration factors (Lockett 1970), the transverse compressive and tensile strengths (Adams and Tsai 1969), and finally the shear strength. The strength-of-materials method is founded on the assumptions that the total load is divided between constituents based on their relative stiffness and the ply is in a state of constant stress–strain, in an average sense (Zweben 1969). The assumptions that constitute the basis of the statistical methods are the fact that the fiber strength is length-dependent, and that the final fracture of a ply occurs when a critical number of cracks have occurred at some section along the axis of ply (Tsai and Schulman 1968; Zweben and Rosen 1969; Armenakas et al. 1970). The advanced methods are primarily classical solid mechanics methods of analysis, such as finite difference, point matching (Miller and Adams 1977; Schaffer and Adams 1980), or the finite element (Saliba et al. 1991).

Both the rule-of-mixtures and most statistical methods do not include any local geometric considerations. Also fiber/fiber interactions are normally not included in the analytical models, nor is the effect of fiber shape included. Other inadequacies include the inability to take into consideration the temperature and environmental dependency of the constituents.

2 Finite Element Method

The finite element method is currently the most commonly used method for existing micromechanics analysis. These finite element approaches rely on the material properties of the constituents to predict the microstress and microstrain distributions inside typical models. A general state of strain is typically assumed, mainly because of simplicity, and the lower computer time and memory needed. However, three-dimensional finite element modeling is sometimes needed.

To demonstrate the use of finite element method, a single ply containing five fibers will be considered (Fig. 1). Since geometry, constraints, material properties and loading are symmetric, a quarter panel will be all that is needed to completely describe the state of stresses and strains in the whole ply. In analyzing such a model the stresses and strains can be proved not to vary along the length of the fiber thus

Fig. 1. Cross-section of a single ply

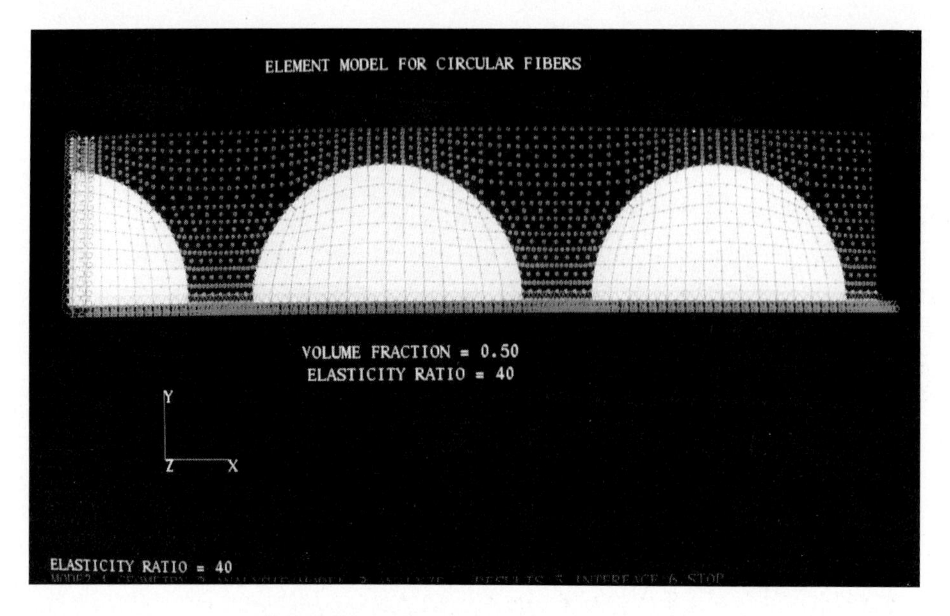

Fig. 2. Typical mesh for plane strain analysis

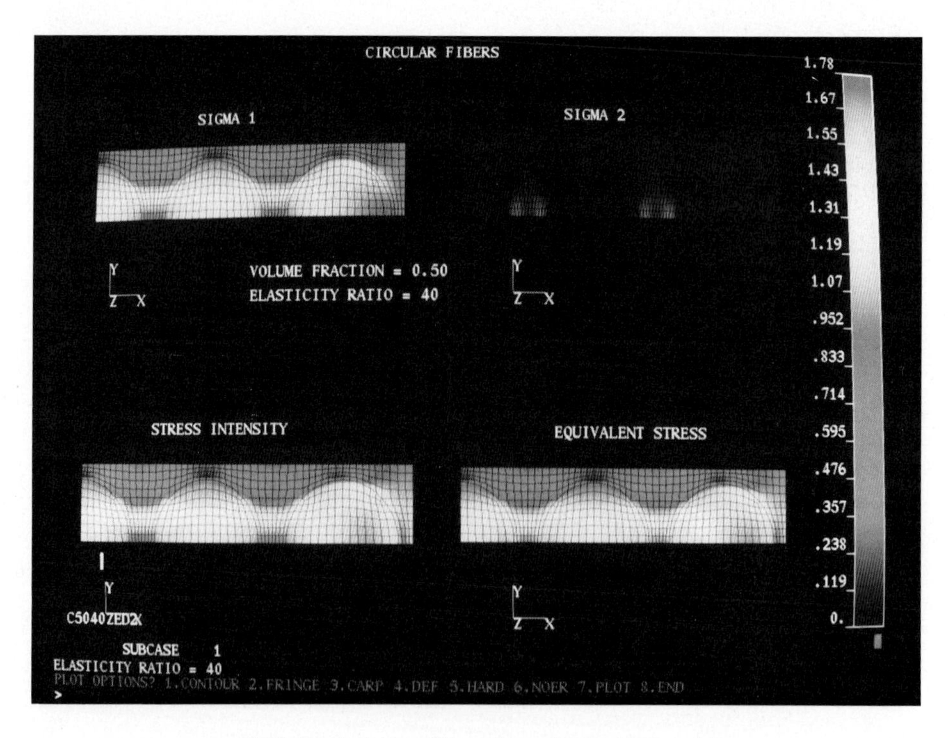

Fig. 3. Typical stress fringes for a circular fiber model

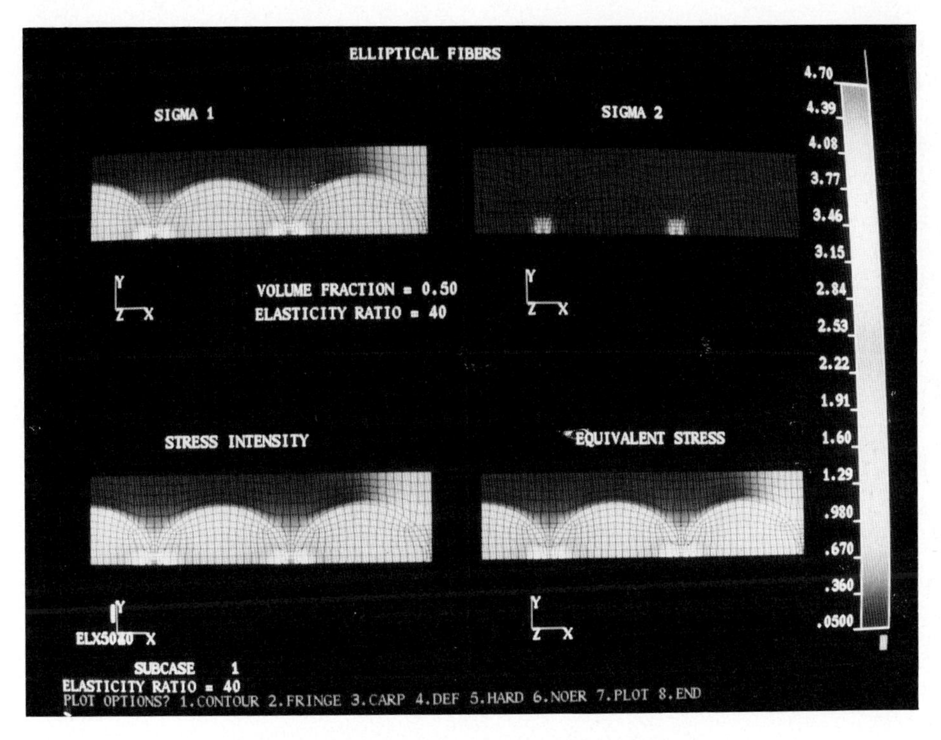

Fig. 4. Typical stress fringes for an elliptical fiber model

plane strain can be assumed. In such a case, a finite element model such as the one seen in Fig. 2 can be justified. It is important to use an effective kind of finite element if good geometric compliance and accuracy is to be obtained. For example, triangular elements are to be avoided when stress concentration factors are to be computed, otherwise an extremely fine mesh will be needed. Similarly, if four-noded elements are to be used the mesh must be kept fine because curves are truly being modeled by straight lines. Thus the preferred element is the eight-noded element because of its ability to use higher-order shape functions which normally results in better modeling. Typical results of a plane strain model are shown in Figs. 3 and 4.

3 Transverse Modulus

To compute the transverse modulus of a ply such as the one in Fig. 1, a typical finite element like that of Fig. 2 can be used. This transverse modulus is, in fact, the ratio of the average transverse stress to the average transverse strain on that same section. By applying a unit load on a transverse edge, the average stress becomes equal to unity. To estimate the average strain, the transverse displacement is integrated along the loading edge and later divided by the length of that edge. Thus the average strain is obtained by dividing the average displacement by the

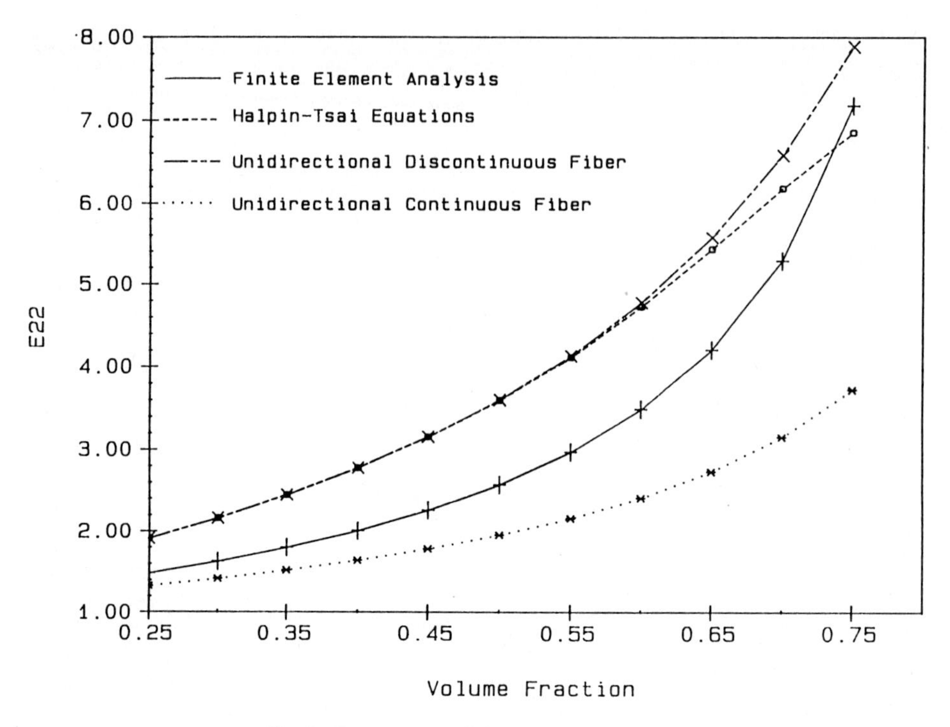

Fig. 5. Transverse modulus vs volume fraction

transverse dimension of the model. Once both the average stress and strain are computed, the transverse modulus is calculated by considering their ratio.

Another way to estimate the transverse modulus is to force equal or constant displacement along the loading edge, and compute an average stress across that same edge. Such a procedure should be accomplished to verify or validate the first approach.

To demonstrate this process, a sensitivity study is undertaken for a ply of circular fibers. The above procedure is repeated for different values of fiber volume fraction and the results shown in Fig. 5.

4 Stress Concentration

The stress concentration value is defined as the ratio of the overall maximum principal stress divided by the average applied load or stress on that section. If the applied load is constant on an edge and is equal to one, then the stress concentration value is simply the largest maximum principal stress. Stress concentration is often needed to determine crack initiation, failure capacity, and many other phenomena. Figure 6 illustrates the results of three finite element models, similar to Fig. 2, where the ratio of the fiber modulus to that of the matrix is set to a constant.

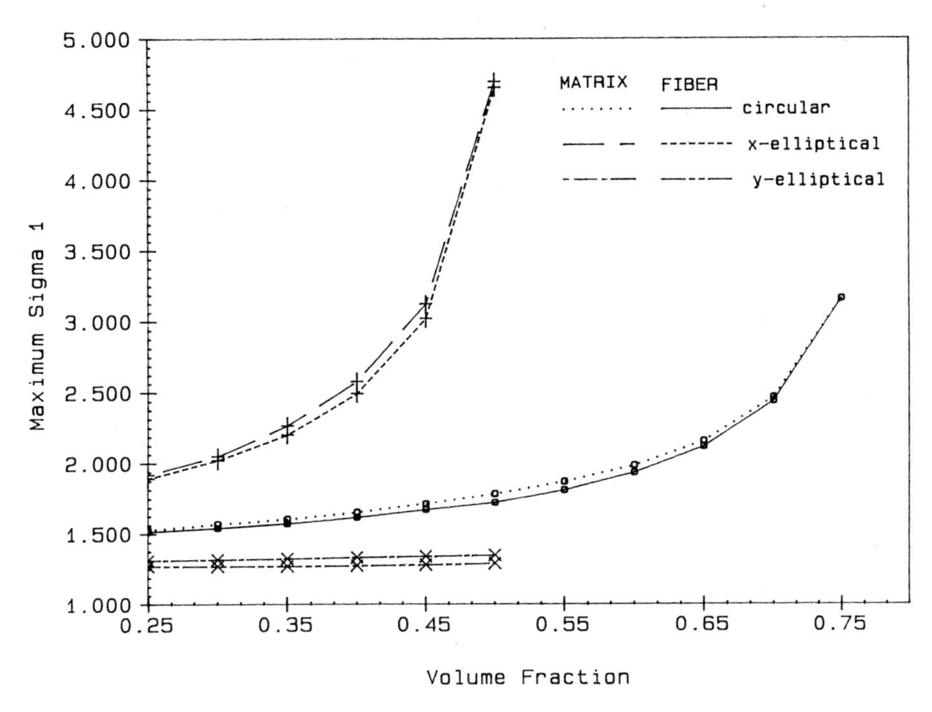

Fig. 6. Stress concentration factor vs volume fraction

The first model has circular fibers while the others have elliptical ones with an axis ratio of 1.5. The two elliptical models are oriented so that the major axis is parallel to the x and y axis respectively. Then the fiber volume fraction is varied from 25 to 75% for circular fibers and between 25 and 50% for the elliptical. A unit load is applied in the transverse direction and the maximum principal stresses are then recorded for each value of the volume fraction. By applying a unit load, the principal stress thus becomes equal to the stress concentration value.

5 Delamination and Crack Propagation

A full understanding of the physical mechanisms of fracture requires a complete characterization of the type of bonding between fiber and matrix, and between plies. Understanding of the interface failure mechanisms and their effect on the mechanical properties is also of importance. Experimental data can shed a light on the above phenomena, but they are difficult if not impossible to obtain, and are in many instances unreliable at best.

Micromechanical studies using the finite element method are used to investigate the stress and strain distribution and can clarify the above concerns. If interfacial bond strength, fiber shapes and plastic flow modeling of the matrix are known, they can easily be embedded in the analysis or the finite element to predict the behavior of the single ply or the whole composite. In fact, the finite element method is by far the most powerful technique available to investigate bonding, delamination, and crack propagation.

References

Adams DF (1974) Practical problems associated with the application of the finite element method to composite material micromechanical analyses. Fibre Sci Tech 7:111–112
Adams DF, Tsai SW (1969) The influence of random filament packing on the transverse stiffness of unidirectional composites. J Comp Mat 3:368–381
Allan GG, Carroll JP, Hirabayashi Y, Muvundamina M, Winterowd JG (1989) In: Skjak-Braek G, Anthonsen T, Sandford P (ed) Chitin and chitosan, sources, chemistry, biochemistry, physical properties and applications. Elsevier, London, p 765
Allan GG, Carroll JP, Hirabayashi Y, Muvundamina M, Winterowd JG (1990) Chitosan-coated fibers for use in papermaking. 1990 Materials Research Society Spring Meet, San Francisco
Armenakas AE, Garg SK, Sciammarella CA, Svalbonas VS (1970) Statistical theories of strength of bundles and fiber-reinforced composites. AFML-TR-70-3 Wright-Patterson Air Force Base
Chamis CG (1967) Micro and structural mechanics and structural synthesis of multilayered filamentary composite panels. DMSMD Rep No 9 Case Western University, Ohio
Chamis CG (1974) Micromechanics of strength theories. In: Broutman LJ (ed) Composite materials. Academic Press, New York, p 93
Christensen RM (1979) Mechanics of composite materials. Wiley, New York
Ekvall JC (1956) Elastic properties of orthotropic monofilament laminates. Am Soc Mech Eng Paper 61–AV–56
Fichter WB (1969) Stress concentration around broken filaments in a filament-stiffened sheet. NASA TN D-5453, Natl Aeronautics Space Adm, Washington DC
Foye RL (1966) Compression strength of unidirectional composites. AIAA Pap 66–143. AIAA 3rd Aerospace Sci Meet, New York
Foye RL (1970) Stress concentrations and stiffness estimates for rectangular reinforcing arrays. J Comp Mat 4:562–566
Foye RL, Baker DJ (1971) Design/analysis methods for advanced composite structures. AFML-TR-70-299, vols I and II
Greszczuk LB (1971) Micromechanics failure criteria for composites subjected to transverse normal loading. AIAA Pap 71–355, AIAA/ASME 12th Struct Dynam Mat Conf, Anaheim, CA
Gunderson SL, Saliba JE, Taylor DG (1991) Circular vs. elliptical cross-sectional geometries for composite fiber reinforcement. AIAA 17th Annu Regional Symp, Dayton (unpubl)
Halpin JC, Tsai SW (1969) Effects of environmental factors on composite materials. Air Force Tech Rep AFML-TR-67-423 In: Primer of composite materials: analysis. Technomic Publ Co, Westport, CT, p 77
Hashin A, Rosen BW (1964) The elastic moduli of fiber-reinforced materials. J Appl Mech 31E:223–232
Hayashi T (1970) Proc Br Plast Fed, 7th Int Reinforced Plast Conf, Brighton, England
Hedgepeth JM (1961) Stress concentrations in filamentary structures. NASA TN D-882, Nat Aeronautics Space Adm, Washington DC
Jones RM (1975) Mechanics of composite materials. Scripta, Washington DC
Lifshitz JM, Rotem A (1970) Longitudinal strength of unidirectional fibrous composites. AFML-TR-70–194
Lockett FJ (1970) Fiber stress-concentrations in an imperfect two-dimensional elastic composite. Nat Phys Lab Per Math 88, Ministry of Technology
Miller AK, Adams DF (1977) Micromechanical aspects of the environmental behavior of composite materials. UWME-DR-701–111–1 University of Wyoming, Wyoming
Pickett G (1965) Analytical procedures for predicting the mechanical properties of fiber-reinforced composites. Air Force Mat Lab Contractor Rep AFML-TR-65–220
Piehler HR (1965) The interior elastic stress field in a continuous close-packed filamentary composite material under uniaxial tension. Mass Inst Tech Rep ASRL TR 132–1
Porter MC (1966) Effect of filament geometry on reinforced composite strength. AIAA Pap 66–142, 3rd Aerospace Sci Meet, New York

Saliba JE, Bogner FK, Taylor DG (1990) A sensitivity study of the mechanical behavior of fiber-reinforced composite materials. University of Dayton, Ohio
Saliba JE, Gunderson SL, Taylor DG (1991) Finite element micromechanics of fiber-reinforced composite materials. AIAA 17th Annu Regional Symp, Dayton, Ohio
Schaffer BG, Adams DF (1980) Nonlinear viscoelastic behavior of a composite material using a finite element micromechanical analysis. UWME-DR-001–101–1 University of Wyoming, Wyoming
Shaffer BW (1964) Material properties of reinforced plastics. Soc Plastic Eng 4:67–77
Shibley AM (1969) In: Lubin G (ed) Handbook of fiberglass and advanced plastics composites. Van Nostrand Reinhold, Princeton, p 438
Stevenson JF (1970) Discrete-element microstress analysis of unidirectional fiber composites. Thesis, Case Western Reserve Univ, Ohio
Tsai SW, Schulman S (1968) A statistical analysis of bundle tests. AFML-TR-67–351 Wright-Patterson AFB, Ohio
Tsai SW, Adams DF, Doner DR (1966a) Effect of constituent material properties on the strength of fiber-reinforced composite materials. Air Force Mat Lab Contractors Rep AFML-TR-66–190
Tsai SW, Adams DF, Doner DR (1966b) Analysis of composite structures. NASA Contractor Rep, NASA CR-620
Vincent JFV (1991) Parallel fibres control fracture in biological systems. Mat Res Soc Symp Proc, vol 218, 221–224
Vincent JFV, Gravell K (1986) J Mat Sci Lett 3:310
Ward WJ, McCarthy TJ (1989) In: Encyclopedia of polymer science and engineering, 2nd edn supplement. Wiley, New York, p 674
Zweben C (1969) Tensile strength of fiber-reinforced composites basic concepts and recent developments. ASTM STP-460, p 528
Zweben C, Rosen WB (1969) A statistical theory of material strength with application to composite materials. AIAA Pap 69–123

Subject Index